PLANT SCIENCE BOOKS

PLANTS *and* VITAMINS

PLANTS *and* VITAMINS

BY

W. H. SCHOPFER
Director of the Botanical Institute University of Bern

authorized translation by NORBERT L. NOECKER

foreword by W. J. ROBBINS
Director of the New York Botanical Garden

ASIATIC PUBLISHING HOUSE
B-181/4, J-EXTN, STREET NO-5 LAXMI NAGAR
DELHI-110092 **INDIA** PH: (011) 22449904

ISBN 81-87067-49-7

Indian Print Edition 2005

Laser Typesetting
Delite Printers
Tri Nagar
Delhi

Printed by :
Ruchika Printers
shahdara, Delhi-110032

Printed and Bound in India

FOREWORD

Thirty years ago when I first became interested in the nutrition of the fungi, poor growth or lack of growth in a particular medium was generally ascribed to some unsuitability of the mineral nutrients, to the presence of some toxic agent, to excess acidity or alkalinity, to the absence of a particular carbohydrate, to the need for some special source of organic nitrogen, to the water relations or to some more esoteric cause. Mycologists recognized that many fungi required special media containing some material of natural origin; and oatmeal, corn meal, potatoes, bean pods, malt extract, peptone, wood, dung and many other natural products were frequently incorporated in the culture medium or used as such. Generally speaking, an effort was made to supply the material on which the organism grew in nature. We know now that the growth of many fungi is conditioned by the presence in the medium of minute traces of specific organic compounds some of them identical with the known vitamins and the presence of these growth substances in products of natural origin largely accounts for their advantages as culture media. This was a possibility seriously considered by few if any of those concerned with the cultivation of fungi thirty years ago and our knowledge of the phenomena involved has come largely from the basic experiments and discoveries made by Professor SCHOPFER.

WILDIERS in 1901 had briefly presented evidence on the importance for the growth of yeast of minute amounts of bios, a concentrate of unknown chemical composition; but his suggestion was roughly handled by many of his contemporaries, and in 1912 physiologists interested in these problems were still mainly concerned in juggling the minerals, the distilled water, the carbon and the nitrogen sources in attempts to determine a medium of known composition suitable for the more discriminating organisms. In that year CASIMIR FUNK drew together his own observations and experiments and those of others and proposed that traces of definite and specific organic substances in addition to minerals, sugars, fats and proteins were essential constituents of the food of animals. He named these substances vitamin(e)s, a term which immediately caught the public fancy. FUNK believed that vitamins were important for plants as well as for animals and during the next twenty years while the animal physiologist proceeded to demonstrate the importance of vitamins for the growth and well-being of animals, and to explore their multiplicity, functions, sources and chemistry, evidences were offered from time to time by various investigators that these substances were important for plants also. BOTTOMLEY's auximones attracted attention for a period; BACHMAN and WILLIAMS published on the vitamin requirements of yeast, a subject which was discussed pro and con by numerous investigators for son.. rs. LINNOSIER contributed an article on vitamins and the fungi; WILLAMAN in 1920 reported on the function of vitamins in the metabolism of *Sclerotinia cineria;* LEPESCHKIN in offered evidence for the importance of vitamins for some yeasts and filamentous fungi; MCCORMICK in made some intriguing observations in the same general direction on *Thielavia basicola;* LASH MILLER and associates pursued the bios problem. But the evidence offered to prove the importance of vitamins for plants was received by the majority of scientists with scepticism. The substances used as supplements by most of these investigators were not pure chemical compounds and it was possible to conceive of some explanation for the results other than one based on a need for vitamins; or, as was true of the influence of m-inositol, the substances used were not recognized as vitamins. It was generally agreed that plants synthesized vitamins and animals used them, but those who considered vitamins to be of importance to a plant were few indeed. In fact up to eight years ago not a single completely convincing example of the importance of a vitamin for a plant could be cited.

In 1934 Professor SCHOPFER published his results on the necessity of an external supply of thiamine for the growth of *Phycomyces.* His experiments grew out of his curiosity about a phenomenon which on the surface appeared to be of little importance; he was interested in determining why *Phycomyces* grew on media prepared with some samples of maltose and not on those prepared with other samples. As so frequently happens in science the answer to this problem which might seem to be trivial, proved to be of fundamental significance. He found the presence of thiamine as a contaminant in the "good" samples of maltose and its absence in the "poor" ones to be the answer. His demonstration of the effect of crystalline thiamine on *Phycomyces* was so clear and so convincing that it opened what may well be considered a new era in our conceptions of plant nutrition.

The test of the importance and validity of any contribution to science is to be judged chiefly by what grows out of it, by its influence. The significance of Professor SCHOPFER's discovery may be judged by the contents of this book because a large part of the material discussed in it is based on his own contributions, on those initiated by others because of his discoveries or on those influenced by his findings.

This is not the place to discuss further the history of the development of our knowledge of the relation of vitamins and similar substances to the growth of plants, nor to evaluate the contributions made by many notable investigators to this and related fields.

Astonishing progress in clarifying many puzzling problems has been made since 1934 and the importance of the results extends far beyond their application to the particular organisms concerned. The use of *Phycomyces* for bio-assay is a minor example of this extension. Nevertheless, the exploration of the importance of vitamins for plant development has merely begun, and we may anticipate with confidence future discoveries of great significance. This book in which the present status of the field is so ably summarized will lead to further contributions.

Professor SCHOPFER in Yverdon, Switzerland) was educated at the University of Geneva under the stimulating supervision of R. F. CHODAT. He was a student with KNIEP at Berlin under a fellowship of the Rockefeller Foundation, at the Pasteur Institute in Paris, and elsewhere, and has been Professor of Botany and Director of the Botanical Institute and Garden of the University of Bern since

PREFACE

This concise volume has the ambitious goal of presenting the present status of the problem of vitamins as related to plants. Only about twelve years ago a botanist interested in these problems was considered unique. Despite the fundamental research on vitamins since the beginning of the present century the problem has not yet reached maturity.

During the past few years extraordinary progress in the study of vitamins has taken place, in fact, it has completely changed the status of the problem. The enormous number of publications justifies this critical review. It does not by any means represent finality, but rather a stage of progress in research which continues to advance. Its purpose is to crystallize our knowledge in this domain, to show the relations of this subject to very diverse sciences, and above all, to point out and prepare the course to be followed. The bibliographies in this book are not complete; only the references pertinent to the reader's orientation are given.

American science has contributed much of importance in the solution of numerous problems concerning the rôle of vitamins in plants. These investigations will be given ample space here.

The literature has been reviewed more or less completely up till 1941. Since that time, during the course of translation and editing, many of the important works of 1941 and 1942 have been added, but omissions exist without doubt (the author has been on active service for many months every year since 1940). The author believes, however, that the reader will profit mainly from the general ideas expressed which have merely been confirmed by the evidence presented in the most recent works.

The author wishes to express his thanks to his assistants, V. KOCHER *and* Dr. W. F. MÜLLER *for the execution of numerous figures.*

He also wishes to give acknowledgment to the Board of Directors of Messrs. Wander (Bern) (Dr. A. *and* G. WANDER), *and to Messrs. Hoffmann-La Roche,* Basel, *for the aid which they have given during the course of the investigations cited in this book.*

Acknowledgments go particularly to Dr. NORBERT L. NOECKER *of the Department of Biology, University of Notre Dame, for his careful translation and invaluable help in editing my French manuscript, without which publication of this volume would hardly have been possible.* Dr. NOECKER *is not only responsible for editing my manuscript. He also made several timely additions and probably not a few corrections. I am very happy to have had the help of such an understanding collaborator!* Dr. NOECKER *has asked me to thank* Dr. W. LAWRENCE POWERS *of the University of Notre Dame and* Mrs. J. G. VERDOORN *for reading portions of the English translation and for helping with the preparation of the indices. Special thanks go to* Dr. THEODOR K. JUST, *also of the University of Notre Dame, for his careful reading and constructive criticism of all the material appearing in this book.*

THE AUTHOR

CONTENTS

PART 1.—*Synthesis of Vitamins in Plants. Auxo-Autotrophic Plants. Research Methods.*

PART 2. — *Vitamins in Relation to Plants unable to synthesize them. Growth factors of Microorganisms.*

Part 3. — *General Problems involving Vitamins.*

Part 1.

SYNTHESIS OF VITAMINS IN PLANTS
AUXO-AUTOTROPHIC PLANTS
RESEARCH METHODS

Chapter I.

GENERAL INTRODUCTION.

Ten years ago a botanist asked to discuss vitamins would have been faced by a hopeless task. He could have done no more than prepare a list showing the vitamin content of various plants. Such a list would have been of importance, in the final analysis, only to animal physiologists.

Modern vitaminology began a few years before the opening of the present century with the important observations of EIJKMAN (1897) and GRIJNS (1901) concerning the cause of beri-beri and polyneuritis (B_1 avitaminosis). Soon it was discovered that plants constituted for the most part the only important source of vitamins. They alone are capable of carrying on a large scale synthesis of these food constituents indispensable to man and animals. However it was still impossible to venture an opinion concerning the rôle of these substances in the vegetable organism. By virtue of the principle of biochemical finality, so clearly formulated by CLAUDE BERNARD, one might well expect them to play a rôle in the plant cell.

Originally a vitamin was described as a substance of unknown chemical composition, active in weak dosage, and different from the ordinary constituents of the diet. This definition was valid when applied to animals, having been made from that point of view. Accordingly vitamins were defined as exogenous substances, whereas hormones were regarded as endogenous. From the zoological point of view this was a very convenient distinction. However, it was soon observed that plants seemed to need analogous substances which were not comparable with any of the plastic or energizing foods required by them. BOTTOMLEY (1914-1917) and MOCKERIDGE (1917) were the first to call attention to certain special activators required by the higher green plants when cultivated under special conditions. These substances were present in a bacterial infusion of peat and were more or less effective in accelerating the growth of aquatic plants such as *Lemna minor, Azolla filiculoïdes, Salvinia natans,* and *Limnobium stoloniferum.* BOTTOMLEY (*cf.* FUNK, 1924) observed that the fixation of nitrogen by *Bacillus radicicola* (*Rhizobium* spp.) and *Azotobacter chroococcum* was favored by this same bacterial infected peat, as well as by its extracts. Attempts to concentrate and purify the active principle by means of the methods used at that time for the study of vitamins (alcoholic extraction of the fermented peat, precipitation by phosphotungstic acid and silver nitrate) led to the production of concentrates of unquestionable activity. A concentrate obtained by precipitation with silver nitrate had a definite action even at a dilution of

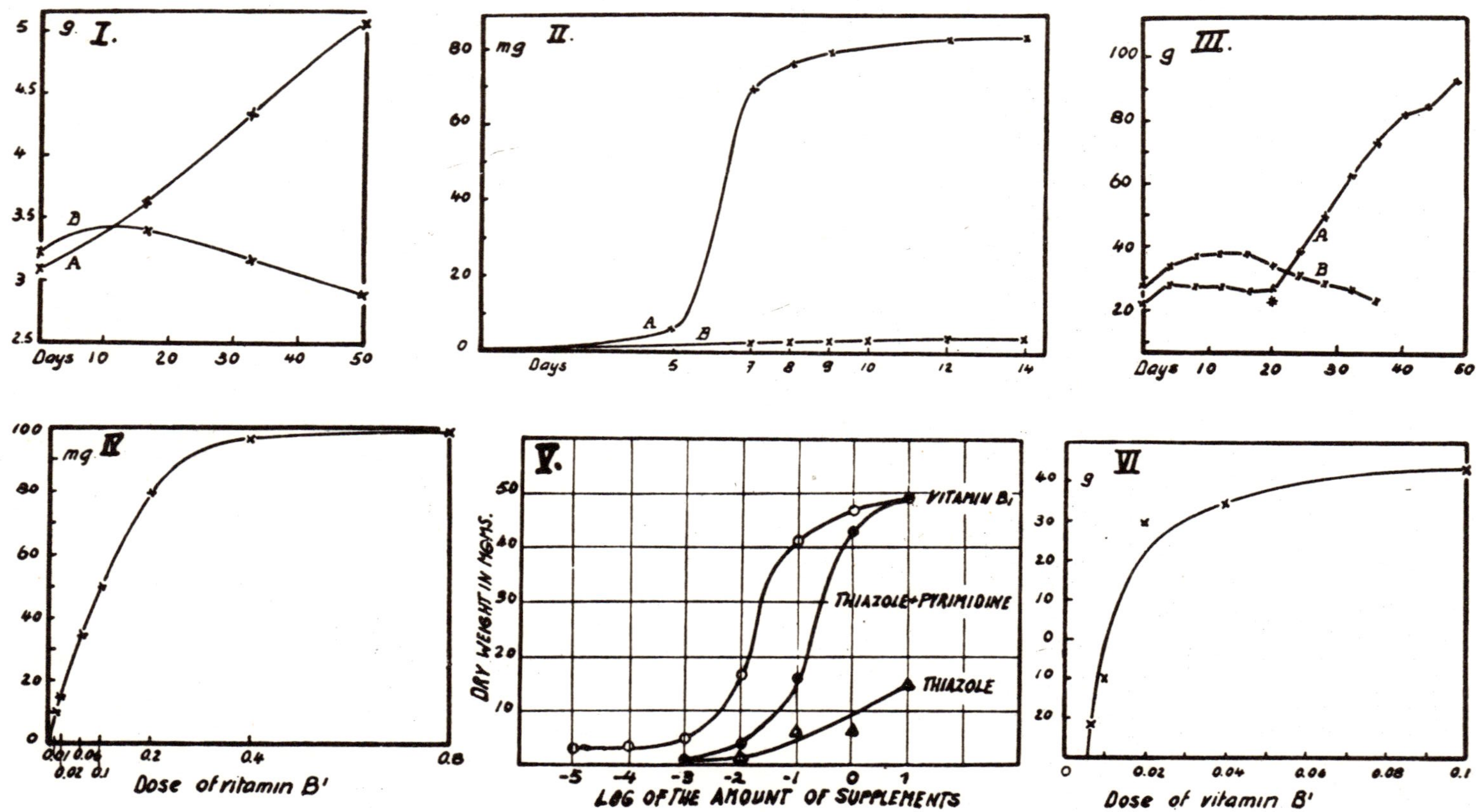
g I.
A
B
Days 10 20 30 40 50
mg II.
Days 5 7 8 9 10 12 14
g III.
Days 10 20 30 40 50
mg IV
Dose of vitamin B'
V.
DRY WEIGHT IN MGMS.
VITAMIN B1
THIAZOLE+PYRIMIDINE
THIAZOLE
LOG OF THE AMOUNT OF SUPPLEMENTS
g VI
Dose of vitamin B'

35:100,000,000. These unknown substances, analogous to vitamins, were called auximones.

In going back still further we find the observations of PASTEUR (1860) concerning yeast. He noticed that the development of yeast cultivated in a synthetic medium was markedly hastened by the addition of small quantities of organic substances, of unknown nature, present in natural products. He also observed that the growth of lactic acid bacteria (1858) was favored by the addition of onion juice to the culture medium. The concept of vitamins embraces also the early observations of WILDIERS (1901) concerning bios, which was at that time a mysterious substance indispensable to certain races of yeasts.

Thus we see that the stage had been set at a relatively early date. The actual investigations responsible for the remarkable progress in plant physiology were based on work whose importance cannot be ignored. Although the early work employed rather crude methods, it none the less represents the starting point in the development of the vast subject treated in this book.

The point of view from which this is written is that of *general physiology*. The author's aim is to point out the importance of vitamins in plants to botanists, particularly plant physiologists, and animal physiologists. We hope to acquaint the latter with the manner in which the biosynthesis of vitamins takes place in plants. On this occasion it is particularly opportune to present evidence showing that certain fundamental phenomena are identical in the two kingdoms. (Fig. 1).

The discovery of a vitamin in the field of animal physiology is based on the following observations: a diet composed of ordinary (unpurified) ingredients consisting of water, mineral salts, carbohydrates, protein materials, and fats in sufficient quantities to meet the plastic and energy requirements of an animal, permits normal development. But if the constituents of the diet are highly purified, deficiency phenomena appear which lead to avitaminosis characteristics. This diet lacks something which is different from all the other constituents in that it is effective in traces — that something is a vitamin.

FIG. 1. — Action of vitamins on various organisms:

as a function of time:

I. *Primula malacoïdes*, with and without growth factors (auximones); from BOTTOMLEY, 1914.

II. *Phycomyces Blakesleeanus*, with and without thiamin; from SCHOPFER, 1935.

III. Rat, with and without vitamin B; from FUNK and MACALLUM, in FUNK, 1924.

as a function of quantity:

IV. *Phycomyces* (from SCHOPFER).

V. Tomato root (from ROBBINS and BARTLEY-SCHMIDT).

VI. Rat (from COWARD, The Biological Standardization of the Vitamins, William Wood & Co., Baltimore, 1938. On the abscissa the dosages are in mg. of internationally standardized vitamin B_1).

The situation is exactly the same with respect to plants.

Let us take a microorganism, *Aspergillus niger*, cultivated in a synthetic medium containing water, a nitrogen source, a carbon source (sugar), and mineral salts, with all the conditions being optimum. The *balance of materials* for the metabolism of this organism can be established with as great exactness as it can for animals.

If we consider the balance of nitrogen alone (in a state of development in which the loss of nitrogen in the form of ammonium is negligible) we obtain, according to TERROINE, Mlle. TRAUTMANN, BONNET and JACQUOT (1925), the following figures:

Duration of the experiment, 4 days

Quantity of N supplied	0.1635 g.
Total N in the mycelium	0.0123 g.
Total N remaining in the liquid, not utilized	0.1498 g.
Total N recovered from the culture	0.1621 g.
Difference	0.0014 g.
Difference in per cent of the quantity of N supplied	0.8%

Likewise, the *energy balance* leads to the same results: the transformations take place without gain or loss.

Energy of the foods supplied	3.8823 cal.
Energy of the unused foods	2.7260 cal.
Energy of the mycelium harvested	0.3830 cal.
Energy liberated in the form of CO_2	0.5408 cal.
Total energy recovered	3.6498 cal.
Difference (energy supplied — energy recovered)	0.1325 cal.
Difference in per cent of the quantity of the energy supplied	4.5%

A second example presented here is taken from the work of ALLEGRA.

Heat of combustion of the mycelium	5.606 cal.
Heat produced during the course of development	3.229 cal.
Heat of combustion remaining in the nutrient solution	10.750 cal.
Total	19.655 cal.
Heat of combustion of the nutrient solution supplied	19.560 cal.

The difference, being only about 0.5 per cent, is within the limits of the experimental error.

Aspergillus niger, as well as *A. Oryzeae,* studied from this point of view by TAMIYA, has all its plastic and energy requirements covered by this synthetic nutrient solution containing only water, a carbon source, a nitrogen source, and mineral salts. In contrast, another fungus such as *Phycomyces Blakesleeanus* which might be expected to get along with the same nutrient solution, actually does not grow at all in it. The latter fungus is like an animal in that its nutrient supply lacks something which likewise is found to be a growth factor—a vitamin.

Let us now consider the higher plants. They are now cultivated in nutrient solutions in much the same manner as are microorganisms. The medium employed contains the necessary mineral salts with a nitrate salt as the nitrogen source. Standard formulas are those of SACHS, HILTNER, and others. If the culture is started, not from the seed, but from the embryo with the cotyledons removed, the solution must contain not only glucose but also essential growth factors to nourish the embryo during its heterotrophic period. Let us follow the development of seedlings grown from the seed (intact), having at their disposal only the reserves of the latter for their necessary energy supply. The energy balance will give the following results (Terroine, Bonnet, and Joëssel, 1924):

	Lentils	Rice	Flax
Energy of the seeds at the beginning (calculated from the weight)	4.2016	2.0205	1.0195 cal.
Energy remaining in the seed	2.6320	0.5014	0.8371 cal. (seed + seedling)
Energy of the seedling	0.9553	1.1081	
Energy corresponding to the CO_2 liberated	0.6025	0.3854	0.1665 cal.
Total energy recovered (remaining seed + seedling + CO_2)	4.1898	1.9949	1.0036 cal.
Difference between the energy at the beginning and at the end (energy lost)	0.0118	0.0256	0.0159 cal.
Difference (energy lost) in per cent of the energy at the beginning	0.28	1.25	1.50

The higher plants studied here (with the methods and according to the viewpoint of the authors) have in the reserves of their seeds everything necessary to cover their plastic and energy requirements. However, if the embryos are separated from their reserves and are cultivated aseptically on a synthetic medium like that used for the seeds, they either fail to develop or grow very little. They lack something which is found in the seed, something which the embryo is unable to synthesize, and which is not present in a strictly synthetic medium. Again we will be able to define this unknown substance as a growth factor, *i.e.*, one or several vitamins.

These experiments of Terroine *et al.*, together with the earlier work by Rodewald on apples, place plant physiology in the same position as animal physiology (Rubner's observations concerning dogs, Atwater and Rona's observations concerning man). The principle of the conservation of energy is expressed in the same manner in both branches of physiology: the organism does not create any new energy, it only uses and transforms the energy which is supplied to it.

The work of the plant physiologist will therefore be exactly comparable to that of the animal physiologist. It has gone through the same stages:

(1) Observation that an unpurified natural medium is likely to assure success in the culture of a microorganism or a higher plant.

(2) Demonstration that a purification of the medium prevents certain organisms from growing.

(3) Presentation of evidence of an indefinite nutrient, different from all the other constituents of the culture medium.

(4) Identification of this indefinite substance as a growth factor, as a vitamin in the broad sense of the term.

One might wonder why plant physiology has been so slow in progressing. The explanation is found in the following analogy. A person who has enough money for a livelihood is perhaps not preoccupied by the manner in which it is acquired, nor does he, in his comfort, dwell upon the many advantages which he enjoys. On the other hand, if he has little or no money he soon acquires some very definite views concerning the manner in which it is obtained and also soon becomes aware of the effects of an insufficiency. The plant finds itself in the situation of the fortunate person. For a long time the plant physiologist did not appreciate the value of his fortune.

The purpose of this book is to present the problem of vitamins as it appears today in the light of general physiology. Up till now no complete exposition of this kind has been made. The great number of publications which have appeared during the past several years justifies this comprehensive review of the subject.

Part one will be concerned with the capacity for synthesis in cells of green autotrophic plants. We shall point out that the capacity for synthesis finds its complete expression in the elaboration of almost all the vitamins. We shall study the rôle of vitamins and their functions in the plants which synthesize them.

Part two will deal with the other aspect of the problem, the inability to synthesize one or several vitamins (loss of growth factors) with the result that these substances must be supplied to the plant as exogenous growth factors. The principal constellations of growth factors will be studied by means of the organisms which have served to demonstrate them.

Part three will present some general phenomena which are wholly or partially explained on the basis of the vitamin concept.

The entire book must appear as a demonstration of the idea that the loss of capacity to synthesize vitamins by a plant leads it to the same level as that occupied by an animal. From this point of view there is one fundamental similarity between the higher animals and microorganisms; both are heterotrophic with respect to growth factors.

Such a problem cuts across numerous branches of science: human physiology, animal physiology, plant physiology, microbiology in general, biochemistry, especially enzymology, and pure organic chemistry. It is impossible to treat the problem of vitamins in plants as if it were completely separated from the general subject to which it belongs. The starting point of the problem is found in animal and human physiology. Microbiology soon made its contribution and gave birth to the well known problem of growth factors. The principal result of modern research is the demonstration that vitamins in the older sense of the term are exactly the same as growth factors. Because of this identity it has been possible to approach the problem of vitamins in reference to higher plants. The study of vitamins in the higher plants has opened the way to one of the most important aspects, namely, the biosynthesis of vitamins. In fact, it is by virtue of the higher autotrophic plants that the chemical mechanisms of biosynthesis could in part be demonstrated.

The necessity of treating the subject on a very comprehensive basis requires the consultation of an extremely extensive bibliography. The dimensions of this book and the spirit in which it was conceived, as well as the necessity of making it accessible to the general public, make it necessary to reduce the citations to a minimum, and the author asks in advance to be excused for this. An effort will be made to cite general references in order that the reader can readily find an extensive bibliography concerning our discussion.

The general works which already cover this subject in part and which contain references to the older literature are: FUNK (1924), KNORR (1925), SERGENT (1928), RANDOIN and SIMONNET (1927), WINTERSTEIN and FUNK (1933), PESKETT (1933), KNIGHT (1935), J. BONNER (1937), KOSER and SAUNDERS (1928), JANKE (1939), and SCHOPFER (1939). Only the books of FUNK and of RANDOIN and SIMONNET, and the review article of BONNER consider the problem from the standpoint of both microorganisms and higher plants. All the other works consider only the microorganisms.

The books of BOYSEN-JENSEN (1935), translated into English by G. S. AVERY and P. R. BURKHOLDER (1936), and of F. W. WENT and K. V. THIMANN (1937) are concerned primarily with plant hormones (auxins). (See p. 216).

Interesting reviews have appeared in the two first volumes of Ergebnisse der Vitamin- und Hormonforschung (1938 and 1939).

A list of important treatises on vitamins is given in the bibliography on p. 58.

REFERENCES:-

1858: L. PASTEUR. — Mémoire sur la fermentation appelée lactique. Ann. Chim. et Phys., III, 52:404. — Oeuvres, 2:12.

1860: L. PASTEUR. — Mémoire sur la fermentation alcoolique. Ann. Chim. et Phys., III, 58:323. — Oeuvres, 2:94.

1897: C. EIJKMAN. — Ein Versuch zur Bekämpfung des Beri-Beri. Virchows Archiv, 149:187.

1901: G. GRIJNS. — Polyneuritis gallinarum. Tijdschr. Netherlandsch-Indië, 41:3. (in Research on Vitamins, 1910-1911, translated and reedited by a committee of honour on occasion of his 70th birthday. Gorinchem, J. Noorduyn en Zoon, 1935).

1901: E. WILDIERS. — Une nouvelle substance indispensable au développement de la levure. La Cellule, 18:313.

1914: W. P. BOTTOMLEY. — Some accessory factors in plant growth and nutrition. Proc. roy. Soc. London, B. 88:237.

1917: W. P. BOTTOMLEY. — Some effects of organic growth promoting substances (auximones) on the growth of *Lemna minor* in mineral culture solution. Proc. roy. Soc. London, B. 89:481.

1917: F. A. MOCKERIDGE. — Some effects of growth promoting substances (auximones) on the soil organisms concerned in a nitrogen cycle. Proc. roy. Soc. London, B. 89:508.

1924: C. FUNK. — Die Vitamine. J. F. Bergmann, München. (Contains an excellent discussion concerning the rôle of vitamins in plants).

1924: E. F. TERROINE, R. BONNET & P. H. JOËSSEL. — L'énergie de croissance, II: La germination. Bull. Soc. Chim. biol. Paris, 6:357.

1925: E. F. TERROINE, S. TRAHTMANN, R. BONNET & R. JACQUOT. — L'énergie de croissance, III: Rendements énergétiques comparés dans le développement de moisissures . . . Bull. Soc. Chim. biol. Paris, 7:351.

1935: W. H. SCHOPFER. — Les vitamines cristallisées B comme hormones de croissance chez un microorganisme (*Phycomyces*). Arch. f. Mikrobiol., 5:513.

1938: K. H. COWARD. — The biological Standardisation of the Vitamins. Baltimore, William Wood & Co., 1938, p. 59.

GENERAL REFERENCES:-

1924: C. FUNK. — Die Vitamine. Ihre Bedeutung für die Physiologie und Pathologie. München, J. F. Bergmann, 522 p.

1925: M. KNORR. — Die Entwicklung des Vitamingedankens in der Bakteriologie. Erg. Hygiene, 7:641.

1928: A. L. SERGENT. — Les facteurs de croissance des microbes sur milieux artificiels. Paris, G. Doin, 182 p.

1927: L. RANDOIN & H. SIMONNET. — Les données et inconnues du problème alimentaire, II: La question des vitamines; Les problèmes biologiques. Paris, Les presses universitaires, 480 p.

1933: A. WINTERSTEIN & C. FUNK. — Vitamine. Handbuch der Pflanzenanalyse, Bd. IV/2, Spezielle Analyse III.

1933: G. PESKETT. — Growth factors for lower organisms. Biol. Rev. Cambridge philos. Soc., 8:1.

1935: B. C. J. G. KNIGHT. — Bacterial Nutrition. Material for a comparative Physiology of Bacteria. London, Medical Research Council, Special Report series no. 210, 182 p.

1937: J. BONNER. — The rôle of vitamins in plant development. Bot. Rev. 3:616.

1938: S. A. KOSER & F. SAUNDERS. — Accessory growth factors for Bacteria and related microorganisms. Bact. Rev. 2:99.

1938: A. LWOFF. — Les facteurs de croissance pour les microorganismes. Rapport au 1er Congrès intern. des microbiologistes de langue française, p. 58.

1939: A. JANKE. — Die Wuchsstoff-Frage in der Mikrobiologie. Centralbl. Bakt. II, 100:409.

1939: W. H. SCHOPFER. — Vitamine und Wachstumsfaktoren bei den Mikroorganismen, mit besonderer Berücksichtigung des Vitamins B_1. Ergebn. d. Biologie, 16:1.

1939: C. H. WERKMAN. — Bacterial dissimilation of carbohydrates. Bact. Rev. 3:187-227.

1941: W. H. EDDY. — What are the Vitamins? New York, Reinhold, 247 p.

1941: W. RUDOLPH. — Die Vitamine der Hefe. Stuttgart, 130 p.

1942: H. R. ROSENBERG. — Chemistry and Physiology of the Vitamins. New York, Interscience Publishers, Inc., 674 p.

1942: E. A. EVANS. — The biological action of the vitamins; a symposium. Chicago, University of Chicago Press, 236 p.

1942: C. H. WERKMAN & H. G. WOOD. — On the metabolism of bacteria. Bot. Rev. 8:1-68.

Chapter II.

THE PLANT CELL AND ITS CAPACITY FOR SYNTHESIS.

Physiologically speaking the plant cell is characterized by an extraordinary capacity for synthesis which is by no means confined to photosynthesis. From elements and mineral substances it builds extremely complex materials and does so with very simple means. Cytologically, the capacity for synthesis is in the main dependent upon the mitochondrial system. The mitochondria are the precursors of various plastids, *e.g.*, leuco-, chromo-, and chloroplasts. From the work of GUILLIERMOND (1933)[1] and his school we know now the genetic relations existing between the various kinds of plastids and the various categories of cellular constituents. The plastids themselves are the most important bases of synthesis. Of the substances synthesized we shall mention, as an example, only the carotinoids since they are of particular interest in the study of biosynthesis.

The plastids, because of their enormous surface area, their chemical composition, and their physico-chemical structure, are adapted to play a very important rôle. Their diastase content and their lipoidal constitution give them an unusual oxidation-reduction capacity. The chondriome has been considered as the possible carrier of glutathione (for example, by JOYET-LAVERGNE and PARAT), whose strong reducing properties and tendency to be oxidized in the air are well known. This property of auto-oxidizability of glutathione lends further support to the possibility that the plastids take part in the oxidation-reduction phenomena accompanying biological syntheses despite the fact that we seldom understand the mechanism of these reactions. The syntheses which yield insoluble final products visible in cytological studies are certainly not the only ones for which the chondriome is responsible. As NAGEOTTE puts it, their chemical activity must likewise permit the formation of substances which diffuse to the place where they accumulate. The recent work of JOYET-LAVERGNE (1935) tends to show that the chondriome of certain animal cells acts as a carrier; for example, red corpuscles act as the carriers of vitamin A, detected histologically by the Carr-Price reaction (antimony trichloride). JOYET-LAVERGNE suspects the existence of an equi-

[1]Recently the Chronica Botanica Company, Waltham, Massachusetts published a book by GUILLIERMOND (1941) concerning the cytoplasm of the plant cell. This book is an international monograph especially written for the "New Series of Plant Science Books", not a translation from one of GUILLIERMOND'S previous publications.

librium between glutathione and vitamin A resulting in the protection of glutathione by vitamin A. He attributes to the chondriome a catalytic oxidizing power of the same order as that required to oxidize a solution of glucose. The second step in cellular respiration occurs supposedly in the chondriome (JOYET-LAVERGNE, 1936). This example shows thát the vitamins, above all other compounds, represent a complete and perfect expression of the capacity for synthesis of the plant cell. The highly complex contents and the intricate structure of the vitamin molecules are the products of intense chemical activity since the initial materials consist of elements or very simple compounds.

A thorough examination of the data concerning the biosynthesis and chemistry of each of the vitamins will be reserved for later chapters. For the present it will suffice to list simply the active substances which are related to the capacity for synthesis in plants and which in most cases take part in the metabolism of the plant.

1. — Vitamin A, anti-xerophthalmia or anti-nyctalopia vitamin; fat-soluble, occurring in plants in the form of a provitamin, *i.e.*, a carotenoid.

Vitamin B group, water soluble:

2. — Antineuritic vitamin, B_1, thiamin (aneurin).

The vitamin B_2 complex:

3. — Growth vitamin, sometimes called B_2 (strict sense of the term) or vitamin G, lactoflavin, riboflavin.

4. — Pellagra-preventative vitamin of humans (P-P factor), canine anti-blacktongue factor of GYÖRGYI - nicotinic acid or its amide.

5. — Pellagra-preventative vitamin of rats, vitamin B_6, "rat acrodynia" factor of GYÖRGYI, adermin, pyridoxine.

The list of "B vitamins" has continued to grow[1], the first addition being the pellagra-preventative vitamin of chicks, also known as the filtrate factor since it is not adsorbed on Fuller's earth with the other B_2 vitamins but remains in the filtrate. It has been identified as pantothenic acid; it is in some cases replaceable by β-alanine. Another member of the "B group" is vitamin W which is necessary for the growth of the rat. It is heat labile but is not the same as vitamin B_1; it is now thought to be identical with biotin.

Other vitamins:

6. — Vitamin C, anti-scorbutic vitamin, water soluble, ascorbic acid.

[1]Since the above was written great progress has been made in the identification of members of the "vitamin B complex". At present nine water soluble "B vitamins" are recognized: thiamin, riboflavin, nicotinic acid, pyridoxine, pantothenic acid, biotin, inositol, *p*-aminobenzoic acid, and "folic acid" (*Cf.* S. ANSBACHER, 1941; R. J. WILLIAMS, 1941; H. K. MITCHELL, E. E. SNELL, and R. J. WILLIAMS, 1941; R. J. WILLIAMS and coworkers, 1941).

7. — Vitamin D_2, anti-rachitic vitamin, fat soluble, calciferol, an isomer of ergosterol.

8. — Vitamin E, anti-sterility vitamin, fat soluble, α-tocopherol.

9. — Vitamin K, anti-hemorrhagic vitamin, fat soluble.

10. — Vitamin P, concerned with permeability, citrin (flavone derivative), the formula of which is approximately but not fully known.

This list includes only those vitamins which have been identified chemically. It cannot be said that the capacity for synthesis of all of these vitamins is confined to plant cells, since animal cells are of primary importance in the synthesis of vitamin D_2. Concerning others, the function of which seemed until recently to be limited to animals, *e.g.*, vitamins E and K, animal physiologists must concede that plants constitute their richest sources. This is indeed comforting to the botanist since it enables him to view the problem with the proper perspective.

This problem concerns not only the vitamins known originally for their action on animals and whose rôle is now sought in plants, but conversely, such factors as those in bios (thiamin, pantothenic acid, pyridoxine, biotin, and inositol), which once seemed to be active only on plants, but which today are known to affect animals as well. These substances will be studied in greater detail in subsequent chapters.

The unusual capacity for synthesis, which is a property of the most primitive plant cells, degenerated during the very early stages of evolution. General autotrophism which is the rule in green plants shows a tendency to decline as evidenced by the complete disappearance of certain substances. The capacity for synthesis, which, when intact, is able to confer complete biological and biochemical independence upon the plant, has evolved by mutation or by slow evolutionary changes of an adaptive nature, leading the plant toward heterotrophism, saprophytism and parasitism. By examining various plant series we are able to follow, step by step, all the stages of this progressive degeneration leading to complete heterotrophism.

In the *Scrophulariaceae* a series of autotrophs and hemiparasites, *e.g.*, *Scrophularia, Pedicularis, Melampyrum,* and *Euphrasia,* is particularly striking. By comparing these species with those having a full complement of chlorophyll it is possible to detect a progressive decrease in their photosynthetic activities (MOLLIARD, 1921). In the case of a *Euphrasia* sp., transpiration and chlorophyll formation are still carried on to some extent, whereas the absorption of water and mineral salts is reduced. It is for this reason that this species might be dependent upon a host since it is not dependent on an outside source of sugars. At any rate, it is a partial heterotroph.

A series of autotrophic and saprophytic *Orchidaceae* is equally instructive. Many of the green orchids are heterotrophic and are dependent upon their mycorrhiza for growth factors of a vitamin nature only during the early stages of their development, since later on they become independent of the symbiotic fungus and can dispense with it (SCHAFFSTEIN, 1938). Others (*Corallorrhiza, Neottia, Limodorum,* and *Epipogon*) remain definitely heterotrophic, but in different degrees. *Neottia,* which grows in shady habitats, has lost almost all of its chlorophyll but still possesses chloroplasts (SENN, 1927), which are lacking in *Corallorrhiza* and *Epipogon.* The causes which have led to partial or complete heterotrophism with all its morphological manifestations need not be discussed here. The causes may be varied in nature and lead to phenomena of convergence. It seems likely that the cause for the chlorophyll deficiency in the humus inhabiting orchids is a lack of light, and that the symbiotic mycorrhiza compensates the plants for this deficiency by supplying carbohydrates.

The origin of heterotrophism can actually be observed. RENNER found wild albino forms of green orchids, *Cephalantera alba* and *Epipactis latifolia,* even though such phenomena had never been described for these orchids. We know, however, that the colorless forms of *Neottia nidus avis* (f. *sulfurea,* f. *pallida,* and f. *nivea*) are found only rarely. The plant probably has suffered no harm in becoming heterotrophic since *Epipactis* has a well developed symbiotic fungus. In the case of *Cephalantera* a large proportion of the root is not infected. We can assume, without definite proof, that the presence of the fungus accounts for the continuation of this albinism. It is hardly probable that these albinos may have been able to give rise to a stable race of definitely heterotrophic forms. Since the leaves remained normal in size in the case cited by RENNER, the loss of chlorophyll represents only one of the characters of the complex of primary and derived characteristics which condition the saprophytism of orchids and which are so difficult to analyze. Likewise in the genus *Pirola,* CAMP (1940) observed the disappearance of the leaves in species which normally possess them. These forms derived from autotrophs survive by virtue of their mycorrhizal fungi.

Experimentally produced heterotrophism can be demonstrated cytologically in all its details in the flagellates. These organisms, which have been studied extensively by LWOFF, are of particular interest because they represent the point of departure of the plant and animal kingdoms, the former containing chlorophyll and being autotrophic, the latter being heterotrophic. The bifurcation involving the mitochondrial system and the plastids begins with this group of organisms. We know green flagellates such as *Euglena gracilis,* others possessing plastids (leucoplasts) but not synthesizing chlorophyll such as *Polytoma uvella,* and still others without plastids and without chlorophyll such as *Glaucoma piriformis.* CHATTON calls the first—chlorophytes, the second—leucophytes, and the third—protozoans.

Protists (*Protista protocarya*)*:* without definite nuclei or mitochrondria (bacteria).

Protists (*Protista eucarya*)*:* with well defined nuclei and mitochrondria.

Protozoans: no plastids, no chlorophyll.
Leucophytes: leucoplasts, no chlorophyll (*Polytoma uvella*).
Chlorophytes: one or several plastids with chlorophyll (*Euglena gracilis*).

Our interest in these organisms pertains particularly to the development of heterotrophism. The physiology of the metabolism

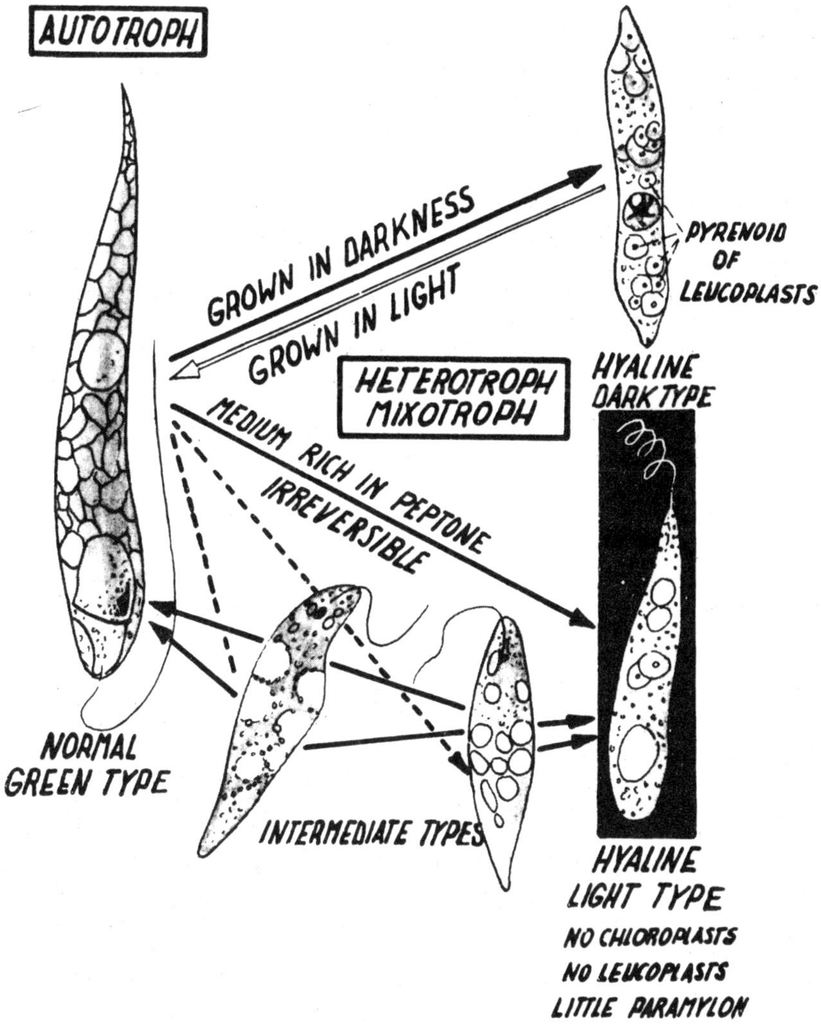

FIG. 2. — Experimental production of reversible and irreversible heterotrophism in *Euglena gracilis* (compiled from data of TERNETZ, 1912).

of these forms is now well known. The leucophytes are heterotrophic with respect to carbohydrates, whereas the chlorophytes are autotrophic.

The case of *Euglena gracilis* is particularly instructive. This green species is autotrophic and grows well in light on a mineral medium based on nitrate or, better yet, ammonium salts. When nitrogen is supplied in the form of potassium nitrate the organism can survive only one transfer, but when this element is supplied by ammonium salts it may be subcultured indefinitely. When cultured in darkness incipient heterotrophism appears and an independent carbon source becomes necessary. Furthermore, it has been found possible to deprive *Euglena gracilis* of its chlorophyll, either by culturing it in the dark, or by culturing it in light on a medium containing a complex organic nitrogen source. This is not applicable to all strains of *Euglena gracilis.* The decoloration in the dark produces reversible effects. The chlorophyll disappears but the plastids remain and, when exposed to light, become green again. On the other hand, when the organism is decolorized in light by means of an organic nitrogen substance such as peptone it not only loses its chlorophyll but also suffers an irreversible loss of its plastids. These forms have become definitively heterotrophic (ZUMSTEIN, 1900; TERNETZ, 1912). Thus the chlorophyte may reversibly become a leucophyte, and the leucophyte may irreversibly become a protozoan (Fig. 2). The chlorophyte, according to the terminology of A. LWOFF, is a protoautotroph (it assimilates CO_2); the leucophyte is an oxytroph (it requires an exogenous source of carbohydrates); the protozoan is a haplotroph (a complex nitrogen source can serve as the source of carbon). Moreover, an organism may be heterotrophic in its nitrogen nutrition but autotrophic in respect to carbohydrates.

We shall not discuss the relations between nitrogen and carbon nutrition which have been well established by the work of A. LWOFF. It seems that protein synthesis is connected with the activity of the plastids. Apparently they play a fundamental rôle in this process since they disappear when heterotrophism is produced experimentally under certain cultural conditions. This heterotrophism, which involves a total change in the metabolism, remains genetically fixed.

The following table demonstrates that the evolution of heterotrophism may also involve nitrogen nutrition since not all green *Euglena* species have the same nitrogen requirements.

TABLE I. — Value of various nitrogen sources to *Euglena gracilis* (after H. DUSI, *in* LWOFF, 1932).

	nitrates	NH_4 salts	amino acids	peptones
Euglena stellata	++	++	++	++
gracilis	+	++	++	++
*Klebsii**	+	++	++	++
*anaboena**	0	+	++	++
*deses**	0	0	++	++
*pisciformis**	0	0	0	++

(*can live only in light!)

The ability to reduce and assimilate nitrates is primitive and characterizes the autotrophs. It has already decreased somewhat in *Euglena gracilis*. *E. Klebsii* and *E. anaboena* are still able to utilize ammonium salts in the synthesis of their amino acids. *E.*

deses requires an amino acid, and *E. pisciformis* has obviously become completely heterotrophic in respect to nitrogen nutrition since it requires a peptone for its growth. The evolutionary change in these species of *Euglena* is complete since it is not only morphological, visibly affecting the plastids, but is also physiological. It is easy to follow the stepwise evolution of carbon and nitrogen nutrition. This physiological degeneration is certainly caused by the loss of some enzyme.

LWOFF and others who have studied this problem have never observed a reappearance of the pigment in those forms of leucophytes which are genetically fixed. This loss of function, namely the abolition of the capacity for synthesis, is therefore definitive and irreversible. Its phylogenetic progress can be traced along orthogenetic evolutionary lines but cannot be reversed either experimentally or genetically. This loss of function gives rise to new needs, thereby causing the organism to change to a new mode of life characterized by a progressive loss of physiological independence.

Such a loss of function is obviously related to saprophytism and parasitism. As already pointed out in connection with the higher plants, the morphological expressions of parasitism are diverse. Among the numerous characters found in parasitic organisms it is difficult to distinguish readily between primary and secondary ones, and between those which appeared abruptly by mutation and those which were perfected by adaptation. The unicellular organisms (flagellates) permit an experimental approach to the problem, whereas the higher plants do not. In the flagellates the loss of function seems to be primitive. This loss must have led the organism to parasitism, and once this relationship was established, the line was able to maintain itself throughout its phylogeny. The initial phenomenon is of a physiological nature since it is the chemistry of the cell which is attacked and modified first. In the case of *Euglena gracilis*, the origin of the definitely decolorized form without plastids occurs in forms with chloroplasts under the influence of organic nutrients. One might be inclined to attribute this change to a direct action of the medium, although such an assumption is illogical. An action of this kind presupposes a genetically determined constitution receptive to the influence of the medium. In reality the organic nitrogen acts upon the genetic mechanism of the cell which in turn conditions the formation of plastids.

From the above we may conclude that physiological evolution by loss of function involves a transposition of action, a phenomenon well known in morphological evolution in which reductions play a primary rôle (WETTSTEIN).

We can well imagine what such a loss of function means to an organism. The performance of all essential functions by a living cell involves a multitude of reactions all of which are interdependent. They have a complex substratum as a base and they are effected by numerous catalytic agents. In this chain of reactions, if one of the links is missing and a method of replacing it can not be found, the whole procedure is interrupted and the products accumulate. Living matter is characterized by the fact that it produces its own instruments which are adapted to and necessary for its conservation. If only one of these instruments is missing, the development is inhibited or undergoes profound modifications.

Let us now return to vitamins and growth factors and consider them in the light of the discussion just presented.

When we compare animals and plants in a general way and note that the former are no longer able to carry on the synthesis of substances they must now obtain from plants, we are evidently viewing the problem in its proper perspective, namely, from the standpoint of loss of function. The problem was considered from this angle at a time when vitaminology was still in its infancy. Today, thanks to the progress in plant physiology, it is possible to attack the problem experimentally.

The first step in the experimental procedure is to find a plant which reacts in the same way as animals, *viz.*, one which is unable to synthesize a vitamin or a similar substance and which therefore requires an exogenous source. Such a plant is a suitable experimental object. The fungus *Phycomyces Blakesleeanus* is ideal for this purpose (SCHOPFER, 1934).

This phycomycete grows well on a medium composed of natural products but does not grow on a strictly synthetic one (glucose 2%, asparagine 0.1%, $MgSO_4$ 0.05%, KH_2PO_4 0.15%). But if a trace of vitamin B_1 (thiamin) is added (0.5γ per 25 cc. of medium) to this synthetic medium, excellent growth is obtained. This plant has become heterotrophic in respect to thiamin.

The proof of this heterotrophism can be demonstrated in the following manner: the spores of the fungus evidently contain a small quantity of the growth factor received from the parent plant. This reserve supply produces no more than a short germ tube. If an extract is prepared from this minute thallus with avitaminosis and is added to a fresh thiamin-free synthetic medium as given above, it is found that this preparation does not support the growth of *Phycomyces*. This certainly is proof that no vitamin is present, that the feeble reserve furnished by the spores has been rapidly depleted, and that the capacity to synthesize a new supply of vitamin has disappeared. It can be proved that the vitamin (or an analogous factor) is actually present in the spores by preparing a *concentrated extract* from a large mass of spores (several millions), and by adding this extract to an inactive synthetic medium and by culturing *Phycomyces* on it. The abundant growth obtained shows that the vitamin is present in the spores, but that the quantity is so small that its action cannot be detected except in a concentrate prepared from the spores.

This experiment proves definitely that *Phycomyces* has lost the capacity to synthesize thiamin. Consequently this plant resembles animals in one of its physiological characteristics. It is not only heterotrophic from the general point of view (a saprophyte), but it is also heterotrophic in respect to vitamin B_1. Other related plants in this group of molds, *Absidia glauca* and *Rhizopus suinus*, are evidently able to live without this vitamin; actually, they are able to synthesize it. Thus they are autotrophic in this respect. Their extracts (obtained from cultures grown on a synthetic medium) when added to a fresh synthetic medium (inactive for *Phycomyces*) permit the growth of *Phycomyces*. The latter received the vitamin (growth factor) synthesized by *Absidia glauca* and *Rhizopus suinus* from the synthetic nutrient material.

This line of reasoning which we have applied to a chemically known vitamin, is equally sound when applied to any of the substances known as growth factors in the field of microbiology (see definitions, p. 195).

FISCHER (1906) established a special category in which he placed the heterotrophic bacteria which require the more complex nutrient materials furnishing special activators analogous to vitamins, and designated this group as *paratrophs*. In this connection we recall the early observations of PASTEUR concerning yeasts and lactic bacteria, and of WILDIERS concerning bios. By virtue of the demonstration that a chemically known vitamin with typical action on animals can also act as a growth factor for plants, the two kingdoms meet again in a definite manner, *i.e.*, the vitamins (in the strict sense) of animals correspond to the growth factors (nutrilites of R. J. WILLIAMS) of microorganisms. Both groups of substances act as catalysts, are effective in very minute quantities, involve the same research techniques, and have the same general characteristics; they are, in fact, one and the same thing. It will be necessary to expand considerably the vitamin concept in accordance with criteria destined to stand the test of time. From now on we shall consider the growth factors of microorganisms as vitamins in the proper sense of the word.

The fact that a loss of capacity for synthesis brings about the need for a special factor has long been known. We find this idea expressed even in some of the early work. TWORT and INGRAM (1913) studying two closely related microbes, KOCH's (human tuberculosis) and JOHNE's bacilli (bovine tuberculosis), made the following observations: JOHNE's bacillus is unable to grow on media which suffice for the development of KOCH's bacillus. On the other hand, JOHNE's bacillus can grow if the medium is supplied with some of KOCH's bacilli or simply a glycerine extract of these bacilli. This fact, which was at that time considered unique, can be explained in only one way, *viz.*, KOCH's bacillus is able to synthesize one or several substances which are indispensable for but not produced by JOHNE's bacillus. It is now known that the factor required by JOHNE's bacillus is vitamin K (WOOLLEY and MCCARTER, 1940).

The fact that the loss of capacity for a particular synthesis by an organism results in its becoming dependent upon a growth factor is definitely established by the investigations of P. FILDES, KNIGHT, LWOFF, SCHOPFER, and ONDRATSCHEK.

REFERENCES:-

1900: H. ZUMSTEIN. — Zur Morphologie und Physiologie der *Euglena gracilis*. Jahrb. Bot. 34:149.

1903: A. FISCHER. — Vorlesungen über Bakterien. Jena, G. Fischer.

1912: CH. TERNETZ. — Beiträge zur Morphologie und Physiologie der *Euglena gracilis* Klebs. Jahrb. Bot. 51:435.

1913: F. W. TWORT and G. L. Y. INGRAM. — A monograph on JOHNE's diseases (Enteritis chronica pseudotuberculosis bovis). London, Baillière, Tyndall & Co.

1921: M. MOLLIARD. — Nutrition de la plante. Formation des substances ternaires. Encycl. scientif., Gaston Doin, Paris, 438 pp.

1927: G. SENN. — Die Chlorophyll-Armut saprophytischer Orchideen . . . Verhandl. Naturforsch. Ges. Basel 38:516.

1928: P. JOYET-LAVERGNE. — Sur les rapports entre le nucléole, le chondriome et le glutathion. C. r. Soc. biol. Paris, 98:567.

1932: A. LWOFF. — Recherches biochimiques sur la nutrition des Protozoaires. Monographies de l'Institut Pasteur, Masson & Cie, Paris, 158 pp.

1933: A GUILLIERMOND, G. MANAGENOT et L. PLANTEFOL. — Traité de Cytologie végétale. Paris, Le François, 1195 pp.

1934: W. H. SCHOPFER. — Versuche über die Wirkung von reinen kristallisierten Vitaminen B auf *Phycomyces*. Ber. d. d. bot. Ges. 52:308.

1935: P. JOYET-LAVERGNE. — Contribution à la recherche de la vitamine A dans les cellules animales et végétales. C. r. Acad. Sc. Paris, 200:346.

1936: P. JOYET-LAVERGNE. — La démonstration expérimentale du rôle du chondriome dans la respiration. C. r. Acad. Sc. Paris, 203:1291.

1938: G. SCHAFFSTEIN. — Untersuchungen über die Avitaminose der Orchideenkeimlinge. Jahrb. Bot. 86:720.

1940: W. H. CAMP. — Aphyllous forms in *Pyrola*. Bull. Torrey Bot. Club 67:453.

1940: D. W. WOOLLEY and JANET MCCARTER. — Antihemorrhagic compounds as growth factors for the JOHNE's bacillus. Proc. Soc. Exptl. Biol. Med. 45:357-360.

1941: S. ANSBACHER. — *p*-Aminobenzoic acid, a vitamin. Science, 93:164.

1941: R. J. WILLIAMS. — The importance of microorganisms in vitamin research. Science, 93:412.

1941: H. K. MITCHELL, E. E. SNELL and R. J. WILLIAMS. — The concentration of "folic acid". Jour. Amer. Chem. Soc. 63:2284.

1941: R. J. WILLIAMS and coworkers. — Studies on the vitamin content of tissues I. The University of Texas Publication No. 4137. (Published by the University of Texas, Austin, Texas).

1941: A. GUILLIERMOND. — The Cytoplasm of the Plant Cell. Waltham, Chronica Botanica Co., 247 pp.

Chapter III.

THE EXPERIMENTAL STUDY OF GROWTH FACTORS AND THE SELECTION OF TEST PLANTS.

The methods employed in the study of vitamins and their actions in plants are, in the main, identical with those used in animal physiology.

Microbiology entered into an active period as soon as it was discovered that microorganisms can be grown on media in the absence of living hosts. The introduction of these methods into biology constitute the main contribution of BREFELD. This period was probably dominated by the misapprehension that an organism cultured on a medium composed of natural products was being supplied with known substances only and that the entire metabolism of the organism could therefore be controlled.

Further progress of unusual importance was made by the introduction of synthetic media prepared from known substances. This advance in technique is due chiefly to PASTEUR who employed a synthetic medium in the culture of yeast. This medium — synthetic in the sense used at that time — was very simple. Already at that time PASTEUR observed that a synthetic medium did not give as good results as one composed of natural products, and that the addition of small amounts of natural products to the synthetic medium resulted in a marked increase in growth. Thus it became evident from the beginning that synthetic media lacked something which could be supplied in small amounts.

The classic work of RAULIN (1870) on the culture of *Aspergillus niger* on a synthetic medium seemed to give the problem a substantial foundation. By supplying a carbon source (sugar or an organic acid), a nitrogen source, and mineral salts containing all the necessary elements, RAULIN succeeded in obtaining excellent growth of this fungus. But he also observed that the addition of yeast extract to the synthetic medium made it possible to obtain even better growth. Thus RAULIN had a vague conception of nutrients of unknown composition about a quarter of a century before animal and human physiologists became aware of such substances.

RAULIN's work probably suggested the possibility of culturing all microorganisms on synthetic media. But when attempts were made to extend the methods of this student of PASTEUR to other microorganisms it soon became obvious that this could not be done. On the contrary, as the constituents of synthetic media were purified more and more the resulting growth decreased correspondingly. In other words the medium employed by RAULIN was not synthetic in the modern sense of this term.

Ingredients of media such as mineral salts and organic substances (sugar), although considered chemically pure, contain a much larger quantity of essential growth substances than one might suspect. In this connection it might be well to share the opinion of STEINBERG, a leading specialist on the action of metallic catalyzers on microorganisms, that the use of actually pure media would result in no growth whatever. Thus the study of highly purified media led to the concept of indefinite nutrients which is still in need of further clarification.

What has just been said concerning microorganisms can likewise be applied to the higher plants. We have already analyzed the results of BOTTOMLEY and MOCKERIDGE according to which the growth of green plants is apparently accelerated by organic substances effective in small dosages and thus different from the nutrients ordinarily supplied.

Let us now examine briefly the precautions necessary in demonstrating a growth factor of vitamin nature as defined by the author. All sources of error must be eliminated. This can be done by observing a very simple rule: the medium supplied to the plant must be truly optimal and of such a composition that *only* the deficient factor can be responsible for the absence of growth. Failure to follow this rule will often lead to false assumptions. For example, the failure of an organism to grow might be attributed to a lack of growth factors when the real cause is of an entirely different nature (physico-chemical properties of the medium or mineral catalysts).

Physico-chemical properties of the medium. — A culture medium for microorganisms or higher plants must not only contain all essential ingredients in correct amounts and proportions, but it must also possess certain physico-chemical properties necessary for the growth of the particular organism, *i.e.*, *p*H, *r*H, viscosity, and surface tension.

*p*H. — A discussion of the effect of the *p*H of the medium is unnecessary in view of the numerous well known cases, *viz.*, an unfavorable *p*H, too acid or too alkaline, prevents growth even if the medium is favorable in other respects.

*r*H. — KNIGHT and FILDES (1930) showed that the germination of spores of *Bacillus tetani* can be inhibited completely by an unfavorable redox-potential. One might be tempted to attribute this inhibition to the absence of a growth factor, but actually it is due entirely to an unfavorable redox-potential (*cf.* HEWITT, 1937).

The *r*H can be adjusted by substances which have the same characteristics as vitamins, or even by vitamins themselves. In such redox systems vitamins can act exclusively on the *r*H of the culture medium, affecting the organism externally without taking part in its metabolism. The following list gives a scale of rH_2 values obtained by means of various vitamins.

TABLE II. Growth factors as redox systems (from JANKE, 1939).

substance	*p*H	T	Eo^1	rH_2	active group
codehydrogenase (factor V)	7.0	30	—315	3.5	nicotinic acid*
riboflavin	7.0	30	—210	7	isoalloxazine**
ascorbic acid	7.0	30	—54	12.2	dienol group
glutathione	7.0	30	?	?	SH— group
cytochrome	7.0	30	+123	18.1	hemin
adrenaline	7.0	30	+380	26.7	dioxyphenyl

*constituent of a codehydrogenase
**constituent of the "yellow enzyme"

These actions in which the effect of a vitamin is confined strictly to the adjustment of the *r*H simulate the action of a true growth factor. Actually it is correct to speak of the action of a growth factor only if the vitamin is absorbed and participates in the metabolism of microorganisms or higher plants. In some cases it is difficult to distinguish between the action of a true growth factor and that which is strictly physico-chemical and external.

Viscosity and surface tension. — These characteristics are much more important than they might appear to be at first. In their work on *Azotobacter,* RIPPEL and B. LEHMANN (1936) observed that an addition of 0.1 per cent agar to a mineral medium visibly favors the fixation of nitrogen and also the production of living material (determined on the basis of dry weight). It was found that the agar acted, neither by furnishing a specific substance to the medium nor by adsorbing toxic inhibiting substances, but by making oxygen more accessible through its colloidal properties.

Agar has a similar effect on *Aspergillus niger.* When grown without agar, the mycelium of this fungus is restricted to the margin of the medium in contact with the culture flask, but when grown with agar, a thick fungus mat is produced over the entire surface of the medium, indicating that surface tension plays an important part.

All experiments concerning extracts of agar obtained by means of various solvents led to negative results, hence the action of the agar in these cases must have been due to its physico-chemical properties.

There are many other cases of similar nature, *e.g., Trichophyton (Oospora) albicans* is favored by a reduction of the surface tension of the medium (VON HAHN, 1926).

Surface tension and viscosity, being related to the degree of dispersion of the medium, are therefore of fundamental importance.

Action of mineral substances. — Mineral substances act not only as plastic materials for the formation of living matter, but they also play a catalytic rôle, acting in very small dosages. The latter action is still erroneously called oligodynamic, since NÄGELI, who discovered the phenomenon, intended this term for an inhibitory or toxic action produced by a trace of metal.

The action of metallic catalyzers has become well known since the observations of RAULIN concerning the effect of zinc on *Aspergillus*, BERTRAND and JAVILLIER — manganese on *Aspergillus*, RICHARDS — thallium on yeast, GOLLMICK — copper on *Aspergillus*, BORTELS — molybdenum, vanadium, and tungsten on *Azotobacter*, STEINBERG (1936) — molybdenum on *Aspergillus*, and of many other investigators. [Concerning the action of mineral elements, *cf.* STEINBERG (1936), PIRSCHLE (1938-39), and NEIPP (1937)]. The account by STEINBERG concerning molybdenum and its action on *Aspergillus* is particularly instructive.

STEINBERG (1936) observed that the growth of *Aspergillus niger* was favored by certain samples of impure sucrose. It was logical to suspect an active substance in the sugar. The peculiar action of the impure sucrose disappeared upon extraction by alcohol, as if the latter had removed a growth factor. However, in place of an organic growth factor (vitamin), a metallic element was found responsible. Of 55 elements studied, STEINBERG found that molybdenum acted as the growth stimulating substance when supplied at the rate of 10γ per liter of solution.

This case is typical because of the fact that all extraction methods employed might have led to the belief that the action was due to a vitamin extracted by alcohol. It shows that one can not be too cautious in dealing with this subject.

Furthermore no misapprehensions should be entertained regarding the purity of nutrient media, since they always contain many more ingredients than expected. Impurities are present in traces even in doubly distilled water. G. LOHMANN has shown that an animal charcoal treatment of the distilled water used in the preparation of media resulted in poorer cultures than those grown with untreated water (*Aspergillus niger*). Distilled water fresh from the still contains traces of zinc which produce excessive development in the control cultures. Furthermore, glass containers liberate elements during the course of sterilization (LASSEUR and GIRARDET, and ONDRATSCHEK). Such substances may also be introduced during inoculation. The most important sources of trace elements are the impurities of the ingredients of the medium. STEINBERG has made detailed spectroscopic studies of the mineral substances present in the well known Pfeffer medium. It contains, in addition to glucose, 7 mineral salts: NH_4NO_3, KH_2PO_4, Mg SO_4, $ZnSO_4$, $CuSO_4$, $MnSO_4$, Na_2MoO_4. The following cations were detected spectroscopically:

in NH_4NO_3: Na, Mg, Ca, K(?).
in KH_2PO_4: Al, Pb, Na, Cu, Mg, Ag.
in $MgSO_4$: Na, Cu.
in $ZnSO_4$: Fe, B(?), As, Mg, Sn(?), Cu, Si, Na, Mn.
in $CuSO_4$: Fe, Mn, Si, Mg, Cu, Pb.
in $MnSO_4$: Na, Fe, Cu, Al, V, Cr, Si, Mg, Ca.
in Na_2MoO_4: Cu, Mn, Fe, Al, Ni, Mg, Ca, K, Na, Mn, Si, Li, V(?).
in glucose: Li, Na, Sr, Ca, Rb, K, Mn, Al, Fe, Rh, Ni, Ag, Cu, Mg, Sn, B, Si.

A certain number of these cations is definitely known to have a favorable action upon the growth of several microorganisms. The importance of mineral substances with catalytic action should not be underestimated. It is known that they play an important rôle in metabolism, *viz.*, as coenzymes, taking part in a fermentation reaction. Zinc is known to raise the economic coefficient of *Aspergillus niger* and to lead to a better utilization of sugars. It must therefore participate in the enzymatic synthesis of polysaccharides (PORGES). The investigations of WAKSMAN and FORSTER have shown that in *Rhizopus nigricans* zinc participates in the production of organic acids.

The importance of mineral elements in culturing microorganisms and higher plants is evident also from a great number of other investigations, not all of which can be cited here. Nevertheless, attention should be called to the brilliant investigations of NIELSEN and his collaborators on *Aspergillus niger*. They found that growth factor B requires a mineral co-factor present in the ash of filter paper and other substances. This co-factor can be replaced by a mixture of 20 metals, in the form of salts, in precise and optimal proportions. The recent investigations of UTIGER and SCHOPFER (Act. Soc. Helv. Sc. Nat. 1941) have demonstrated the great importance of mineral catalyzers (present as impurities in the organic constituents of the medium) in the synthesis of vitamins by *Rhodotorula rubra* and *Mucor Ramannianus*. The author groups the mineral substances together and calls them pseudo-growth-factors (see p. 38). They act alone, or as catalyzers of a growth factor, or as coenzymes (detailed bibliography in SCHOPFER, 1939).

Various actions simulating those of growth factors, or interfering in their study. — The substances which fall in this category are not to be underestimated. It has been known since the work of NIELSEN and HARTELIUS that the heat used in autoclaving a solution of ammonium tartrate and glucose leads to the formation of a synthetic factor of the "B group" (B_2) having the same action as that of the natural B_2 factor on *Aspergillus niger*. Thus in heating a medium containing a natural substance in which ammonium tartrate and glucose are present, one unwittingly can produce a substance acting as a growth factor of the type found by NIELSEN. Recently DAGYS (1940) observed the production of the same factor upon heating ammonium malate and fructose. This factor was also produced by heating sap from birch wood which undoubtedly contains malic acid, ammonium, and fructose. Similar observations have also been made by FULMER, WILLIAMS and WERKMAN, FULMER and HUESSELMANN, KOSER, FINKLE, DORFMAN and SAUNDERS.

It is likewise necessary to consider the possibility of destroying thermolabile growth factors by sterilization, and also the possibility of forming inhibiting substances as demonstrated by MIRIMANOFF in his work on yeast. The latter substances can lead to the belief that a growth factor is required although this may not be the case at all.

Such inhibiting substances can also be introduced into the medium by natural substances used in making the medium. The author has frequently observed cases in which the treatment of a medium by an adsorbant eliminated the inhibitor and restored the favorable qualities of the medium. In this case the improvement in the medium was accomplished without adding any growth factor. If one is working with a medium which has not been treated by an adsorbant and is dealing with an auxo-heterotrophic (heterotrophic for growth factors) organism, it is obvious that no effect can be produced by the addition of growth factors.

It has also been observed that growth factors may be masked by being bound in such a manner that their effects do not manifest themselves. A case of this kind involving a combination of factors was demonstrated by using yeast as the test object. The factors were combined with proteinaceous material, and had to be liberated by enzymatic splitting before their presence could be demonstrated.

Aside from growth factor deficiencies, the causes which are responsible for failures to obtain growth may be summarized as follows:

In the case of auxo-autotrophic organisms:

1) a medium sufficient in all respects containing:
   - a) preformed inhibiting substances.
   - b) inhibiting substances formed during the preparation of the medium.
2) a medium insufficient because of conditions preventing the synthesis of growth factors. In this case the organism becomes conditionally auxo-heterotrophic.

In the case of auxo-heterotrophic organisms:

- a) growth factor masked by being adsorbed or bound.
- b) inhibiting substances, preformed and not removed.
- c) inhibiting substances formed during the preparation of the medium. The inhibiting substances referred to under b) and c) interfere with metabolism in a general manner.
- d) substances which inhibit specifically the effect of the growth factor.
- e) growth factor destroyed during the preparation of the medium.

This list of causes for lack of growth is applicable to both microorganisms and higher plants growing on either synthetic or natural media.

In order to identify a substance as a true growth factor it is necessary to proceed as follows:

1) Observe that a given organism grows well on a medium composed of natural products.
2) Observe that the organism does not grow on a synthetic medium purified as much as possible.
3) Observe that the organism can be made to grow satisfactorily on such a highly purified medium by adding a very small quantity of a natural substance such as fruit, milk, extracts of plant or animal organs, or by adding some product such as sugar containing an impurity despite its

reputation to the contrary. The amount of active material should apparently be too small to furnish energy or ordinary plastic material.

4) Concentrate the active factor contained in the natural or contaminated substance, eliminating such products as fats, sugars and proteins and thereby ruling out the possibility of their action.

5) Eliminate carefully all physico-chemical factors (*p*H, *r*H, surface tension, viscosity, etc.) capable of inhibiting the culture and likely to lead to the false assumption of the absence of a true growth factor.

6) Devote particular attention to the mineral substances. The ash of the concentrate must be proved inactive. This is important but not conclusive since the mineral element might act only in conjunction with an organic material destroyed during incineration. Only after the above steps have been completed is it possible to proceed with the purification and isolation of the active substance present in the concentrate. The final step required to complete the investigation is to determine the mechanism of action of the active substance.

It must be proven conclusively that the organism is completely and definitely auxo-heterotrophic. As already seen (p. 20), it is possible to have conditions which make an organism appear completely auxo-heterotrophic although it is actually able to synthesize its own growth factor or factors if properly cultured. In this connection the possibilities for research are unlimited. In the case of an organism which can not be cultivated under any of a thousand different conditions, it is difficult to determine whether there might still be one set of conditions under which the synthesis of growth factors might be possible.

Sources of growth factors. — Several methods of approach are possible in studying the sources of growth factors.

Chemically impure products. — In 1930-31 the author found that the only sugar capable of producing growth of *Phycomyces Blakesleeanus* on a synthetic medium was a certain brand of maltose (Kahlbaum, Berlin). There was no reason to suspect maltose itself of being endowed with any unusual properties since glucose was known to be ineffective. More intensive investigations (SCHOPFER 1931-32) clearly demonstrated that this sugar, although sold as chemically pure, contained an impurity of organic nature which could be removed by adsorption with animal charcoal. This affinity for animal charcoal was responsible for the identification of the substance as thiamin (vitamin B_1). It was logical to suspect impure saccharose of containing the same substance. By studying all the stages of preparation of this sugar, beginning with the sugar beet, the author, working with *Phycomyces,* was able to determine the exact stage at which the sugar no longer exhibited activity through impurities. Similar observations have been made concerning the growth of yeast (WILLAMAN and OLSEN). In the latter case the commercial sugar sold in grocery stores proved to be very pure. However, successive purifications by means of alcohol brought about a progressive decrease in the size of the crop as long as any bios factors remained. When these factors had all been removed, no further decrease in the crop was obtained.

It is necessary to consider impurities accompanying other ingredients used in making up media. Commercial peptone (ORLA-JENSEN, OTTE and SNOG-KJAER, 1936) contains growth factors for bacteria. The same is true of gelatin (KOSER and SAUNDERS, 1938), agar (ROBBINS, 1939, ROBBINS and MA, 1941), and even crude cotton frequently used as plugs for culture flasks (SCHOPFER and RYTZ, 1937). Ordinary asparagine contains impurities (bios?) favoring the growth of flagellates (LWOFF and DUSI).

Substances furnished by another organism. — Cases of unilateral stimulation of one organism upon another, or of reciprocal stimulation of two organisms (artificial symbiosis, MUELLER and SCHOPFER, KÖGL and FRIES) have frequently led to the discovery of growth factors of vitamin nature (see p. 245, Chapter XXIII). The *Polyporaceae,* in which growth is accelerated by or is dependent upon bacterial products, present a typical example of this phenomenon. This observation led N. FRIES (1938) to investigate their growth factor requirements which were found to be thiamin and bios. This phenomenon is explained by the fact that the products of an auxo-autotrophic organism contain the factors required by an auxo-heterotrophic one. Cases of this kind are frequent in the older literature but they could not be explained at the time they were reported. Innumerable cases have been reported concerning the acceleration or initiation of growth of microorganisms and higher plants by crude yeast extract.

Extracts of plant and animal organs. — Microorganisms and their diffusion products are not the only causes responsible for accelerating the growth of auxo-heterotrophic organisms since various organs and their extracts are likewise effective. Extracts from liver and other animal glands have a marked growth promoting action on *Phycomyces* grown on a synthetic medium. In general the growth factors of the *Mucoraceae* can be demonstrated in a great number of leaves, stems, and roots, in fact, they appear to be ubiquitous and are found in both kingdoms.

Bios II (biotin), which is so important for the growth of yeast, was originally extracted from egg yolk and not from yeast. It is also present in animal tissues.

Liver extracts furnish *Corynebacterium diphtheriae* with an indispensable growth factor which has been identified as pimelic acid. It can also be extracted from cows' urine (MUELLER, 1937).

The situation is similar in respect to factors X and V which are required by the hemophilic bacteria and are normally supplied by blood. These factors were first found in plant and animal extracts.

In later chapters in which the various groups of organisms will be discussed, there will be occasion to point out the original as well as other sources of the growth factor involved. The abundant information now available on the vitamin requirements of plants permits a detailed treatment of the subject. This discussion will be followed by a consideration of the mechanism of the action of growth factors. An understanding of the mechanism will permit a more precise definition of these substances.

The preparation of concentrates and the isolation of growth factors. — After the plant or animal material has been chosen and the extraction has been completed, the next step is to concentrate the active substance into smaller and smaller volumes. The final step is the isolation of the growth factor in chemically pure crystalline form. Although this is the work of a chemist, the biologist can proceed until a highly active concentrate is obtained.

It is impossible to summarize here the methods of preparation for all growth factors. Certain general methods of separation frequently employed in a number of techniques may, however, be discussed at this time. A very serviceable method is that of adsorption. Adsorption by animal charcoal when applied to a crude aqueous extract permits the separation of various water soluble factors of the "vitamin B group". Likewise, adsorption with Fuller's earth at a *p*H of 4.62 leads to the separation of thiamin. These techniques permit the biologist to prepare concentrates with which he can work (see STRAIN, 1942).

Precipitation reactions permit the separation of active substances from a large mass of inactive material.

The work must be quantitative in every respect. If possible, each step in the preparation of the concentrate should be accompanied by an assay of the exact amount of growth factor contained in the extract or the concentrate. These operations always result in a loss of the growth factor and the final yield is often as low as one or two percent.

Even when working with an impure substance its general properties can be determined: various solubilities, ultrafiltrability, diffusion, and resistance to dry or moist heat, autoclaving, acids, bases, and oxidizing agents. For a long while these general properties were used to distinguish the groups of active substances. We still speak of the "B substances" of NIELSEN which are distinguished from the "A substances" (auxins) by means of their solubilities, the former being soluble in water, the latter soluble in ether and water. The "B substances" can be subdivided into two groups, B_1 and B_2, on the basis of oxidation and adsorption. These general characteristics usually do not permit the identification of growth substances, but they are helpful in guiding the investigator.

One method by which a growth factor can be identified is to concentrate and isolate the active substance. To date this method has resulted in the identification of only a few growth factors. For example biotin was concentrated and isolated in crystalline form by KÖGL and TÖNNIS (1936). In a similar manner nicotinic acid and its amide were identified as growth factors by KNIGHT by the action of their concentrates on *Staphylococcus aureus*. In most cases growth factors cannot be readily isolated and identified in this manner, and an indirect method must be employed. The active substance is concentrated and the characteristics of the concentrate are compared with those of chemically known substances. The latter with characteristics similar to those of the concentrate are tested at random to determine whether any of them have the same growth promoting activity as the concentrate. Sometimes a known substance is found with the same activity as that of the unknown factor and thereby its identity is established. The latter method was employed in the identification of thiamin as the active factor for *Phycomyces Blakesleeanus.* The characteristics of the substance in Kahlbaum's maltose (p. 25) was such that it led unmistakably to the group of "vitamin B substances" which were still incompletely known in 1931. When crystalline thiamin (WINDAUS *et al.*) became available the first tests with it showed very definite activity on *Phycomyces.* In this case the method of procedure was purely intuitive and the pure active substance (thiamin), which was used by chance, was not proven to be present in the active concentrate. Even today there is no definite proof that thiamin or its components were present in the samples of Kahlbaum's maltose. The important thing is the finding of a growth factor of vitamin nature with the same activity as that of the concentrate and the ability *to replace it completely.*

We have already mentioned the fact that most organisms require several growth factors. In a few cases, however, a *single* growth factor suffices and its action is able to replace completely that of a natural substance (milk, extracts of organs, of fruit, etc.) or a concentrate thereof. Sometimes a factor permits only a very weak growth as compared with that produced by a natural substance indicating that other essential growth factors or nutrients are present in the natural product. Suppose we consider a theoretical case involving such an organism on which a single isolated growth factor has only a limited effect. The cultural conditions are optimal, hence it is certain that the substances which are lacking are growth factors. Yeast is a typical example of an organism illustrating this phenomenon. In order to study such an organism the products eliminated during the isolation of the first factor must be investigated. If this factor was removed by precipitation it is necessary to study the other factors remaining in the liquid in which the crystallization occurred; if the first factor was removed by adsorption the investigations must be continued with the filtrate of the adsorbate. Before attempting to identify the factors present in the filtrate the activity of the latter must be verified. This may be done as follows: the amount of growth produced by the natural substance added to the basal medium is observed and designated as 10; the amount of growth produced by the basal medium plus the first factor isolated from the adsorbate is given a comparative rating, for example 2; the activity of the first isolated factor plus the filtrate of the adsorbate is then ascertained. If these two fractions have the same activity as the natural substance their rating becomes 10, proving that other factors are present in the filtrate. Yeast extract is a good example of a natural substance containing a multiplicity of factors. The standard solution (impure yeast extract), added to the basal medium, and supplied to a culture of yeast produces a growth curve representing the maximal development of this organism. When inositol alone is used a much poorer growth is produced. Other factors when used singly likewise have a very low growth promoting activity. On the other hand, the combination, inositol + biotin, gives a curve approaching that of the standard solution. Upon the addition of a third factor, thiamin, the curve obtained becomes more nearly like, but not identical with, that of the standard. In this particular case it can be assumed that almost all of the growth factors required by yeast have been ascertained (see p. 132). The first factor is only partially active, being limited by a deficiency of the second factor. If the second factor is added in sufficient quantity growth is somewhat increased. At this point the third factor, when deficient, becomes a limiting factor. If this factor is supplied in sufficient quantity growth is increased still further. Thus it is possible to go as far as a fourth or fifth factor. A difficult case is represented by the formation of roots on etiolated cuttings of *Pisum* on which the number of roots produced increases as the complexity of the constellation of growth factors is increased: saccharose < saccharose + auxin < saccharose + auxin + biotin < saccharose + auxin + biotin + factor X (*Cf.* WENT and THIMANN, 1937). Factor X represents the undetermined substance required for the maximum production of roots.

The two cases described above are relatively simple in that the individual factors of the constellations have an additive effect. In other cases, the factors when used singly have practically no effect, and it is necessary to use combinations of two or three factors in order to obtain any appreciable amount of growth. A case somewhat similar to this is illustrated by *Phycomyces* in its vegetative development and formation of zygotes, the former requiring only one growth factor, the latter requiring a combination of factors. The vegetative development produced by thiamin is almost as good as that produced by a natural substance such as yeast extract, but thiamin is unable to provide for the maturation of zygotes. It lacks something. ROBBINS has shown that this "something" is found in agar or in potato tubers from which it can be extracted by pyridine or methyl alcohol. The methyl alcohol extract alone is ineffective in producing zygotes; it must be used in conjunction with thiamin in optimal dosage. The author has confirmed ROBBINS' results.

Nematospora Gossypii, studied by KÖGL and FRIES, is another example of an organism on which growth factors are only slightly effective when used singly. Inositol, biotin or thiamin alone have only a very weak effect. On the other hand, inositol and biotin when used together, are far more effective than when they are added separately. Additional activity is produced by adding thiamin to the combination inositol + biotin (p. 134).

All degrees of intergradation occur between the two cases, *i.e.,* factors having strictly additive action and those having no activity except when used in combination.

In closing this technical and didactic discussion, it is well to remember that a mere tabulation of results is unsatisfactory unless the reader learns how the data were obtained.

Measurement of the action of growth factors. — For preliminary work a qualitative estimate of the activity of a growth factor made by observation is sufficient; later, however, it must be supplemented by an accurate quantitative measurement. It is important to establish the ratio between the dry weight of the crop and the weight of the vitamin. When the investigation concerns the growth of higher plants the *lengths* of shoots, roots and leaves serve as useful criteria. However, the better criterion of growth is always the *dry weight* obtained by desiccation at 105°C. Of course, one must be sure that the desiccation does not entail any destruction of material in the organisms studied. For the unicellular microorganisms and particularly the coenocytic organisms such as the mucoraceous fungi, the determination of the dry weight is absolutely necessary. This means of measuring growth is often difficult in the case of unicellular microorganisms and it is permissible to measure their growth by means of a nephelometer or by *counting the number of cells formed.* Various types of nephelometers have been proposed. LANGE's model, employing a photoelectric cell, is very serviceable and can be used to measure the growth of cultures in small quantities of media (*cf.* MILATZ and ROTTIER). It must be remembered that the relationship between the nephelometric values and the weight of the living matter conditioning the opacity of the medium is not always exact. The absorption of light depends not only on the quantity of cells present, but also on their color, their opacity, and the reserve material which they contain. The author, working with *Rhodotorula rubra,* has found good agreement between the weight of cultures and the nephelometric values obtained by various dosages of growth factors. Although the nephelometric values are not always absolutely quantitative they still have relative value and can be used to follow the growth of a culture. In coenocytic organisms such as the mucoraceous fungi with their compact structure, the only reliable criterion for growth is the *dry weight* of the culture. Unicellular microorganisms present two distinct phenomena, *i.e.,* an increase in the amount of dry matter and an increase in the number of cells. Sometimes these two phenomena proceed simultaneously but need not do so necessarily since they are conditioned by different factors. Consequently it seems desirable in microorganisms to determine both the dry weight and the number of cells produced.

Growth curves. — It is possible to determine both the amount and the rate of growth and to construct respective growth curves: 1) One is a growth curve based on time. The same amount of vitamin is used in all of the experimental flasks; it may be optimal or otherwise. At regular intervals a culture is harvested and weighed. 2) The other type of curve shows the relation between the amount of growth produced and the amount of vitamin supplied. Each experimental flask contains a different amount of vitamin. After the cultures have completed their development their weights are determined. This type of curve demonstrates the optimal vitamin dosage, *i.e.,* the least amount for maximal growth. It is represented on the curve as a maximum beyond which further additions of the vitamin are ineffective. In all experiments which are conducted properly both types of curves must be prepared. It is evident that these curves are also functions of the composition of the medium. A complete study implies that they have been constructed on the basis of several different cultural conditions. The optimal dosage of a growth factor naturally depends on numerous chemical and physical factors: composition of the medium, depth and surface area of the medium, the shape of the vessel, etc. This optimum is of value only when the conditioning factors are known as far as possible. The principal task of the biologist is the determination of the optimal dosage under conditions which are such that the growth factor is utilized to the fullest extent. In other words, this optimum must be such that it is the quantity of the growth factor which plays the rôle of the limiting factor (see *Phycomyces*, p. 102). Whenever possible, it is desirable to construct a three-dimensional graph showing the weights obtained at the ends of various growth periods with various dosages of the growth factor. Such graphs are highly instructive. They have been prepared by VAN HASSELT for yeast, by SCHOPFER for *Phycomyces*, and by SCHOPFER and BLUMER for *Ustilago violacea*. They are valid for any given set of conditions.

Other methods of determination. — Other criteria of growth can be utilized. The rate of respiration determined by the evolution of CO_2 has been used for yeast. The nitrogen assay has been employed by J. H. MUELLER *et al.* to measure the growth of cultures of *Corynebacterium diphtheriae*. Such methods must, however, be used with discretion. They take into consideration only one function, which often has only an indirect relationship to the growth factor, resulting in misleading figures. The ideal method of determining growth would be to measure only the function directly related to growth. Since the function of a growth factor is that of a coenzyme the ideal method would necessitate the accurate measurement of the enzymatic reaction in relation to this coenzyme. Since this is often difficult or impossible it is preferable to use the dry weight as an expression of the rate of all vital functions.

In closing this chapter it is well to point out that the determination of the relationship, *weight of crop / weight of growth factor*, should be carried out under as precise a set of conditions as possible (see p. 111).

REFERENCES:-

1870: J. RAULIN. — Etudes chimiques sur la végétation. — Thèse sciences, Paris.

1926: F. V. VON HAHN. — Über die Einwirkung oberflächenaktiver Substanzen auf die Wachstumsformen des Soor neonatorum. Kolloidzeitschr. 40: 321.

1930: B. C. J. G. KNIGHT. — Oxydation-reduction studies in relation to bacterial growth, III. The positive limit of O-R. potential required for the germination of *B. tetani* spores in vitro. Biochem. J. 24:1496.

1931: W. H. SCHOPFER. — Recherches expérimentales sur la formation des zygotes chez *Phycomyces Blakesleeanus.* Influence des substances vitaminiques. Bull. Soc. bot. suisse 40:87.

1932: *Idem* 41:73.

1936: A. RIPPEL und B. LEHMANN. — Über die Wirkung von geringen Mengen von Agar auf Wachstum und Stickstoffbindung von *Azotobacter* und auf andere mikrobiologische Vorgänge. Arch. f. Mikrobiol. 7:210.

1936: F. KÖGL und B. TÖNNIS. — Über das Biosproblem. Darstellung von kristallisiertem Biotin aus Eigelb. Zeitschr. physiol. Chem. 242:43.

1936: R. A. STEINBERG. — Relation of accessory growth substances to heavy metals, including Molybdenum, in the nutrition of *Aspergillus niger.* J. Agric. Res. 52:439.

1936: S. ORLA-JENSEN, N. C. OTTE und A. SNOG-KJAER. — Über Wuchsstoffe der Peptone. Centralbl. Bakt. II, 94:452.

1937: W. H. SCHOPFER et W. RYTZ, JR. — La ouate comme source de facteur de croissance de microorganisme. Arch. f. Mikrobiol. 8:244.

1937: F. W. WENT and K. V. THIMANN. — Phytohormones. New York, Macmillan Company, 294 pp.

1937: L. NEIPP. — De l'influence de divers cations sur le croît microbien. Paris, Masson, 430 pp.

1937: L. F. HEWITT. — Oxydation-reduction potentials in Bacteriology and Biochemistry. London County Council.

1937: B. C. J. G. KNIGHT. — The nutrition of *Staphylococcus aureus*: Nicotinic acid and vitamin B_1. Biochem. J. 31:731.

1937: J. H. MUELLER. — Pimelic acid as a growth stimulant for *C. diphtheriae.* J. Bacter. 34:164.

1938: S. A. KOSER and F. SAUNDERS. — Gelatin as a source of growth-promoting substances for bacteria. J. Bact. 36:57.

1938: N. FRIES. — Über die Bedeutung von Wuchsstoffen für das Wachstum verschiedener Pilze. Symbolae botanicae Upsalienses, III, 2.

1938: K. PIRSCHLE. — Die Bedeutung der Spurenelemente für Ernährung, Wachstum und Stoffwechsel der Pflanzen. Erg. Biol. 15:69.

1939: *Idem* 17:255.

1939: A. JANKE. — Die Wuchsstoffrage in der Mikrobiologie. Centralbl. Bakt. II, 100:409.

1939: W. H. SCHOPFER. — Vitamine und Wachstumsfaktoren bei den Mikroorganismen. Erg. Biol. 16:1.

1939: W. J. ROBBINS. — Growth substances in agar. Amer. J. Bot. 26: 772.

1940: J. DAGYS. — Untersuchungen über die Wuchsstoffe des Birkensaftes. C. r. Trav. Lab. Carlsberg, série physiologique, 23:1.

1941: W. J. ROBBINS and ROBERTA MA. — Biotin and the growth of *Fusarium avenaceum.* Bull. Torrey Bot. Club 68:446.

1942: H. STRAIN. — Chromatographic Adsorption Analysis. New York, Interscience Publishers, Inc., 222 pp.

Chapter IV.

CLASSIFICATION AND DEFINITION OF ACTIVE SUBSTANCES.

Classification, terminology, and synonymy. — The classification and nomenclature of growth factors are indeed difficult subjects to handle. At present it is impossible to construct a tenable system of classification of active substances since our information concerning their chemical nature, function, and mechanism of action is still quite incomplete. Despite this lack of knowledge, various authors have attempted to arrange the active substances in a more or less logical and valid system. Hence it is desirable to bring these classifications together, to define the various terms and establish their synonymy.

Vitamins and Hormones. — According to the classic definitions, a vitamin is a substance with catalytic action which is indispensable to animals. Since it cannot be synthesized by animals and must be furnished by plants, its origin is exogenous as far as animals are concerned. By comparison, a hormone is a product of animal metabolism, also effective in very small amounts, produced by special endocrine glands and active on the organism itself in a very specific manner. In its relation to animals a hormone is thus endogenous in origin. Since specific animal hormones, *e.g.*, estrone (theelin), are synthesized in appreciable quantities in plant tissues where they certainly must play a rôle, and vitamins likewise participate in the metabolism of those plants in which they are formed, this distinction is obviously untenable. These definitions which originated in the field of animal physiology can not be used in the field of general physiology.

All definitions proposed so far have been designed principally to distinguish the active substances ("Wirkstoffe") from ordinary nutrient materials. This distinction is not always easy to make, because the mineral elements are, in the last analysis, comparable with the active substances of organic nature. They (Zn, Mn, Mo) activate the metabolism of aerobic and anaerobic bacteria and likewise participate in the metabolism of higher plants. Manganese by itself acts as a coenzyme and, as we know now, the mineral substances which are active in small dosages are undoubtedly related physiologically to vitamins. Zinc is a co-factor of certain organic growth factors. VON EULER (1930-32) and MITTASCH (1936) place all of the highly active substances, organic and inorganic, in one group and call them *biocatalyzers*. This term is very appropriate, but its scope is too wide. More recently VON EULER (1938) proposed for vitamins and hormones the term "ergone" to be used in all cases in which it is not easy to distinguish between these two categories of substances. In order to understand the nature of "ergones" it is necessary to consider the constitution of enzymes.

An enzyme is now known as a complex substance made up of a colloidal protein "carrier" ("apoenzyme", "Träger") and an active group (the "prosthetic" group, "Wirkgruppe", or coenzyme):

$$\text{Holoenzyme} \rightleftarrows \text{coenzyme} + \text{apoenzyme.}$$

A certain degree of dissociation and equilibrium exists between the coenzyme and the protein "carrier". The active group has been identified chemically for a few enzymes, *e.g.*, thiamin is the active group in cocarboxylase. The coenzymes must therefore be considered as equivalent to "ergones", which will be designated as "*ergozymes*" because of their enzymatic nature. The "ergozymes" include the *vitazymes* and *hormozymes.* These terms designate the vitamins and hormones which function as coenzymes and must be combined with the proper "carrier" in order to become active. The main object of this classification is to connote the dynamic nature of these substances. Vitazymes and hormozymes move in a cycle in the organism, that is, they are liberated from the "carrier" and are recombined with it whenever necessary. AMMON and DIRSCHEL (1938) used the term *ergines* to include the sum total of the active substances and subdivided them into two groups: 1) the *ergones* (vitamins and hormones of VON EULER), acting on living systems, and 2) the *enzymes,* activating chemical reactions and acting on non-living systems. The ergones are stimulants (*Reizstoffe*), while the enzymes are catalyzers. Typical vitamins act as coenzymes and are effective only *in vivo,* whereas the catalytic action of an enzyme occurs *in vitro.* This system of classification is purely artificial and is not very satisfactory since it is difficult to distinguish between living and non-living systems and to separate stimulants (*Reizstoffe*) from enzymes (catalyzers) on that basis. MITTASCH (1936) examined the nature of catalyzers and stimulants (*Reizstoffe*). The distinction could be maintained only as long as "Reizstoffe" were considered as having certain physiological characteristics, *i.e.,* a "Reizstoff" was the unknown substance, acting in small amounts and initiating or conditioning a physiological process in a living system. Many of the substances formerly known as Reizstoffe are now actually chemically known. Some years ago the most that could be said was that a microorganism did not thrive on a synthetic medium unless a specific stimulant was present, which alone was capable of initiating growth. This stimulant lost its mysterious nature when it was learned that thiamin, for example, could function in that capacity, *i.e.,* as the coenzyme of carboxylase, and now we simply say that an organism which has lost its capacity to synthesize this coenzyme must have it supplied as a growth factor. Coenzymes, which according to the definition must act on non-living enzymatic systems, are indispensable to living organisms. JANKE (1939) made the same mistake as AMMON and DIRSCHEL in attempting to distinguish between ergones and enzymes by claiming that the former act on living protoplasm. This distinction is entirely erroneous. On the basis of the above analysis we arrive at the following outline:

- biocatalyzers
  - inorganic: Mn, Zn, Mo, etc.
  - organic (ergines):
    - enzymes
    - ergones
      - vitamins
      - hormones[1]

This classification has the advantage of including all of the phenomena of biological catalysis. We shall return to it later.

Other classifications. — G. K. LINK, H. A. WILCOX and A. D. LINK (1937) proposed the term auxone. Thus we would have *autoauxones,* if the substances are synthesized by the organism (auxoautotroph), and *heteroauxones,* if the substances must be supplied because the organism is unable to synthesize them (auxoheterotroph). In contrast to the preceding classification, this one is strictly physiological and based solely on the capacity of the living cell for synthesis. So far this classification has been used very little.

VON EULER (1932) proposed the symbol BP for all water soluble factors required by plants (vitamins of the B group) on the assumption that the growth factors of microorganisms and plants are related to vitamin B_1. This symbol was never used extensively and lost its significance when it was found that the factors called BP were not specific for plants but were, for the most part, ubiquitous and universal catalyzers.

In 1937 NIELSEN and HARTELIUS proposed the following classification, based on both chemical and physiological characteristics:

1) the ether soluble substances acting as hormones on the *Avena* coleoptile, *i.e.*, the auxin group — the "A substances";
2) the water soluble substances acting as growth factors for microorganisms — the "B substances";
   a) substances destroyed by oxidation (under the conditions described by the authors), capable of adsorption by animal charcoal, and acting in the absence of co-factors — "B_1 substances";
   b) substances not destroyed by oxidation (under the conditions described by the authors), incapable of adsorption by animal charcoal and requiring a mineral co-factor (Zn), growth factor for *Aspergillus niger* and other fungi — "B_2 substances".

The choice of the symbols B_1 and B_2 was rather unfortunate since these letters had already been used for vitamins and were introduced in this new sense without discussion. Furthermore, no clear distinction was made between true growth factors and products acting in relatively high dosages like the B_2 factors. For *Aspergillus* the B_2 factor can be replaced by 1 mg. of pyruvic acid + 0.3 mg. glycolic acid + 10 mg. glyoxalic acid. This classification summarizes very nicely the numerous investigations of NIELSEN and

[1]Ergozymes, vitazymes, and hormozymes in case they are considered in accordance with their functions and the relationships between them.

his school as well as those of DAGYS but it is not suitable for general application.

The first International Conference on Phytohormones (1938) realized the need of a standard system of classification and submitted the following:

A. The auxin group
  1) Phytohormones
    auxin *a*
    auxin *b*
    auxin *a* lactone
    heteroauxin
  2) Substances with similar activity but not found in plants (*e.g.*, 3-indol acetic acid)

B. The bios group
  1) Phytohormones
    biotin
    thiamin
    estrone (theelin)
    factor Z of fermentation
  2) Specific nutritive substances
    i-inositol
    β-alanine
    pyruvic acid
    glyconic acid

The major divisions (A and B) which correspond to those of NIELSEN and his collaborators, are suitable for continued usage. On the other hand, the term phytohormone which had already been proposed by WENT, KÖGL and KONINGSBERGER is not a logical one. The intention was that the term phytohormone should designate the endogenous substances acting on plants and help distinguish them from the exogenous substances with similar activity. Moreover, since biotin and auxin are found in animal tissues (their function in animal metabolism is still unknown) it would be necessary in these cases to speak of zoohormones. According to this usage the term hormone is employed for both thiamin and auxin although their modes of action and physiological functions are entirely different. It is impossible to give the term hormone such a broad meaning without excluding the vitamin concept from this classification.

It should be noted that this classification makes no mention of any substances acting in minute dosages by which vitamins might be distinguished from nutrients with specific action which act in relatively large amounts such as pyruvic, glycolic, and other acids, as well as *i*-inositol.

It is surprising to find substances such as β-alanine which, according to LASH MILLER, functions as bios IIa, and is effective in small dosages (a few gammas), and also acts in conjunction with other bios substances. Since all of its characteristics are those of a true growth factor, it is impossible to consider this substance as a specific nutrient.

Factor Z of VON EULER and PHILIPSON, which accelerates fermentation and which is distinct from cozymase, although enzymatic in nature, is classified here with the phytohormones. This also seems untenable.

Thus the classification proposed in 1938 by the International Conference on Phytohormones, represents the status of the problem of growth substances at the time it was made. The terminology does not seem logical and apparently it will have to be abandoned.

MASON (1939), adopting the concept of HUXLEY, published the following classification of active substances:

Activators

A noncommital term for those chemical substances produced by the organism, which exert specific functions in regard to correlation or differentiation.

A. Local Activators

1. *Intracellular activators*
   a. Substances produced by genes and causing the cells which produce them to differentiate in a specific manner.
   b. Substances responsible for sex characters, eye color, etc., in certain insects.
2. *Regional activators*
   a. Substances which predetermine specific regions in the embryo; *e.g.*, limb disc, parts of eye rudiments.
   b. Substances responsible for growth gradients.

B. Distance Activators

3. *Diffusion activators*
   a. Direction of diffusion unrestricted—
      e.g., Numerous growth factors necessary for yeasts, molds, protozoa, and bacteria.
      e.g., Molting hormones of insects.
      e.g., Pigmento-motor neurohormones of fishes.
   b. Direction of transport restricted by structural organization—
      e.g., Growth hormones of higher plants.
   c. Action restricted to contiguous tissues—
      e.g., Organizer substances of vertebrate embryo.
      e.g., Lens-inducing substance of optic cup.
   d. Action restricted by specific means—
      e.g., Neurohormones of higher vertebrates.
4. *Circulatory activators*
   a. Hormones as originally defined.
   b. Vitamins.

Para-Activators

By-products of metabolism, with effects on correlation or differentiation, *e.g.*, carbon dioxide, histamin, etc.

The above classification is essentially that proposed by HUXLEY (1935). It does not employ fundamentally important criteria (*e.g.*, the mechanism of action) in distinguishing the major groups, hence it is highly artificial and is not likely to enlist many followers.

This review covers the principal classifications and concepts employed or proposed up to the present.

The author would like to eliminate the term phytohormone but would like to conserve the concepts of hormones, vitamins, and growth factors, since these terms have become well established through usage. The term growth factor is of value in its original sense as an accessory growth factor because it preceded the term vitamin. Either term can be used regardless of whether the source of the substances is endogenous or exogenous. Actually the source of a substance is of no value whatever as a criterion for classification. A better criterion, and the principal one used in the author's system of classification, is the general mode of action. Thus the fundamental criterion is a physiological one. We shall no longer consider a substance by itself but instead its action and the nature of the effect produced (Schopfer, 1937).

Any action will be considered as *hormonal* if the substance in question acts, in the *final analysis,* specifically on the morphology of the organism. The term hormone is therefore applicable to auxins and other substances with similar action.

An action of *vitamin nature* is, in the main, concerned with intermediate metabolism. Vitamins favor or condition an action. They are concerned with the nutrition of the cytoplasm and participate in its growth. They do not act, in the final analysis, in a specific manner on the form of the plant.

The author proposes the following scheme of classification:

			Example
Active substances ("Wirkstoffe", biocatalyzers)	I.	Pseudo-growth-factors of mineral nature. Various physico-chemical actions	Zinc
	II.	Growth factors (in the strict sense of the term) of vitamin nature, acting on metabolism and on the growth of cytoplasm (vitamins and vitazymes)	Thiamin
	III.	Factors of hormonal nature, in the final analysis acting in a specific manner on the morphology of the organism (hormones and hormozymes)	Auxin

Group I includes all the mineral substances with catalytic and "oligodynamic" action such as Zn, Mn, Mo, and Cu. Some of these are able to function as co-factors of growth or act in conjunction with co-factors, thereby participating in enzyme systems.

Group II includes all the vitamins whose action has been demonstrated in plants. This would include the "B substances" of Nielsen and some of the "BP substances" of von Euler. The vitazymes should be listed here if it is desirable to emphasize their function as coenzymes. Other factors belonging in this group are factor Z of fermentation and the bios substances which cannot be designated as formative substances.

Group III includes chiefly the auxin group of substances.

Certain substances are hard to classify in this system. Estrone, for example, is of doubtful position since little information is available concerning the mode of action of this animal hormone. Should it be found to have a general action on assimilation it would belong in group II with the vitamins. On the other hand, should its specific morphogenic action, *e.g.,* on the formation of flowers, be confirmed it obviously would belong in group III with the hormones. Thus it is difficult to present a definitive classification of these substances since our knowledge concerning them is still incomplete. The author, however, believes that it is advantageous to retain the classical terms hormone and vitamin with slightly modified meanings by placing the emphasis on the physiological criterion.

The substances to be discussed in this book are mainly growth factors of vitamin nature.

Definition and characteristics of growth factors. — The study of the capacity for synthesis and its loss served to introduce the problem of vitamins in relation to plants, whereas the discussion of the various systems of classification familiarized us with the different concepts and definitions. The author will now attempt to define the term growth factor from his point of view (SCHOPFER, 1939 *a* and *b*).

First characteristic: The need of a growth factor arose from a loss of capacity for synthesis as amply discussed in the preceding chapter. This character, however, does not suffice to define satisfactorily a growth factor of vitamin nature since the loss of capacity for synthesis can affect substances other than vitamins, *e.g.*, specific nutrients or even ordinary nutrients. Thus a completely heterotrophic plant has lost its capacity to synthesize its sugar and also its growth factors. A. LWOFF (1938) proposed the loss of capacity for synthesis as the sole criterion for a growth factor, but the author does not agree with him on this point, since this criterion, although essential, is insufficient for a complete definition.

Second characteristic: A true growth factor of vitamin nature is organic. Thus chemical elements and substances of mineral nature are excluded (see p. 21).

Third characteristic: A characteristic essential to our concept is the so-called "catalytic" action of very small dosages, improperly referred to as "oligodynamic" action. It is difficult to set a definite upper limit for the dosage of a growth factor beyond which it becomes a nutrient; it is sufficient to say that the maximal dosage is small. At any rate this characteristic is a justifiable one. It can be expressed by what we call the "economic coefficient" of a growth factor, *i.e.*, the ratio between the weight of the crop obtained by the use of a definite quantity of growth factor and the weight of the growth factor (weight of crop / weight of growth factor). Examples of "economic coefficients" for thiamin are as follows:

Phycomyces Blakesleeanus	200,000/1	(from SCHOPFER)
Rhodotorula rubra	200,000/1	(*id.*)
Ustilago violacea	2,500,000/1	(from SCHOPFER and BLUMER)
Polyporus adustus	380,000/1	(calculated from the results of KÖGL and FRIES)
Cercosporella herpotrichoïdes	433,000/1	(from DEFAGO)

(For the algae see ONDRATSCHEK, 1940).

Although these figures give a certain impression, the determination of this coefficient is not as easy as it would appear since it involves certain theoretical difficulties. The "economic coefficient" is of value only when it is obtained under a well established set of conditions. Failure to recognize this fact has resulted in the publication of ratios which are not comparable. This subject will be discussed at greater length later on after the reader has acquired some additional information pertinent to a better understanding of it.

The "economic coefficient", *e.g.*, weight of culture / weight of vitamin, is particularly difficult to interpret when based on a limited number of cases. This is especially true when the substance in question is intermediate between a true growth factor and a specific nutrient required in relatively large dosage, *e.g.*, inositol. In order to classify such a substance it is necessary to resort to other characteristics of growth factors. JANKE (1939) introduced the concept of *micronutrients* for substances which act in small dosages as plastic material in the construction of cytoplasm. These are likely to be mistaken for true growth factors because of their activity when present in traces. The term is a very fortunate one; however, it merely conceals and does not explain a difficulty which such substances present. Nevertheless, it is generally conceded that certain growth factors are able to function as *micronutrients* ("Spurenbausteine", trace nutrients).

Fourth characteristic: Quantitative action. The activity curve of a growth factor begins by rising uniformly (almost a straight line) as the dosage is increased until an optimum is reached; at this point there is a sharp bend, beyond which there is no rise, at least theoretically, with increased dosage.

Fifth characteristic: Marked specificity of action tied up with the chemical constitution. This is one of the most interesting aspects and one of the most important characteristics of growth factors. In some cases the specificity of action is so pronounced that a single substitution in one of the important positions completely nullifies the growth promoting (auxogenic) action of the growth factor. Such marked specificity of action is explained by the fact that a growth factor is actually a constituent of an enzyme, *i.e.*, a coenzyme or a fraction of a coenzyme.

THIAMIN

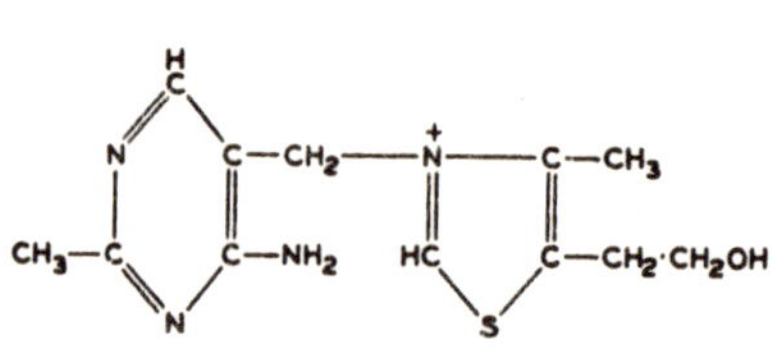

ACTIVITY ON PHYCOMYCES=100%

ANALOGUE OF THIAMIN

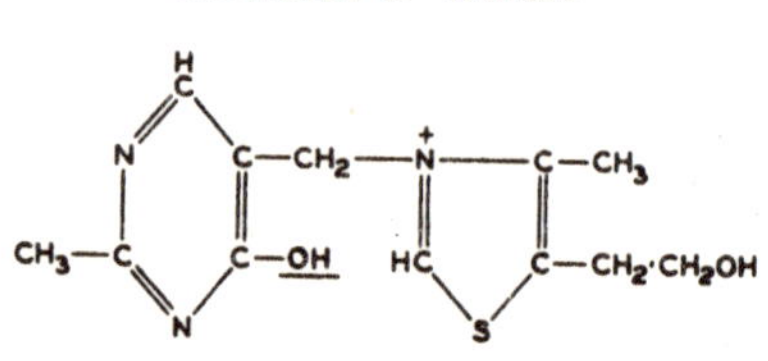

ACTIVITY ON PHYCOMYCES=0

This specificity of action is subject to variation, as will be seen during the course of this study.

Other characteristics will be brought to the attention of the reader later. Meanwhile, it is sufficient to know that a growth factor of vitamin nature such as will be studied in this book is an *organic substance, the need for which results from the loss of the capacity for its synthesis, whose action is catalytic (active in small amounts), quantitative, and markedly specific.*

REFERENCES:-

1930: H. VON EULER. — Biocatalysatoren. Sammlung chemischer und chemisch-technischer Vorträge, Neue Folge, Heft 4, 68 S.

1932: H. VON EULER und T. PHILIPSON. — Wasserlösliche Wachstumsfaktoren. Biochem. Zeitschr. 245:418.

1936. A. MITTASCH. — Über Katalyse und Katalysatoren in Chemie und Biologie. Berlin, J. Springer, 65 S.

1937: G. K. LINK, H. A. WILCOX and A. D. LINK. — Response of bean and tomato to *Phytomonas tumefaciens, P. tumefaciens* extracts . . . Bot. Gaz. 98:816.

1937: N. NIELSEN und V. HARTELIUS. — Über die Trennung der auf die Stoffproduktion der Hefe und Schimmelpilze einwirkenden Wuchsstoffe. C. r. Trav. Lab. Carlsberg, série physiologique, 22, no. 1.

1937: W. H. SCHOPFER. — La forme de la plante. Quelques facteurs chimiques de sa genèse et de son déterminisme. Actes Soc. helv. Sc. nat. Genève, 118th session, p. 62.

1938: H. VON EULER. — Bedeutung der Wirkstoffe (Ergone), Enzyme und Hilfsstoffe im Zellenleben. Erg. Vitamin- und Hormonforschung, 1:159.

1938: First phytohormone conference. — Chronica Botanica 4:48-49.

1938: R. AMMON und W. DIRSCHEL. — Fermente, Hormone, Vitamine und die Beziehungen dieser Wirkstoffe zueinander. Leipzig, G. Thieme, 451 S.

1938: A. LWOFF. — Les facteurs de croissance pour les microorganismes. Rapport au 1er Congrès international de l'association des microbiologistes de langue française, p. 58.

1939: A. JANKE. — Die Wuchsstoff-Frage in der Mikrobiologie. Centralbl. Bakt. II, 100:409.

1939: KARL E. MASON. — Vitamins and hormones (see p. 135 in Science in Progress). Yale University Press, New Haven, Conn.

1939*a*: W. H. SCHOPFER. — Vitamine und Wachstumsfaktoren bei der Mikroorganismen. Erg. Biol. 16:1.

1939*b*: W. H. SCHOPFER. — Définition du facteur de croissance de micro organisme. Third Intern. Congress f. Microbiology, New York, Report of Proceedings, p. 486.

1940: K. ONDRATSCHEK. — Experimentelle Untersuchungen über den Einfluss von Wirkstoffen auf die Vermehrung einiger mixotropher Algen. Arch f. Mikrobiol. 11:89.

Chapter V.

THE PRINCIPAL VITAMINS SYNTHESIZED BY PLANTS.

This chapter is designed to orient the reader and prepare him for what is to follow. It consists of a brief enumeration of the principal vitamins synthesized by plants and is in no way a duplication of the detailed treatment on vitaminology found in subsequent chapters since it gives only the chemical structure and distribution of these vitamins. (*Cf.* VOGEL, 1940).

Fat soluble vitamins.

Unsaturated alcohols:-

Vitamins A_1 and A_2. Polyene derivatives. — These vitamins are of interest to botanists for the present only because of their provitamin, carotene. Although plants do not seem to be able to synthesize the vitamin itself, they are able to synthesize the provitamin. The carotinoids correspond to the former lipochromes of KRUKENBERG, so named because of their solubility in fats and because of their yellow color. The great majority of the yellow pigments of the plant world are carotinoids, and in this respect there is no difference between the cryptogams and the phanerogams or between plants and animals. Provitamin A is an unsaturated hydrocarbon, carotene. It occurs in the form of α-, β-, γ-carotene with the empirical formula $C_{40}H_{56}$. β-carotene contains two β-ionone groups and is a symmetrical compound, whereas the α-, and γ-carotenes are asymmetrical and contain one β-ionone group and an isoprene chain. The activity as provitamin is dependent upon the β-ionone group, *i.e.*, the position of the double bond. The splitting of the molecule of β-carotene is effected by animals according to the following equation:

$$C_{40}H_{56} + 2H_2O = 2\,C_{20}H_{30}O$$

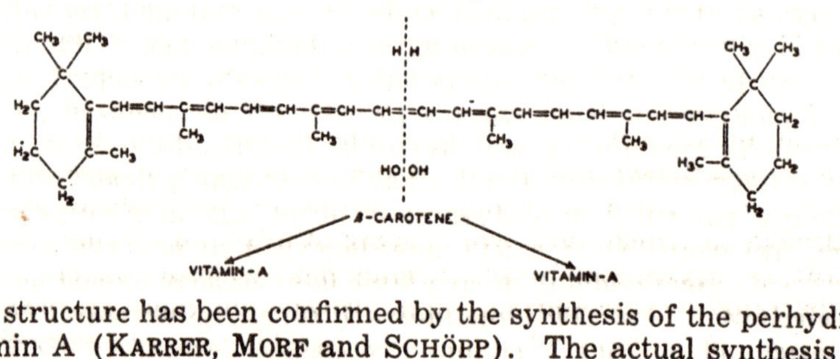

This structure has been confirmed by the synthesis of the perhydrovitamin A (KARRER, MORF and SCHÖPP). The actual synthesis of vitamin A itself has also been announced (KUHN and collaborators).

Distribution of carotinoids. — The plant cell has unusual capacity for the synthesis of carotinoids. A tissue generally contains several carotinoids, which can be separated by the chromatographic adsorption method. Animal tests are of no value in distinguishing carotene from preformed vitamin A. The Carr-Price reaction (antimony trichloride in chloroform solution) is not specific for vitamin A since it gives an intense blue color when applied to vitamin A, all carotinoids, and polyenes in general. Under certain conditions, however, it can be rendered specific for this vitamin. Direct spectroscopic analysis permits the determination of total carotinoid content or of the individual carotinoids after they have been separated by the chromatographic methods of analysis.

Practically all groups of plants contain carotinoids. They are present in the root, shoots, leaves, flowers and fruits of phanerogams. They also occur in abundance in the various groups of cryptogams: myxomycetes, phycomycetes, ascomycetes, basidiomycetes, bacteria, *Cyanophyceae, Diatomeae, Flagellatae, Chlorophyceae, Phaeophyceae, Rhodophyceae,* and *Characeae.* The microreaction of MOLISCH (saponification of fats in which the carotinoids are dissolved, and crystallization of the pigment) permits the detection of the pigment inside the cell where a chemical analysis would be impossible. The quantities vary but may be remarkably high in some cases.

TABLE III. Carotene content of plants (from WALKER, Diss. Zürich 1935, modified):-

	Carotin, g. per 100 g. dry matter	*Ratio of xanthophyll/carotene*
Spinach (dry leaves)	0.006	1:3.58
Beech *do.*	0.00604	1:1.72
Asparagus *do.*	0.0168-0.0533	1:1.12 , 1:1.22
Lettuce *do.*	0.00311	1:1.12
Nettle *do.*	0.00928	1:3
Caltha palustris (flower)	0.00322	1:8
Maize	0.0002	1:6.5
Pisum (max.)[1]	0.02643	
Triticum (max.)[1]	0.01814	

Among the fungi, the phycomycetes have an unusually high carotene content. This has been known since the time of ZOPF (1902). *Phycomyces Blakesleeanus,* cultivated under a given set of conditions, produces carotene in quantities reaching 0.1 percent of the dry weight. The carotene is almost exclusively of the beta form as shown by the strong vitamin A activity of the thallus of this fungus. Likewise one of the red yeasts, *Rhodotorula Sanniei,* is rich in various carotinoids, and also possesses a strong vitamin A activity on animals (TCHANG and P. CHAIX).

Vitamin A activity has been found in various microorganisms and their extracts: *Chlorella* (COWARD), *Nitschia closterium* (di-

[1]From VIRTANEN.

atom) (JAMESON, DRUMMOND and COWARD, AHMAD), phytoplankton in general, *Chlorococcus* (GUNDERSON and SKINNER), *Trentepohlia* (LEDERER), lichens (ELLIS, PALMER, and BARNUM). It is more than probable that these organisms, like *Phycomyces* and *Rhodotorula,* are rich in carotinoids and do not contain preformed vitamin A.

Sterol derivatives:-

D Vitamins (*antirachitic vitamins*). — This group of vitamins, although very important for animal organisms, does not, according to present knowledge, play an important rôle in plant physiology. Plants contain certain sterols (phytosterols, mycosterols), the most important mycosterol being *ergosterol,* $C_{28}H_{43}OH$, which occurs in yeast. Upon irradiation it furnishes vitamin D_2 (calciferol). Ergosterol occurs in nature and the artificial vitamin D_2 produced upon its irradiation was for a long while thought to be identical with the highly active natural vitamin D found in animals. However, it seems quite probable that the only strongly active natural vitamin D is vitamin D_3, the provitamin of which is 7-dehydrocholesterol. This vitamin is found only in animal tissues where it accompanies cholesterol, $C_{25}H_{47}OH$. The provitamin D_4, 22-dihydrocholesterol is not found in nature. On the other hand, provitamin D_5 which is represented by 7-dehydrositosterol, $C_{29}H_{47}OH$, is found in plants, particularly in soybean oil. Upon irradiation it yields vitamin D_5 with a slight antirachitic activity.

D vitamins in the active form do not occur in plants, but they can be formed there by irradiation, *e.g.*, in *Azotobacter chroococcum.* The characteristic of plants rests in the fact that they can synthesize the provitamins, particularly ergosterol. Certain microorganisms, such as *Azotobacter chroococcum,* are able to synthesize provitamin D from a mineral medium (GREAVES).

This capacity for synthesis is lost in certain flagellates, *e.g.*, *Trichomonas* sp., for whom the sterols must be supplied as indispensable growth factors (see p. 181).

Provitamin D_2, occurring in yeast as ergosterol, certainly plays an important rôle as a constituent of the plasma membrane but it is beyond the scope of this book to consider the function of this substance.

Phenols of the chroman series:-

Vitamin E (antisterility vitamin, alpha-tocopherol). — This fat soluble vitamin has long been known from its action on animals. It has also been known that certain plant organs are extremely rich sources of this vitamin. In spite of these facts chemical studies of it are very recent and it is just now beginning to receive the attention of plant physiologists. In 1936 EVANS *et al.* isolated this vitamin in its crystalline form and gave it the name α- and β-tocopherol ($C_{29}H_{50}O_2$). According to FERNHOLTZ the structure of vitamin E is as follows:

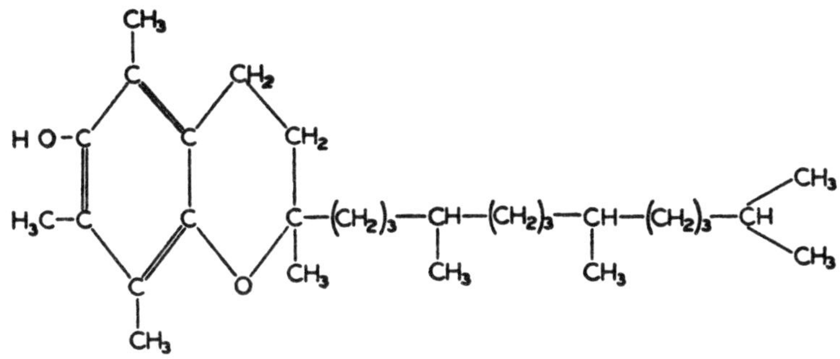

This structure has been confirmed by synthesis, carried out first by KARRER and his collaborators. Starting with trimethyl hydroquinone and phytol (phytylbromide) they were able to obtain alpha-tocopherol by a condensation process in the presence of zinc chloride as catalyzer, the yield being almost quantitative.

Vitamin E content and distribution. — Vitamin E is widely distributed in plants. The content of a few plants is given here in milligrams per 100 grams of substance:

Germ of wheat	30
Germ of maize	16.4
Oil of wheat germ	520
Green cabbage	6
Lettuce	6
Banana	0.5
Yeast (dry)	0.3

The germ of wheat seed is the best source for the extraction of vitamin E.

Specificity of action of vitamin E. — β-tocopherol and γ-tocopherol are slightly less active than α-tocopherol. The natural E vitamins are by no means specific in their action on animals. There is a series of simpler compounds which are active when used in relatively large dosages.

Derivatives of naphthoquinone:-

Vitamin K (antihemorrhagic vitamin). — This vitamin which is very widely distributed in plants is one of the most recently studied. ALMQUIST and KLOSE have reported that vitamin K activity is present in a series of compounds of which the simplest is phtiocol (2-methyl-3-hydroxy-1, 4-naphthoquinone). This compound, however, is only slightly active.

The constitution of vitamin K_1, verified by synthesis, confirms these views. The structure of this vitamin has been studied especially by DOISY and collaborators who have found that it is 2-methyl-3-phytyl-1, 4-naphthoquinone in the presence of $ZnCl_2$.

It is synthesized by condensing 2-methyl-1, 4-naphthoquinone with phytol (phytylbromide). Phytol is one of the precursors of vitamin K and also of vitamin E.

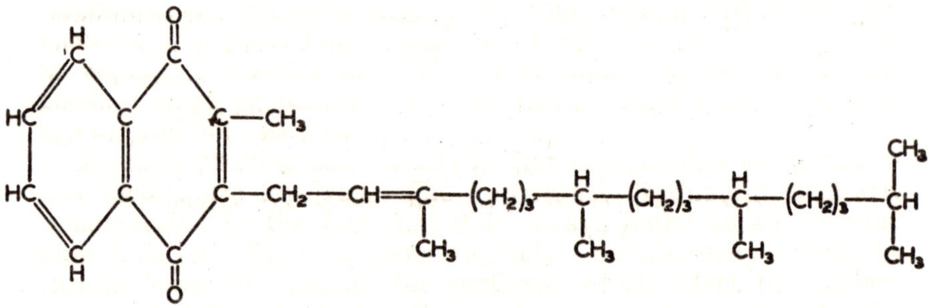

VITAMIN K_1 ($C_{31}H_{46}O_2$)

A second K vitamin is known, vitamin K_2, with less activity than vitamin K, and differing from it in certain properties. The structure established by DOISY and co-workers (BINKLEY, MCKEE, THAYER, and DOISY, 1940) is as follows:

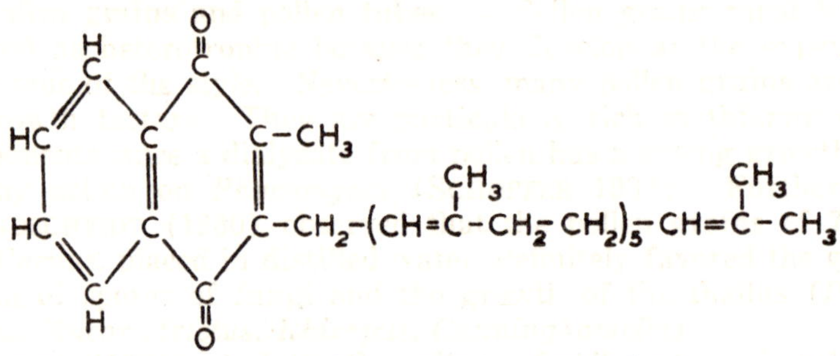

VITAMIN K_2 ($C_{41}H_{56}O_2$)

It occurs in bacteria.

Vitamin K content and distribution. — Plant tissues contain much larger amounts of this vitamin than animal tissues do. The content of some plants is as follows (expressed in Dam units - 1 unit corresponds to 0.08 gamma of pure vitamin K (phylloquinone), after DAM, 1938) :-

Lucerne	200–400	Soybean	25
Cauliflower	400	Pea	15
Spinach	500	Wheat	3
Nettle	400	Wheat germ	3
Strawberry	15	Carrot	10
Tomato	50	Potato	8

It is interesting to note that the germ of wheat seed contains much less vitamin K than vitamin E.

The *specificity of action* is rather marked. In addition to vitamin K_1 and vitamin K_2, another compound, 2-methyl-1, 4-naphthoquinone, has vitamin K activity, approximately three times as great as that of vitamin K_1. Vitamin K_1 in turn has twice the activity of vitamin K_2. There are other synthetic substances possessing vitamin K activity but the dosages required are relatively large.

Water soluble vitamins.

Heterocyclic nitrogen ring compounds:-

Pyrimidine ring:-

Vitamin B_1 (Antineuritic vitamin, thiamin, aneurine). — This is one of the most widely distributed vitamins in the plant kingdom. It occurs in the preformed condition in plants which in turn supply it to animals. It was the first vitamin discovered and its discovery marks the beginning of modern vitaminology (EIJKMAN, 1897). After numerous attempts it was finally isolated in the pure condition (JANSEN and DONATH, WINDAUS *et al.* 1932). The first work on the structure of thiamin was that of R. R. WILLIAMS, WATERMAN, KERESZTESY and BUCHMAN (1935). WINDAUS and his collaborators were the first to propose the proper empirical formula, $C_{12}H_{18}ON_4S\ Cl_2$. WILLIAMS *et al.* found that thiamin can be split by sulfite into two components according to the following equation:

$$C_{12}H_{18}ON_4SCl_2 + H_2SO_3 \rightarrow \underset{(1)}{C_6H_9ONS} + \underset{(2)}{C_6H_9O_3N_3S} + 2\ HCL$$

From product 1 a thiazole was isolated and identified (BUCHMAN, WILLIAMS and KERESZTESY). The study of product 2 led to the identification of pyrimidine (CLINE, WILLIAMS and WATERMAN). It is now known that thiamin is actually composed of:

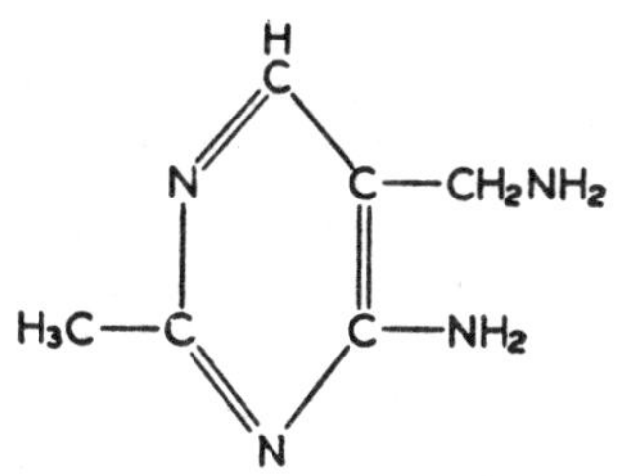

2-METHYL-4-AMINO-5-AMINO-METHYL-PYRIMIDINE

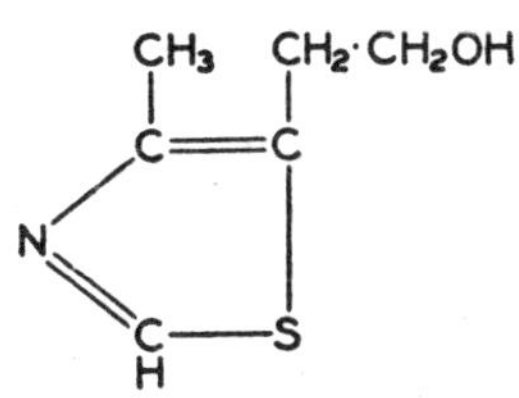

4-METHYL-5-β-HYDROXY-ETHYL-THIAZOLE

In the thiamin molecule, the thiazole and pyrimidine nuclei **are** united by a methylene bond:

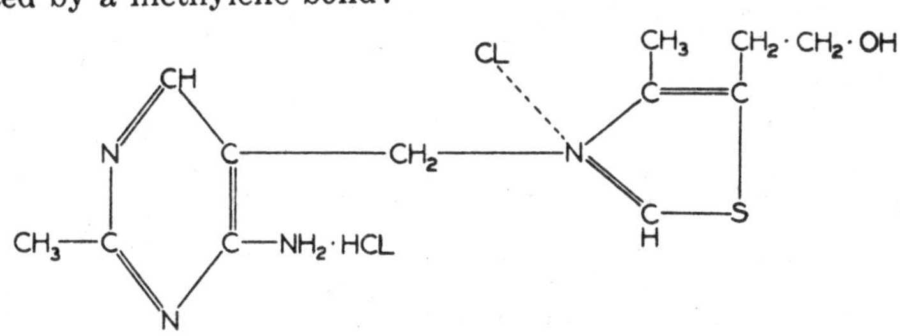

The chemical synthesis of thiamin which we must understand in order to study its biosynthesis later on is now relatively easy to accomplish. The two synthetic components are condensed in the following manner (synthesis of ANDERSAG and WESTPHAL, and of WILLIAMS and CLINE):

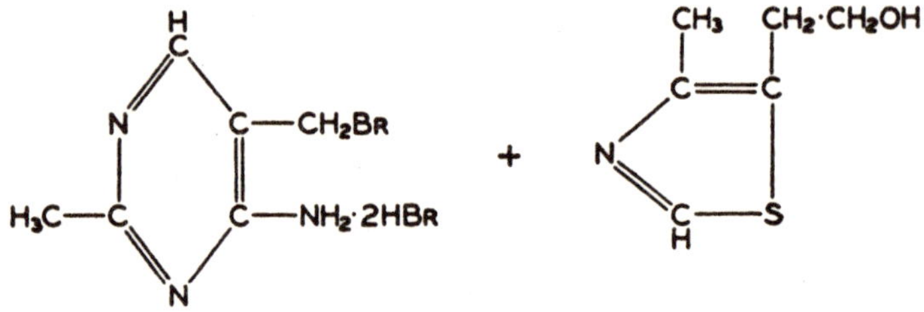

2-METHYL-4-AMINO-5-BROMOMETHYL PYRIMIDINE-DIHYDROBROMIDE

4-METHYL-5-β-HYDROXY ETHYL-THIAZOLE

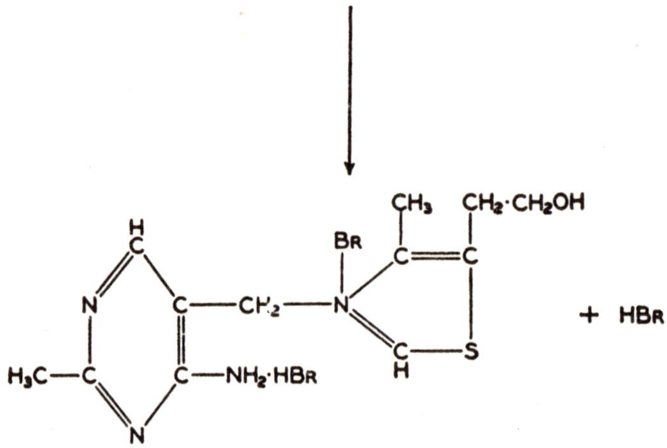

THIAMIN BROMIDE HYDROBROMIDE

The bromide hydrobromide can be converted into the chloride hydrochloride by shaking with silver chloride. Thiamin is therefore chemically known as 4-methyl-5-β-hydroxy-ethyl-N- [(2-methyl-4-amino-pyrimydyl [5]-) methyl]-thiazole.

Pyrimidine has an active group in position 5, *e.g.*, CH_2Br, which makes the condensation possible.

The precursors of thiazole are also of interest to physiologists. Thioformamide and chloracetopropyl alcohol *in vitro* condense to form thiazole according to the following reaction:

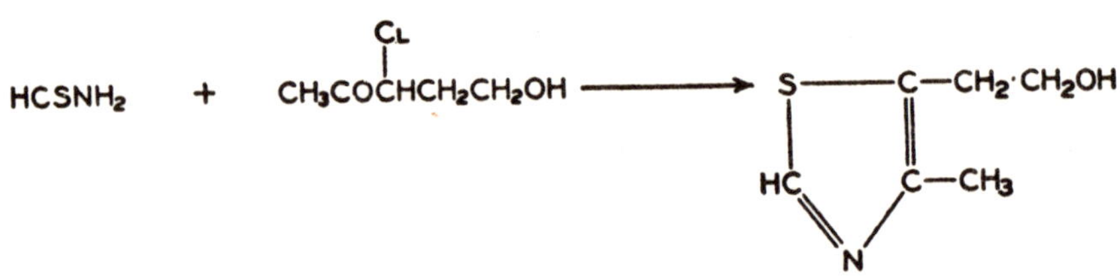

THIOFORMAMIDE

CHLOROACETOPROPYL ALCOHOL

THIAZOLE

Another method has been employed for the chemical synthesis of thiamin (TODD and BERGEL), consisting of a modified synthesis of thiazole, and a progressive union of this with pyrimidine.

Various methods of syntheses on which R. R. WILLIAMS *et al.*, ANDERSAG and WESTPHAL, TODD and BERGEL have worked have led to the complete elucidation of the structure of thiamin. Furthermore an appreciable number of substitution products of thiamin and its components are known.

Thiamin is of particular interest to plant physiologists. For a complete account of this vitamin the excellent monograph of R. R. WILLIAMS and T. D. SPIES should be consulted. This work, the first of its kind, considers the problem in all of its biological and chemical aspects, and constitutes a reference work of the first order.

Distribution of vitamin B_1 in the plant kingdom. — Thiamin is found in almost all groups of plants. In higher plants it is found in all organs, *e.g.* roots, stems, leaves, and fruits. It is particularly abundant in fruits, where it occurs as a reserve material. It is also present in the various organs of the flower, calyx, corolla, and stamens (SCHOPFER). The leaves of 134 species of plants, representing 71 families, studied by means of the *Phycomyces* test, evidently contain one or more of the following: thiamin, its two components (pyrimidine and thiazole), or a substance very closely related to thiamin.

Various assays show that the thiamin content of plants is sometimes very high (gammas per 100 grams of fresh material):

Germ of wheat	1100–4200	Lentil	400
Germ of rye	2500	Spinach	110
Germ of maize	800	Potato	70

Much of the data obtained by animal physiologists is not very reliable because of incorrect identification of the species of plants employed. Furthermore, the figures vary with different methods of analysis. If vitamin B_1 is being considered with reference to its action on man and animals it is obvious that the assay must be made by means of animal tests. Many plant materials have been assayed by this method and, in addition, a considerable number have been analyzed by means of the thiochrome and the *Phycomyces* tests. Consequently we are now well informed concerning the distribution of thiamin (or substances having the same physiological action) in the plant kingdom.

A large number of bacteria, yeasts and other fungi are autotrophic with respect to thiamin. This vitamin is also found in the *Chlorophyceae* (*Spirogyra*), *Phaeophyceae* (various species of *Fucus*), *Rhodophyceae*, Bryophytes (Musci and Hepaticae), and Pteridophytes (*Polypodium*). It is possible to find some well defined species in the same genus which are autotrophic and others which are heterotrophic with respect to thiamin (see p. 104).

Multiplicity of B_1 vitamins — Specificity of action. — It has been customary to consider vitamin B_1 (thiamin) as an individual substance, but it is now well established that certain substitution products of this vitamin exhibit some activity if used in relatively high dosage. It is therefore necessary to explain what is meant by specificity of action. If only the maximal action of a substance like thiamin is considered, there are only a few substitution products which are capable of replacing it quantitatively. If, on the other hand, the degree of activity is not considered there are several products with positive action, the effective dosage being sometimes 5000 times greater than that of ordinary thiamin. Recently SCHULTZ, using an animal test, studied 39 homologues and analogues of thiamin, among which 22 compounds were found with definite vitamin B_1 activity. Furthermore, there are reasons to believe that plants contain substances which are related to thiamin and, although not identical with it, have the same vitamin activity.

In animal metabolism thiamin apparently has a certain number of functions which are absolutely indispensable. The NH_2 group in position 4 of the pyrimidine portion as well as the hydroxyethyl group in position 5 of the thiazole are of prime importance. The absence of these groups markedly diminishes the vitamin activity. The H in position 2 of the thiazole is of similar importance to animals. It should be recalled that nicotinic acid and certain pyrimidine compounds can enter into the composition of heterovitamins of B_1 recently synthesized by BAUMGARTEN and DORNOW. In thiamin the thiazole group which seems indispensable can be replaced by 2-methyl-3-hydroxyethyl-pyrimidine. This pyrimidine when condensed with normal pyrimidine of thiamin (in its bromide form) yields the bromide of 2-methyl-3-hydroxyethyl-N-(2-methyl-4-amino-pyrimidyl-5-methyl-pyridine). Starting with nicotinic acid it is possible to obtain 3-hydroxyethyl-N- (2-methyl-4-amino-pyrimidyl-(5) methyl-pyridine in the form of the bromide hydrobromide:

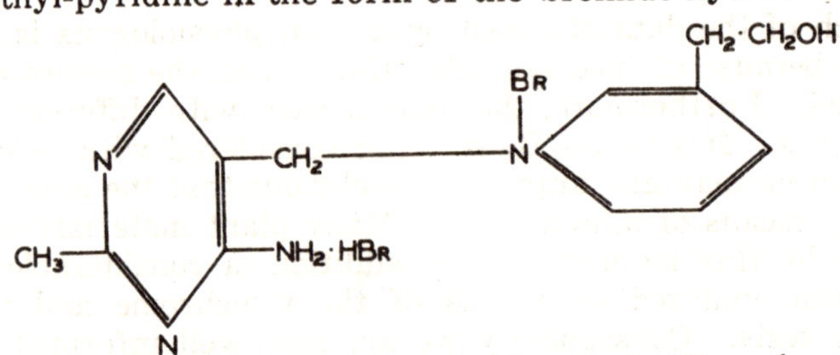

The absence of the methyl group in the pyrimidine ring does not increase the vitamin activity. The above compound has an activity 1/240 of that of thiamin. These heterovitamins have not actually been found in organisms, although their presence is not precluded. These investigations show that an interesting relationship may exist between vitamin B_1 and the antipellagra vitamin (nicotinic acid). This relationship appears to be purely chemical but it may prove to be physiological.

Alloxazine:-

Vitamin B_2, s. str. Riboflavin (Growth vitamin). — Riboflavin has not been investigated as extensively as thiamin. Nevertheless, it is an important vitamin with reference to plants. As in the case of the carotenes, the water soluble yellow pigments which are now classified as flavins have long been recognized. The history of the flavins and of their discovery is intimately tied up with that of the respiratory coenzymes. In 1932 BANGA and SZENT-GYÖRGYI isolated from an extract of pig heart a coenzyme with a golden yellow color, which they called *cytoflav*. The colored constituent was also found by KUHN. It was isolated in its crystalline form by GYÖRGY and WAGNER-JAUREGG. The yellow pigments called flavins (KUHN *et al.*), *lactoflavin* (from milk), *ovoflavin* (from egg white), *hepatoflavin* (from liver), all turned out to be identical and the name which is rapidly coming into general use is *riboflavin*.[1] The name riboflavin indicates that the substance is a ribityl derivative of a flavin. It may be described chemically as 6, 7-dimethyl-9-(*l'-d*-ribityl)-iso-alloxazine ($C_{17}H_{20}O_6N_4$). Its structure, as shown by various methods of synthesis, is:

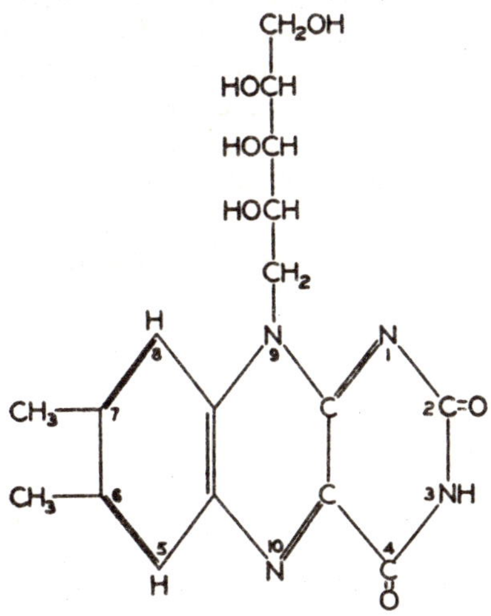

Riboflavin content. — The higher plants are rich in flavins according to the figures presented below (from EULER, KUHN, WAGNER-JAUREGG, WARBURG; chemical test):

Maize	1–5	mg. flavin per kg. of fresh material
Barley germ	0.5–1	*do.*
Peas, Spinach, Tomato	0.5–1	*do.*
Carrot	0.2	*do.*
Wheat germ	0.2	*do.*
Potato	0.07	*do.*

The presence of flavins in bacteria proves that the biosynthesis of these pigments is not directly dependent upon the action of

[1]In April 1937 the Council on Pharmacy and Chemistry of the American Medical Association adopted the name "Riboflavin" for the substance which was originally referred to as lactoflavin (Council on Pharmacy and Chemistry. — Riboflavin, the accepted name for vitamin B_2. Jour. Amer. Med. Assoc. 108:1340-1341, 1937). Their action was followed by the Society of Biological Chemists, the American Institute of Nutrition, and the Federal Food and Drug Administration.

chlorophyll. The flavin content of microorganisms is remarkably high as shown by the following figures (from WARBURG and CHRISTIAN):

	mg. per kg. dry wt.
Butyric bacteria (*Clostridium butyricum*)	136
Lactic bacteria (*Bac. Delbrückii*)	115
Baker's yeast	36
Beer yeast	30
Acetic bacteria (*Bac. Pasteurianum*)	15

Riboflavin does not occur in *Bacterium coli* and has not been proven to be present in *Staphylococcus.*

Almost nothing is known concerning the flavin content of the higher fungi, the algae and the higher cryptogams, although these organisms undoubtedly contain this vitamin.

Pyridine ring :-

Nicotinic acid (human pellagra-preventative vitamin, canine anti-blacktongue vitamin). — This factor, belonging to the B_2 vitamins for animals, was discovered in the field of microbiology when it was found to act as a growth factor for *Staphylococcus aureus.* ELVEHJEM and his collaborators established the identity of the pellagra-preventative vitamin with the amide of 3-pyridine-carboxylic acid.

Chemistry. — The structure of the human pellagra-preventative vitamin is as follows:

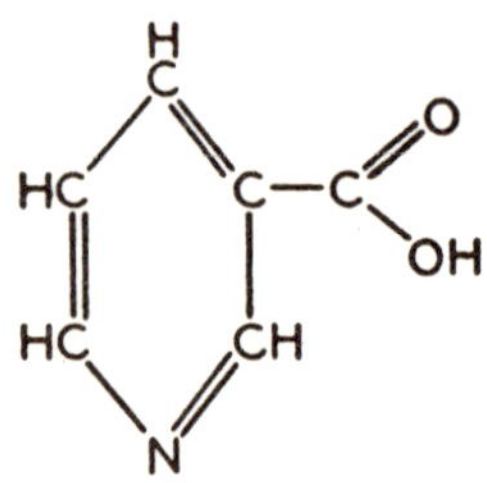

NICOTINIC ACID

$C_6H_5O_2N$

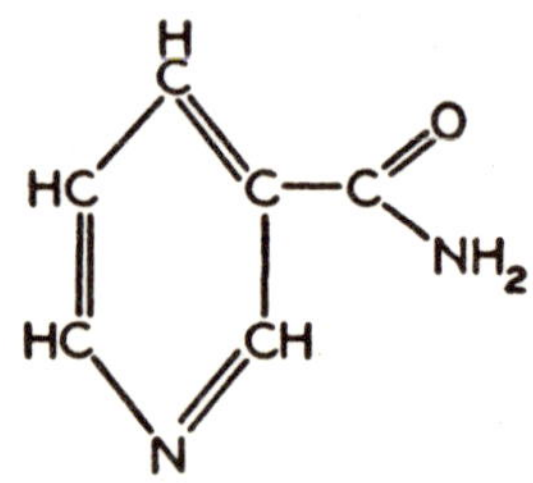

NICOTINAMIDE

$C_6H_6ON_2$

Distribution and content. — A systematic study of the distribution of the nicotinamide in various tissues has not yet been made. There are indications that it occurs in green peas, soybeans, spinach, and dry peas. A small amount is present in wheat, carrots, lettuce, green onions, but almost none in potatoes, apples, and dry prunes. The data are very fragmentary. Since nicotinic acid may occur in the free state or combined as one of the components of codehydrogenase (see later discussion) it is necessary to take these facts into consideration in connection with nicotinic acid assays.

The chief rôle of nicotinic acid lies in the formation of codehydrogenase. This fundamental function makes it very probable that nicotinic acid is widely distributed in animals and plants.

Specificity of action. — The specificity of nicotinic acid is well known with respect to microorganisms, whereas, with respect to animals, little is known. One remarkable fact is that plants contain an alkaloid trigonelline which is capable of completely replacing nicotinamide, although the two are chemically different. The

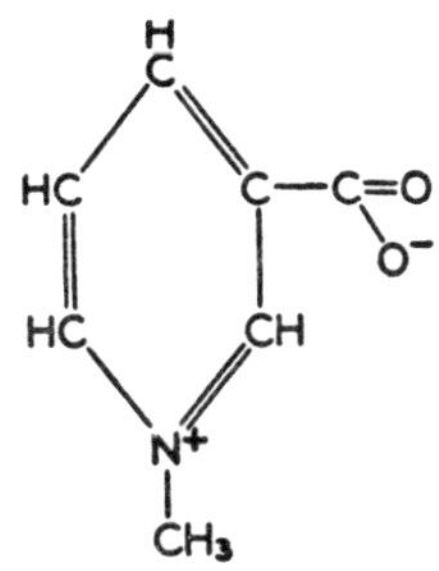

TRIGONELLINE (INTERNAL SALT OF N-METHYL-PYRIDINE-β-CARBOXYLIC ACID, $C_7H_7O_2N \cdot H_2O$).

transformation of nicotinic acid into trigonelline can be brought about by animals in the liver (dog). (For the specificity of action on microorganisms see p. 143).

Codehydrogenases. — The codehydrogenases I and II are able to function as the human pellagra-preventative vitamin (VON EULER, MALMBERG and collaborators). In this case the nicotinamide functions as the precursor of the codehydrogenase and it is to the latter that the pellagra-preventative action must be attributed. Although animals are unable to synthesize the nicotinamide constituents of codehydrogenase, they are able to synthesize the other components of this complex coenzyme. Codehydrogenase I (cozymase of yeast) has the following structure (from SCHLENK and VON EULER):

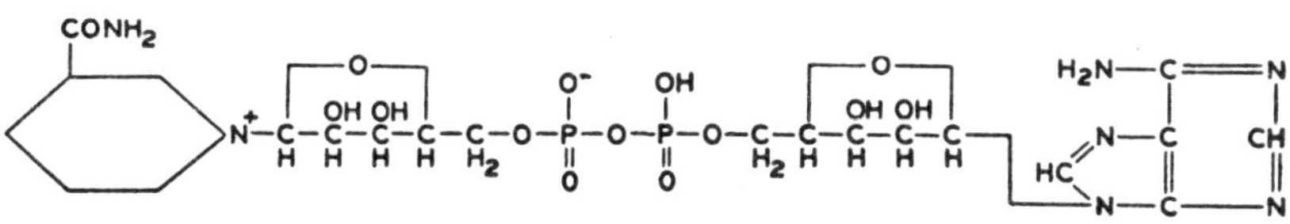

NICOTINAMIDE ADENINE

This codehydrogenase I (of VON EULER, ALBERS and SCHLENK) is diphosphopyridine-nucleotide. The codehydrogenase II (of red blood corpuscles) (WARBURG, CHRISTIAN and GRIESE) differs from codehydrogenase I by having one more molecule of phosphoric acid. A reciprocal transformation of the two codehydrogenases is possible. Codehydrogenase II upon dephosphorylation yields codehydrogenase I. The transformation in the opposite direction can be accomplished enzymatically.

In animals the pellagra-preventative action is dominated not by codehydrogenase in the form of the complete molecule but instead by the nicotinamide. It is correct to consider the nicotinamide as the precursor of codehydrogenase I and II, but it is not certain that the codehydrogenases actually represent the active form of the pellagra-preventative vitamin. It has been demonstrated by animal physiologists that the non-oral administration of codehydrogenase in the form of the complete molecule has only a weak pellagra-preventative activity. Hence, in animals, the nicotinamide must obviously undergo various transformations which still remain undetermined. It is not obligatory that the vitamin action of the nicotinamide is due to its rôle as a constituent of the codehydrogenase molecule. The pyrimidine group of nicotinic acid possibly enters into the composition of other vitamin substances (see p. 47). Among microorganisms one case is known in which codehydrogenase as such (complete molecule) functions as a growth factor. *Hemophilus parainfluenzae* requires the complete molecule of codehydrogenase I or II and cannot synthesize it from the nicotinamide.

Vitamin B_6 (pyridoxine[1], "rat acrodynia" factor of GYÖRGY, factor of elution, adermin). — Pyridoxine represents one of the three antidermititis factors (Hautfaktoren). It is concerned with a dermititis disease of rats (rat pellagra or "achrodynia") and should be distinguished from the other two antidermititis factors, nicotinamide which prevents human pellagra, and the "filtrate factor" (pantothenic acid) which prevents a dermititis of chickens and pigeons.

Pyridoxine is a member of the vitamin B_2 complex. It is indispensable for the maintenance of a healthy skin and for the growth of hair in rats.

Chemistry. — It was first isolated in the crystalline condition from rice polishings by KERESZTESY and STEVENS and has since been isolated from yeast, liver, molasses, and sugar beets. Its chemical structure was established by KUHN and WENDT (1939) following its synthesis by KUHN and collaborators. It is known chemically as 2-methyl-3-hydroxy-4,5-di-(hydroxy-methyl) pyridine, with the formula $C_8H_{12}NO_3.Cl$ and is the hydrochloride of $C_8H_{11}NO_3$:

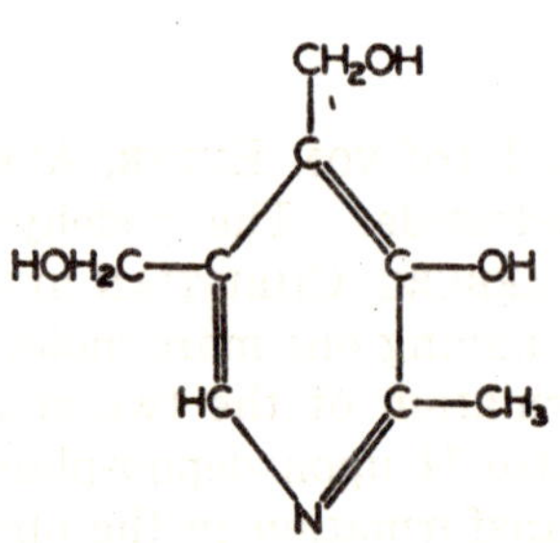

[1]Pyridoxine is the name adopted for the rat antidermititis factor by the American Institute of Nutrition, the American Society of Biological Chemists, and the Council of Pharmacy and Chemistry of the American Medical Association.

Specificity of action. — The specificity of action of pyridoxine has not been investigated extensively with animals, but bacteria have been subjected to tests involving an appreciable number of substitution products (see p. 156).

Keto-lactone group of sugars:-

Vitamin C (ascorbic acid). — This water soluble vitamin, which is now prepared in large quantities, has been studied extensively although its chemical history is recent. Historically it is of interest by virtue of the fact that the avitaminosis associated with it, called scurvy, has been known since ancient times (middle ages, and even the days of antiquity). The first chemical investigations on this vitamin were made simultaneously by SZENT-GYÖRGYI, and by TILLMANS and KING, who extracted a substance from the suprarenal glands of cattle and from oranges with the formula of $C_6H_8O_6$, which was identical with the anti-scorbutic vitamin. Its structure, studied by MICHEEL and KRAFT, HAWORTH and HIRST, KARRER *et al.*, has been confirmed by synthesis by REICHSTEIN (1933-34). Its formula is:

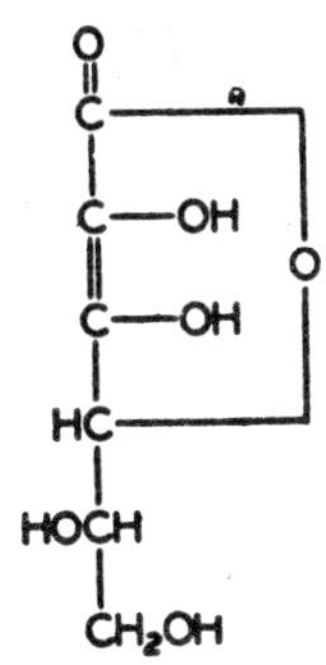

L-ASCORBIC ACID ($C_6H_8O_6$)

Ascorbic acid is known chemically as 2,3 dienol-*l*-gulonic acid lactone. Its synthesis may be accomplished by various methods which fall into two general types: 1) starting with *l*-xylose which is transformed into *l*-xylosone, 2) starting with *d*-glucose, which is transformed by reduction into sorbitol which in turn is transformed by *Bacterium xylinum* into *l*-sorbose. The latter synthesis is of particular interest to botanists engaged in the study of the biosynthesis of ascorbic acid.

Ascorbic acid possesses a strong capacity to reduce a wide variety of organic and inorganic reagents. This capacity and the acidic nature of this vitamin are due to the dienol group,

a structure rarely occurring in nature. Its capacity for reduction is surpassed by only a few natural substances, *e.g.* glutathione (—SH group).

Content and distribution in nature. — Ascorbic acid is widely distributed in all plant organs. Some examples are as follows:

	gm/kg.
fruits:	
lemon	0.5
orange	0.3–0.9
apple - pear	0.04
tomato	0.05
vegetables:	
turnip	0.2
potato	0.1
cabbage	0.25–0.5
leaves:	
iris	6.0
pine	0.25
spruce	0.8

Specificity of action. — A series of homologues of vitamin C (*l*-ascorbic acid) are known but none of them have as great an activity as the vitamin itself. These products are obtained by replacing the pentose by another sugar.

l-ascorbic acid: activity 100
d-ascorbic acid: activity 0
l-rhamnoascorbic acid: activity 1/5
d-araboascorbic acid: activity 1/20
l-glucoascorbic acid: activity 1/40
l-galactoascorbic acid: activity 1/60
d-glucoheptoascorbic acid: activity 1/100

Later we shall have occasion to return to the various homologues of *l*-ascorbic acid when we consider their action on higher plants.

Pantothenic acid (*pantothen*[1], *"filtrate factor", "anti-grey-hair factor", anti-pellagra vitamin of chickens and pigeons*) *and its precursor beta-alanine.* — This vitamin is now included in the list of vitamins required by animals, but it was originally known only as a growth factor for microorganisms. Beta-alanine in some cases and pantothenic acid in others are necessary for the development of yeast and certain bacteria, *e.g. Streptococcus lactis* (see p. 154). R. J. WILLIAMS demonstrated several years ago that pantothenic acid is universal in its distribution. He and his collaborators succeeded in isolating it in its crystalline form and in synthesizing it. Beta-alanine is a growth factor for the diphtheria bacillus (*Corynebacterium diphtheriae*). Its action was first discovered by R. J. WILLIAMS and E. ROHRMAN (1936) working with certain strains of *Saccharomyces cerevisiae*.

Pantothenic acid is identical with the "filtrate factor" and the anti-pellagra factor of pigeons and chickens (JUKES, WOOLLEY, WAISMAN and ELVEHJEM). This vitamin also stimulates the growth of rats (SUBBAROW and HITCHINGS). Moreover, it has been observed that beta-alanine alone stimulates the growth of rats when there is a deficiency of the "filtrate factor" (HOFFER and REICHSTEIN, 1939).

---

[1]The name, pantothen, has been suggested by R. J. WILLIAMS (Science 94:462, 1941) as a substitute or abbreviation for pantothenic acid.

The synthesis of pantothenic acid and the establishment of its structural formula have made it clear that beta-alanine is a precursor of pantothenic acid. The chemistry of pantothenic acid will be considered in detail under the section entitled "bios" (see p. 129). It suffices here to say that pantothenic acid ($C_9H_{17}O_5N$) is a combination of beta-alanine and dioxyvaleric acid. This explains why the one or the other of these factors is required by certain microorganisms. The latter precursor of pantothenic acid is a lactone ($C_6H_{10}O_3$) which has been isolated from a concentrate of liver and shown to have the following structure:

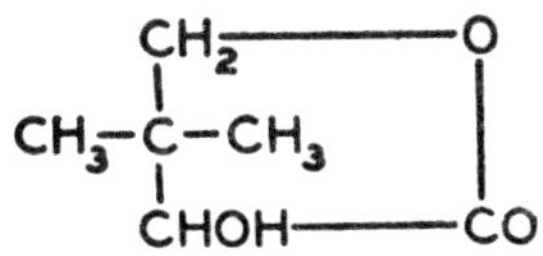

This lactone, known chemically as alpha, gamma-dihydroxy-beta, beta-dimethylbutyrolactone, upon condensation with beta-alanine forms pantothenic acid:

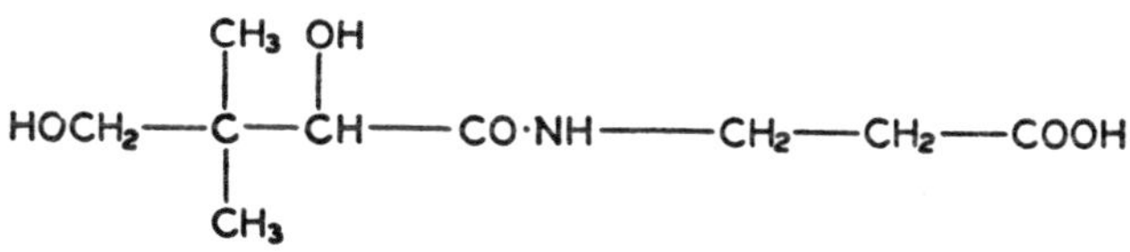

α,γ-DIHYDROXY-β,β-DIMETHYLBUTYRYL-β′-ALANIDE

This substance, once known only as a growth factor of yeast, must now be considered as a typical vitamin with the same general importance and with a distribution as universal as that of the other known vitamins.

Bios. — Other members of the vitamin B complex with bios activity still remain to be discussed, namely, bios I (*i*-inositol) and bios II (biotin of KÖGL). These substances are widespread in animals but their rôle has been recently established by DU VIGNEAUD *et al.* (1942)[2]. It is noteworthy that biotin whose structure is still unknown ($C_{11}H_{18}O_3N_2S$) is found in the ovarian follicles of chickens where the content increases from 439 US (0.7 gram follicle) to 1562 US (18.1 gram follicle) during the period of maturation.[1]

The "vitamin B_2 complex" is not yet completely understood since it contains other factors which have never been chemically isolated. Some are probably not vitamins at all. The water soluble vitamins B_3 and B_5 which occur in yeast, cereals and malt extract, as well as vitamin B_4 found in yeast have not been identified as to their chemical nature. Another little known factor is the *anti-anemic factor* (hemogene, "extrinsic factor") found in yeast, malt extract, the embryo of wheat seed, and the polishings of rice. Related to this factor is still another for anti-anemia called the tropic anemia-

---

[1]Recent work by GYÖRGY *et al.* has proven the identity of biotin, vitamin H (curative factor for egg-white injury), and coenzyme R, indicating that it plays a fundamental rôle in both plants and animals.

[2]V. DU VIGNEAUD, The structure of biotin. Science 96:455.

preventative factor ("tropiques", "tropenanämieverhütender Faktor"), found in yeast and the germ of wheat seed. Additional factors exhibiting anemia-preventative action are: *vitamin M* which prevents cytopenia in monkeys (in yeast), hallochrome $C_9H_7O_4N$ (in various plants), and xanthopterine (uropterine) $C_{19}H_{20}O_7N_{16}$ (in *Trifolium*). Hallochrome probably plays a rôle as an intermediate product during the course of the formation of melanin. These animal vitamins are incompletely understood and, for the moment, are of no particular interest to botanists. However, they must not be disregarded entirely since their presence in plant materials suggests that they play a rôle there. It is highly significant that many of the typical vitamins required by animals have been found to act as growth factors for microorganisms and green plants and, conversely, that many of the growth factors of microorganisms have turned out to be vitamins required by animals.

REFERENCES AND SUGGESTED READINGS:-

The references given at the close of this chapter consist entirely of review articles and works which are more or less general in nature.

1927: L. RANDOIN & H. SIMONNET. — Les données et les inconnues du problème alimentaire, II. La question des vitamines. *In* Les Problèmes biologiques, Presses universitaires de France, 1927, 480 p.

1927: R. BERG. — Die Vitamine. Leipzig, S. Hirzel, 714 S.

1931: E. BROWNING. — The Vitamins. Monograph of the Pickett-Thompson Research Laboratory, Vol. 1, London, Baillière, Tindall & Cox, 575 p.

1931: H. C. SHERMAN and S. L. SMITH. — The Vitamins. New York, The Chemical Catalog Company, 575 p.

1932: L. K. WOLFF. — Vitaminen. s'Gravenhage, Algemeene Landsdrukkerij, 54 p.

1935: C. BOMSKOV. — Methodik der Vitaminforschung. Leipzig, G. Thieme, 301 S.

1937: R. GREWE. — Das Aneurin (Vitamin B_1). Erg. der Physiologie 39: 252-293.

1937: L. RANDOIN. — Vues actuelles sur le problème de l'alimentation, avec tables de composition des aliments. Actualités scientifiques et industrielles no. 579, Paris, Hermann & Cie, 125 p.

1937: M. MITOLO. — Vitamine, odierni aspetti del problema. Torino, Rosenberg & Sellier, 485 p.

1938: J. SIVADJIAN. — La chimie des vitamines et des hormones. Les monographies de chimie pure et appliquée, Paris, Gauthier-Villars, 239 p.

1938: R. R. WILLIAMS and T. D. SPIES. — Vitamin B_1 (thiamin) and its use in medicine. New York, The Macmillan Co., 411 p.

1938: R. AMMON & W. DIRSCHEL. — Fermente, Hormone, Vitamine und die Beziehungen dieser Wirkstoffe zueinander. Leipzig, G. Thieme, 451 S.

1939: W. STEPP, J. KÜHNAU und H. SCHROEDER. — Die Vitamine and ihre klinische Anwendung. Stuttgart, F. Enke, 283 S.

1939: W. N. HAWORTH and E. L. HIRST. — The chemistry of ascorbic acid (vitamin C) and its analogues. — Erg. der Vitamin- und Hormonforschung 2:160-191.

1939: H. BROCKMANN. — Die Chemie der antirachitischen Vitamine. Erg. der Vitamin- und Hormonforschung 2:55-104.

1939: P. KARRER. — Über Chemie der Flavine. — Erg. der Vitamin- und Hormonforschung 2:381-418.

1939: F. SEITZ. — Vitamine und Hormone und ihre technische Darstellung, II. Darstellung von Vitaminpräparaten. Leipzig, S. Hirzel, 205 S.

1939: C. A. ELVEHJEM, P. W. WILSON *et al.* — Respiratory enzymes. Burgess Publishing Co., Minneapolis, Minn., 236 p.

1940: H. VOGEL. — Chemie und Technik der Vitamine. Sammlung chemischer und chemisch-technischer Vorträge, Heft 45, Stuttgart, F. Enke, 271 S.

1940: F. GSTIRNER. — Chemisch-physikalische Vitaminbestimmungsmethoden. Stuttgart, F. Enke, 226 S.

1940: Anonymous. — Riboflavin, Annotated Bibliography. By Merck & Co., Rahway, N. J., 119 p.

1940: —— ——. — Ascorbic Acid Merck. Merck Service Bulletin, revised. By Merck & Co., Rahway, N. J., 39 p.

1940: —— ——. — Alpha-tocopherol (Vitamin E), Annotated Bibliography. By Merck & Co., Rahway, N. J., 138 p.

1940: —— ——. — Vitamin B_6, Annotated Bibliography. Published by Merck & Co., Rahway, N. J., 46 p.

1940: —— ——. — Vitamin B_6 Hydrochloride (Pyroxidine Hydrochloride). Service Bulletin, by Merck & Co., Rahway, N. J., 22 p.

1940: —— ——. — Nicotinic Acid, Annotated Bibliography. By Merck & Co., Rahway, N. J., 88 p.

1940: —— ——. — Nicotinic Acid, and Nicotinamide. Service Bulletin, by Merck & Co., Rahway, N. J., 22 p.

1940: —— ——. — The Treatment of Prothrombin Deficiency with 2-methyl-naphthoquinone. Merck Service Bulletin, by Merck & Co., Rahway, N. J., 20 p.

1940: —— ——. — Vitamin K, Annotated Bibliography. By Merck & Co., Rahway, N. J., 91 p.

1940: —— ——. — Vitamin B_1 (Thiamine Hydrochloride), Annotated Bibliography. By Merck & Co., Rahway, N. J., 97 p.

1940: —— ——. — Vitamin B_1 Merck (crystals). Merck Service Bulletin, Revised, by Merck & Co., Rahway, N. J., 31 p.

1940: —— ——. — Riboflavin Merck. Service Bulletin, 3rd rev. By Merck & Co., Rahway, N. J., 39 p.

1940: C. R. ADDINALL. — The Story of Vitamin B_1. Published by Merck & Co., Rahway, N. J., 72 p.

1940: A. R. TODD *et al.* — Vitamin E; a Symposium held under the auspices of The Food Group (Nutrition Panel) of the Society of Chemical Industry. New York, Chemical Publishing Co., 88 p.

1940: S. B. BINKLEY, R. W. MCKEE, S. A. THAYER and E. A. DOISY. — The constitution of vitamin K_2. J. Biol. Chem. 133:721.

1941: H. C. SHERMAN. — Chemistry of Food and Nutrition. 6th ed., New York, The Macmillan Co., 611 p.

1941: H. A. WAISMAN and C. A. ELVEHJEM. — The Vitamin Content of Meat. Minneapolis, Burgess Publishing Co., 210 p.

1941: W. H. EDDY. — What are the Vitamins? New York, Reinhold Publishing Co., 247 p.

1941: Anonymous. — Vitamins in Nutrition. By Merck & Co., Rahway, N. J., 30 p.

1941: E. A. DOISY. — Recent developments in the investigation of vitamin K and other antihemorrhagic compounds. The Chem. Bull., Chicago Sect. Amer. Chem. Soc. 28:199-202, 228.

1941: M. LANDY. — Vitamin H, Biotin and Coenzyme R, a brief review of the literature. Published by the Research Laboratories, S. M. A. Corporation, Chagrin Falls, Ohio.

1941: H. R. BUTT and A. M. SNELL. — Vitamin K. Philadelphia, W. B. Saunders Co., 172 p.

1942: H. R. ROSENBERG. — Chemistry and Physiology of the Vitamins. New York, Interstate Publishers, 700 p.

1942: OTTO MEYERHOF *et al.* — A symposium on respiratory enzymes. The University of Wisconsin Press, Madison, Wis., 281 p.

1942: W. J. ROBBINS and V. KAVANAGH. — Vitamin deficiencies of the filamentous fungi. Bot. Rev. 8:411-471.

1942: E. A. EVANS, Jr. *et al.* — The Biological Action of the Vitamins, a Symposium. Chicago, The University of Chicago Press, 227 p.

Chapter VI.

VITAMINS AND THEIR ACTION ON PLANTS SYNTHESIZING THEM. I. EMBRYOS AND ROOTS.

The fact that vitamins occur in higher plants suggests that they participate in the metabolism. Since vitamins are always present under normal conditions of germination and culture (normal synthesis of vitamins) they are generally overlooked. In order to call attention to the synthesis of these vitamins, it is necessary to find a means whereby the plant can be rendered wholly or partially heterotrophic. This can best be accomplished by the use of tissue cultures. Although this type of culture was attempted long ago, it has met with numerous failures. The first experiments were made with embryonic organs, the cells of which have a considerable capacity for proliferation, *e.g.*, embryos of certain seeds.

Embryo cultures. — During the first stages of its development an embryo is completely heterotrophic and dependent upon the resereve material contained in the cotyledons.

The first attempt to culture embryos is generally but incorrectly credited to VAN TIEGHEM. In 1754 CHARLES BONNET reported in his celebrated work, "L'usage des feuilles", that he had succeeded in separating the embryos from their cotyledons in the seeds of *Phaseolus multiflorus* and *Fagopyrum*. By cultivating the excised embryos in soil he obtained plants which were somewhat dwarfed. The modern method of embryo culture was described in 1890 by BROWN and MORRIS (Jour. Chem. Soc. London). This method was employed by BOTTOMLEY in his experiments on the influence of vitamins on higher plants cultured on synthetic media. In the case of seeds of the *Gramineae*, which have been used by various workers, the endosperm is removed instead of the cotyledon. Numerous experiments have been performed in connection with various physiological problems in which embryos have been grown to a certain stage of development. The methods have been improved so that it is now possible to start with sterilized seeds and grow them to the mature fruiting stage in a sterile synthetic mineral medium. This was made possible by the work of MOLLIARD and COMBES on culture methods for embryos. It is now possible to start with an embryo quickly removed from the seed and to obtain a dwarf plant with flowers (RYTZ 1939). Upon the removal of the cotyledons, the embryos are deprived of their carbon and nitrogen reserves, thus making it necessary to employ a culture medium enriched by those substances which are lacking, particularly sugar-and nitrogenous materials (Fig. 3). Furthermore, as we shall see, the embryos of many plants are dependent upon vitamins and growth factors which are very abundant in the cotyledons.

In the culture of embryos it is important to separate the embryo from the cotyledons as soon as possible, *i.e.*, before any substances are able to pass from the cotyledons into the embryo during the early stages of germination. If possible, it is even advisable to remove the embryo from the seed as soon as the seed reaches maturity and before desiccation takes places. This precaution is necessary in the case of seeds which are able to germinate as soon as they are ripe, *i.e.*, those which do not require a dormant or after-ripening period such as the seeds of the *Gramineae* and *Cruciferae*. Even during maturation the embryo is able to accumulate a small quantity of growth factors, which must be removed by allowing

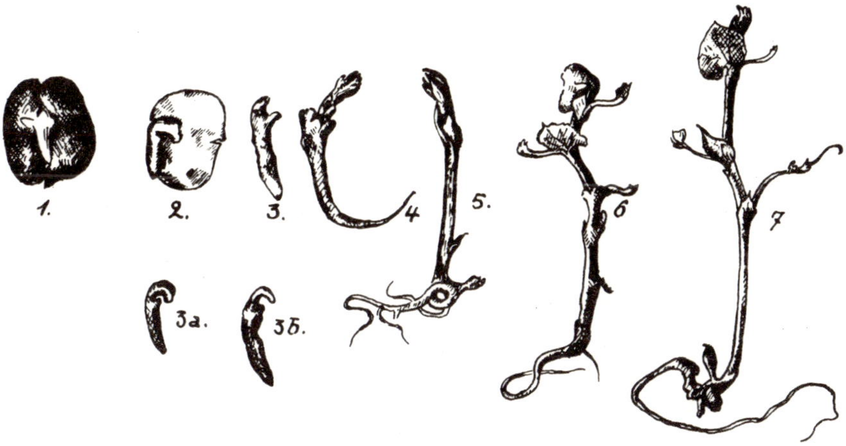

Fig. 3. — Culture of embryos of seeds of *Pisum*, without cotyledons, on solid synthetic medium (from W. Rytz jr., 1939); 1, 2, sketch showing the seed of Pisum; 3, *3a*, *3b*, excised embryos in early stages of development; 4-7, later stages of development.

the embryo to remain for some time in a medium devoid of growth factors. This medium permits growth to start, but soon checks it. At this time the embryo is devoid of all growth factors and is ready for experimental use. The observations of Kögl and Haagen Smit (1935) are very significant in this connection. The embryos of *Pisum* contain more biotin (in the plumule and especially in the radicle) after 48 hours of growth than after six hours of growth. These investigators, working with cultures of embryos from the seeds of *Pisum* ("Kortstroo Schokkers" variety) were the first to demonstrate the necessity of growth factors of vitamin nature during the course of germination. The embryos were cultured in a medium containing: saccharose 250g., KNO_3 2g., KH_2PO_4 2g., $Ca(NO_3)_2$ 9g., $MgSO_4.7H_2O$ 2g., $FePO_4$ 0.015g., $Na_2B_4O_7$ 0.000001g., $ZnSO_4$ 0.000001g. They found that extracts of cotyledons and yeast were very beneficial when added to the synthetic medium. Additions of either asparagine or aspartic acid to the medium had a favorable action. Since biotin is abundant in the seedling during the course of normal germination, it is natural to suspect this substance as an important factor in the culture of embryos. KÖGL and HAAGEN SMIT, working with "Kaapsche Groene" variety, obtained an increase of 63 per cent in dry matter, with 2,000 Saccharomyces units (1 US = 1/25,000 gamma) of biotin. Even better results were obtained with the combination biotin + thiamin. With the "Kaapsche Groene" variety, thiamin alone markedly increased root growth, the action beginning at 0.008γ per 10 cc. of medium and reaching a maximum at 0.4γ. The growth of stems was affected less than that of roots by thiamin alone. The maximal increase in stem growth was obtained with the combination of biotin (0.2γ) + thiamin (2.0γ). In general the increase in growth was found to be less in stems than in roots. The green weight of the stem was increased considerably by biotin alone in a dosage of 0.1γ, and the combination, biotin (0.2γ) + thiamin (2.0γ), had only slightly greater effect despite the fact that thiamin, when used alone, had a marked effect on the green weight of the stem. On the other hand, it was found that the fresh weight of the root was not affected by biotin alone (0.1γ), and that thiamin alone was as effective as the combination, thiamin + biotin. The dry weight of the stem was affected more strongly by biotin than by thiamin but the maximal effect was obtained with the combination, biotin + thiamin. The dry weight of the root was not affected by the addition of biotin alone; the maximal effect was obtained with thiamin alone and was not increased upon the addition of the combination, biotin + thiamin.

From these observations of KÖGL and HAAGEN SMIT it is obvious that biotin and thiamin have about equal importance, but that they behave differently in the different organs of the embryo. In the final analysis, it would appear that stems and roots differ in their ability to synthesize these two factors. The fact that additions of biotin have practically no effect on roots (growth in length, fresh weight and dry weight) indicates that this vitamin is readily synthesized in this organ. Thiamin, on the contrary, markedly affects the growth of roots. In stems the situation is different, both biotin and thiamin being very active when used alone; however, their maximal action occurs when they are used together. The stem, therefore, has considerable difficulty in synthesizing both of these substances.

Experiments concerning the effect of ascorbic acid (vitamin C) on the growth of plants have yielded conflicting results. KÖGL and HAAGEN SMIT (1935) were unable to find any activity, whereas VIRTANEN and VON HAUSEN found that this vitamin definitely promotes the growth of embryos deprived of their cotyledons. VON HAUSEN (1936), in particular, found that ascorbic acid had an extremely marked effect on the growth in length, the increase in dry weight, and the subsequent vitamin C content of seedlings (see Fig. 8). VIRTANEN and VON HAUSEN used a method different from that of KÖGL, according to which they allowed the seeds to germinate for seven days before removing the cotyledons. Furthermore, the dosages of ascorbic acid were rather high, up to 10 mg., whereas those used by KÖGL did not exceed 800γ. The experiments of VON HAUSEN give the impression that ascorbic acid may have functioned otherwise than as a growth factor. The investigations of these men differed still further in that the varieties of peas used were not the same. J. and D. BONNER (1938), who studied the action of ascorbic acid on nine varieties of peas found that they reacted differently. The "Perfection" variety responded to ascorbic acid with a 213 per cent increase in weight, whereas the variety "Little Marvel" did not increase in weight beyond 103 per cent (as compared with the control). This clearly demonstrates the fact that varieties differ in their response to ascorbic acid added to the culture solution. The embryos cultivated in a solution containing thiamin as the only growth factor exhibited a marked increase in growth and also a considerable increase in the ascorbic acid content of the embryo (Perfection variety). In this case the thiamin apparently compensates for the absence of vitamin C by permitting the synthesis of the latter.

Of other substances tested on pea embryos the following are active according to J. BONNER and G. AXTMAN (1937): thiamin *alone*, pantothenic acid *alone*, and estrone (theelin) *alone* (Perfection and Alaska varieties). The best results were obtained with the following combination: thiamin + pantothenic acid + estrone + ascorbic acid. KÖGL and HAAGEN SMIT (1935) found the activity of estrone almost negligible and that of biotin slight. BONNER (1938) found that nicotinic acid possessed strong activity when used alone but that the action of this factor was increased when it was used in conjunction with thiamin (growth of shoots).

The results of these investigations on embryo culture have been confirmed with the exception of those concerning ascorbic acid.

It is now possible to regard an embryo, detached from its cotyledons, as heterotrophic as far as vitamins are concerned (auxoheterotrophic), a condition which can easily be explained by the large growth factor content of the cotyledons which normally supply the embryo. The loss of capacity for synthesis differs markedly in the root and shoot of the embryo. The root still has the ability to synthesize a considerable amount of biotin, but no thiamin. The rudimentary stem, on the other hand, has lost its capacity to synthesize both of these factors. Other factors likewise affected are nicotinic acid, pantothenic acid, estrone and, in some cases, ascorbic acid. The capacity to synthesize ascorbic acid depends largely on the variety of plant. The synthesis of biotin differs to a lesser extent with the variety, whereas that of thiamin does not differ at all. Biotin is the principal factor by virtue of its unusually great activity.

One remarkable discovery resulting from these studies was that the needs vary with the state of development of the embryo and with the preliminary treatment received.

It is difficult to compare the data obtained by different investigators because the media employed by them were not the same.

The carbon source is of considerable importance in its effect upon the synthesis of ascorbic acid. Various other factors influencing the biosynthesis of this factor in higher plants have been demonstrated, *e.g., temperature, light,* and the *catalytic action of metals* (see p. 95). These factors still remain to be studied in connection with the synthesis of ascorbic acid in the embryo. Such studies might possibly explain certain discrepancies of the results. The knowledge which has already been accumulated concerning the part played by growth factors represents an advance in the study of the physiology of germination.

After having established the fact that embryos are partially heterotrophic we might well ask what happens in the adult plant. Does it become completely auxoautotrophic? Do all the organs become auxoautotrophic? If not, what physiological relationships, relative to growth factors, exist between the various organs? In order to answer these questions it is necessary to investigate individually each organ which can be isolated and cultured apart from the rest of the plant.

Leaf cultures. — Little is known concerning the synthesis of vitamins in excised leaves. As the "chemical laboratory" of the plant, the leaf theoretically should be able to carry on all of the essential syntheses, and should be self-sufficient with respect to vitamins. According to D. BONNER, HAAGEN SMIT and WENT (1939), the leaf requires one or several "hormones". These can be found in the extracts of leaves, in the cotyledons of pea seeds, and in yeast extract. These factors have not been identified and the problem remains unsolved. It is known, however, that adenine, in a concentration of 0.1-1.0 mg. per liter, plays a rôle in the expansion of the leaf.

Root cultures. — The root, which is generally devoid of chlorophyll, is of particular interest with respect to growth factors. This organ may well be expected to be dependent upon organs with a greater ability to synthesize growth factors.

The growing points represent meristems endowed with unusual capacity for regeneration. GAUTHERET (1935) demonstrated this by cutting off the extreme tips of roots of *Zea Mays.* These tips regenerated rapidly and the absence of cell initials did not interfere with this phenomenon.

Numerous attempts to culture roots have been made, particularly in connection with studies concerning the action of hormones of the heteroauxin type (see BURLET for the technique). From the point of view which is of interest here, real progress has been made since the initial observations of ROBBINS (1922). We are indebted to him for some of the best work on this subject, including the successful culture of roots of *Zea Mays* in sterile media. In contrast to the work of KOTTE, who cultivated the very tips of the roots, ROBBINS employed longer portions (1 cm. or more) and succeeded in obtaining excellent cultures in which not only cell elongation occurred but also cell division was very rapid, resulting in marked increase in length of the root accompanied by considerable branching. He was able to maintain cultures over a long period of time by transferring the excised roots periodically to fresh media. Glucose is essential for growth, and peptone and yeast extract are decidedly beneficial. These products, particularly the latter, are recognized as extremely rich sources of growth factors. ROBBINS and MANEVAL (1923) made a systematic study of roots of various species of plants and were able to determine the action of various factors on their growth rate. Mineral elements are also very important for the development of root cultures. Particularly interesting results were obtained with tomato roots which respond nicely to culture *in vitro*. WHITE (1934) obtained an unlimited culture of excised tomato roots in a synthetic mineral medium by supplying glucose and yeast extract. This extract has been shown by numerous investigations to contain substances essential to the growth of roots.

WHITE (1937 *a* and *b*) attempted to separate the active constituents from yeast. He made the interesting observation that a certain number of them is indispensable to excised tomato root cultures, namely, histidine, phenyl-alanine, lysine, leucine, isoleucine, valine, glutamic acid, proline, and serine. But these substances are not the only ones which account for the activity of the extract of autolyzed yeast.

Encouraged by the good results obtained by means of pure vitamins, ROBBINS and BARTLEY (1937) employed thiamin in an experiment and succeeded in obtaining excellent growth of their root cultures. This vitamin proved to be an essential factor for the growth of roots. It can not be replaced by any of the following substances: heteroauxin, ash of yeast, ascorbic acid, pantothenic acid, asparagine, cysteine, inositol, urea, pimelic acid or hydrolytic products of nucleic acid. This observation is of prime importance to the problem of tissue culture. It was confirmed by WHITE (1937). ROBBINS and BARTLEY-SCHMIDT (1938) found that thiamin was still active at a dilution as high as 1:40,000,000,000,000, when used with WHITE's solution: $Ca(NO_3)_2$, $MgSO_4$, KNO_3, KCl, KH_2PO_4, $Fe_2(SO_4)_3$ and glucose (Plate I). Another important discovery by ROBBINS and BARTLEY (1937) was that the complete molecule of thiamin is not required, but only the thiazole portion. The combination, pyrimidine + thiazole, is able to replace thiamin. The dry weight increases as a function of the concentration of the thiamin components but the increase is somewhat slower than that obtained with thiamin. At the end of the experiment, however, the weights obtained by the use of pyrimidine + thiazole are practically the same as those obtained by the use of thiamin. Thiazole possesses definite activity when used alone, and even greater activity when used in conjunction with pyrimidine. (See Fig. 1, V).

The importance of thiamin for growing roots *in vitro* has been confirmed by BONNER and ADDICOTT (1937) with peas (*Pisum*). In this case the essential factor, thiamin, can be replaced by the two thiamin components pyrimidine + thiazole, providing both of these components are supplied together. Peas, therefore, have suffered a greater loss in their capacity to synthesize than tomatoes. The former have lost their ability to synthesize both pyrimidine and thiazole, whereas the latter have lost their ability to synthesize only thiazole. Peas are still able to unite the two components of thiamin into a complete molecule.

The synthesis of thiamin in roots of peas, which is comparable with that established for microorganisms (see p. 117), was demonstrated in the following manner: the *Phycomyces* test permits the detection of thiamin or its two components, pyrimidine and thiazole, while the *Phytophthora* test permits the detection of thiamin only (whole molecule). These two tests make it possible to demonstrate that peas growing in a medium containing pyrimidine and thiazole are able to condense these two components into a thiamin molecule. These tests also permit the establishment of the fundamental fact that tomato (*Lycopersicum esculentum*) roots growing in pure and unlimited culture have lost their capacity to synthesize thiazole, but are still able to synthesize pyrimidine. Pea roots have lost the capacity to synthesize both pyrimidine and thiazole. In the roots of both plants the thiamin molecule is resynthesized. The capacity for synthesis in pea roots will be studied in greater detail later on (see p. 89). It is clear that thiamin must not be considered as the only factor involved in the growth of roots, but that this vitamin acts in conjunction with other active substances of the auxin group. The action of thiamin has been carefully studied in regard to its effect upon the cytology of the root (*Pisum*) (ADDICOTT, 1939). Thiamin stimulates primarily the meristematic activity of roots but does not affect cell elongation. Its action is associated with that of auxin.

The work concerning the nutrient requirement of excised roots started with the extract of autolyzed yeast as the source of the active factors. It soon became apparent that yeast extract could not be completely replaced by thiamin, or amino acids alone or in combination with thiamin (WHITE in particular, 1937). From the figures of WHITE (Fig. 4), it is obvious that tomato roots do not grow as rapidly in a synthetic medium supplemented with amino acids, accessory mineral salts, and vitamin B_1, as with yeast extract. Other growth factors might well be suspected since the solution was optimal in all other respects. Further support of this assumption is found in the fact that the behavior of the roots of peas and tomatoes differ.

ROBBINS and BARTLEY-SCHMIDT (1938*a*) found that pyridoxine (vitamin B_6) when added to thiamin or thiazole increased considerably the activity of the latter substances on tomato roots:

	DRY WEIGHT OF CROP IN MG.
Control	0.4
5γ thiamin	3.4
5γ thiamin + 1γ pyridoxine	16.1
1γ pyridoxine	1.8

Pyridoxine exhibits no activity when used alone. It belongs to a constellation based on thiamin.

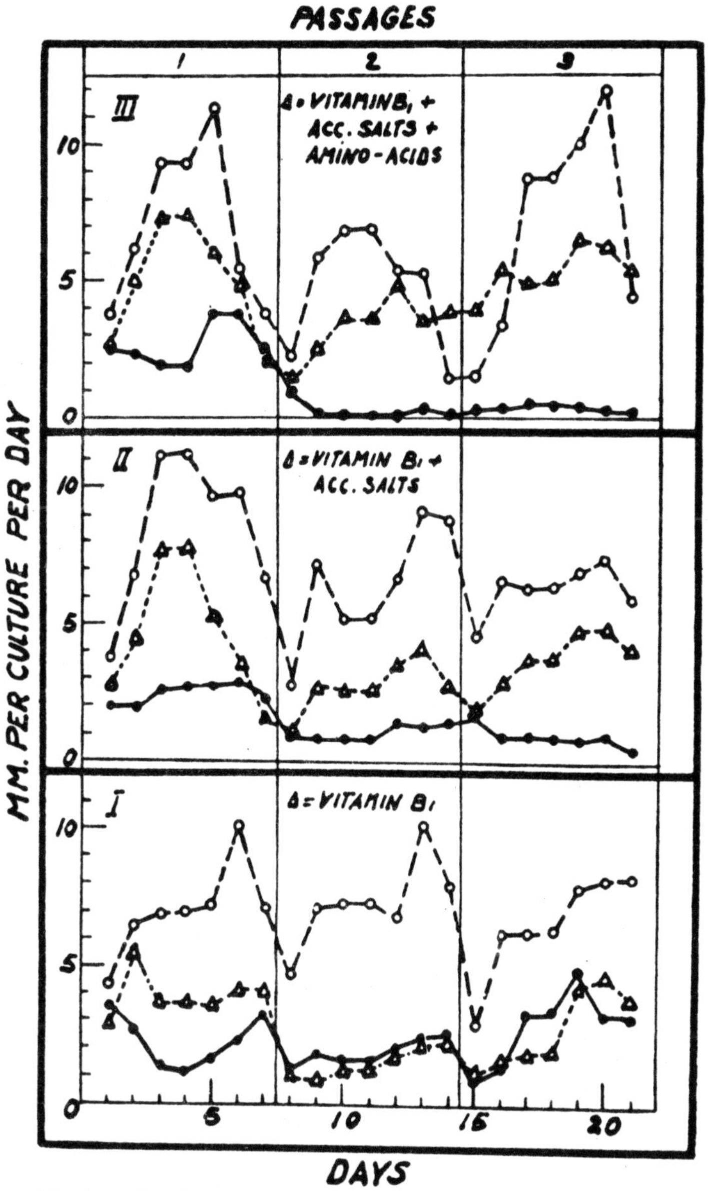

FIG. 4. — Growth of excised tomato roots during the course of 3 passages. Solid circles: standard salts and sugar (basal medium); open circles: basal medium + yeast extract; triangles in I: basal medium + vitamin B_1 (1 mg. per liter); triangles in II: basal medium + vitamin B_1 + accessory salts; triangles in III: basal medium + vitamin B_1 + accessory salts + amino acids (from WHITE, 1937).

The growth of excised pea roots is increased when nicotinic acid is added to thiamin (ADDICOTT and BONNER 1938, ADDICOTT and DEVIRIAN 1939). Pantothenic acid has been credited with a stimulating effect on *Ricciocarpus natans* (R. J. WILLIAMS and ROHRMAN) and on the radicle of the embryo of *Ginkgo* (LI-TSI-TUNG and SHEN). The pantothenic acid which was used was not in crystalline form or sufficiently purified to permit any definite conclusions. Such experiments are of value only for orientation purposes.

The most interesting fact is the specificity of the reactions of roots from various species of plants. BONNER and DEVIRIAN (1939) and BONNER (1940) made comparisons of the growth factor requirements of roots of several species of plants. The results are presented in the following table:

TABLE IV. Vitamin requirement of roots growing in pure culture (from J. BONNER, 1940) :-

	VITAMIN B_1	NICOTINIC ACID	PYRIDOXINE
Flax	+	—	—
Pisum	+	+	—
Radish	+	+	—
Alfalfa	+	+	—
Clover	+	+	—
Cotton	+	+	—
Carrot	+	—	+
Tomato	+	+	+
Datura	+	+	+
Sunflower	+	+	+

For the growth of radish roots, the combination thiamin + nicotinic acid is as favorable as yeast extract. In this case all of the necessary growth factors are known.

WHITE has shown that the growth of wheat seedlings is promoted by the SH radical, the action of which has been known since the time of HAMMET. J. and D. BONNER found that a constellation consisting of vitamin B_1 + vitamin B_6 + nicotinic acid is required by the roots of *Cosmos* and that vitamins E, K, B_2, pantothenic acid, beta-alanine, ascorbic acid, adenine and many of the amino acids have no effect.

From the above studies of the BONNERS it can be concluded that if all the other hormonal factors are present the ten different kinds of roots employed vary in their ability to synthesize vitamins. The least capable in this respect are the roots of tomato, *Datura*, and sunflower, whereas the most capable is the root of flax.

The results of BONNER and DEVIRIAN concerning excised tomato roots agree with those of ROBBINS and BARTLEY-SCHMIDT regarding the action of pyridoxine but disagree regarding the action of nicotinic acid. In an attempt to reexamine the results of BONNER and DEVIRIAN and those of ROBBINS and BARTLEY-SCHMIDT, WHITE (1940) was unable to find any growth-promoting activity in either pyridoxine or nicotinic acid. He also pointed out that the discrepancies between his results and those of the other investigators were probably caused by differences in method and in behavior of different strains of tomato roots (WHITE's strain, Bonnie Best; BONNER and DEVIRIAN's strain, Beefsteak; ROBBINS and BARTLEY-SCHMIDT's strain, pink-fruited Mexican variety). In contrast with the embryo, which is auxoheterotrophic during only a part of its development, the root is always more or less auxoheterotrophic, and must therefore depend on the leaf, which is able to synthesize all the necessary substances.

We still have very little information concerning the underlying reasons for this specificity in the loss of capacity for synthesis. It is very probable that the composition of the medium plays a rôle. It will be interesting to investigate the manner of action of the amino acids, which are indispensable and which must be present along with the growth factors.

Specificity of action of growth factors on roots. — A great deal of information is available concerning the specificity of action of thiamin (pyrimidine + thiazole) on the root of *Pisum*. In this case we are able to describe the manner in which the concept of specificity of action was conceived. In these experiments (BONNER, 1938) pyrimidine and thiazole were supplied to pea roots in the form of various substitution products.

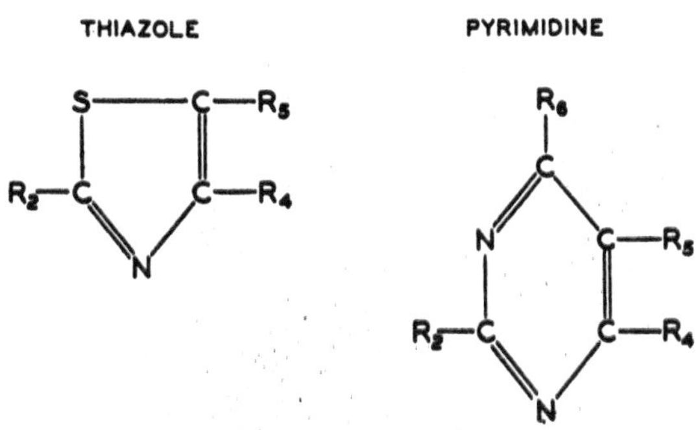

From table V it is obvious that a few substitutions are possible, both for thiazole and pyrimidine. In the case of thiazole, the H in position 2 is of prime importance (see Nos. 9 and 12 in the table). The CH_3 group in position 4 can be substituted without destroying the activity (Nos. 3 and 8). The β-hydroxyethyl in position 5 can also undergo substitutions providing there remains an active group capable of reacting with pyrophosphoric acid for the formation of cocarboxylase (see p. 70). In the case of pyrimidine the NH_2 group in position 4 of prime importance (No. 5). The CH_2NH_2 group or CH_2Br must remain since it must react with the thiazole in order to form the thiamin molecule.

BONNER studied, in addition to the analogues of the components of thiamin, the activity of certain substitution products of thiamin (whole molecule); in this connection he found thiochrome inactive when used in conjunction with pyrimidine or thiazole.

It is interesting to compare this specificity in relation to pea roots with that of thiamin in relation to animals, since there appears to be a difference. The specificity with relation to pea

TABLE V. Specificity of action of the components of thiamin on isolated pea roots growing in pure culture (from J. BONNER, 1938).

Thiazole and its analogues (in the presence of normal pyrimidine)

	R_2	R_4	R_5	PERCENT ACTIVITY
1.	H	$-CH_3$	$-CH_2CH_2OH$	100 (THIAZOLE OF ORDINARY THIAMIN)
2.	H	$-CH_3$	$-CH_2CH_2Cl$	100
3.	H	$-CH_2OH$	$-CH_2CH_3$	100
4.	H	$-CH_3$	$-CH_2CHOH \cdot CH_3$	100
5.	H	$-CH_3$	$-CH{=}CH_2$	100
6.	H	$-CH_3$	$-CH_2Br$	90
7.	H	$-CH_3$	$-(CH_2)_3OH$	75
8.	H	$-CH_2CH_2Cl$	$-CH_3$	75
9.	CH_3	$-CH_3$	$-CH_2CH_2OH$	35
10.	H	$-CH_3$	$-CHOH \cdot CH_3$	30
11.	H	$-CH_3$	H	0
12.	$-NH_2$	$-CH_3$	$-CH_2CH_2OH$	0
13.	H	$-CH_3$	$-COCH_3$	0

Pyrimidine and its analogues (in the presence of normal thiazole)

	R_2	R_4	R_5	R_6	
1.	$-CH_3$	$-NH_2$	$-CH_2Br$	H	100
2.	$-CH_3$	$-NH_2$	$-CH_2NHCSH$	H	100
3.	$-CH_3$	$-NH_2$	$-CH_2NH_2$	H	95
4.	$-CH_3$	$-NH_2$	$-CH_2OC_2H_5$	H	25
5.	$-CH_3$	$-OH$	$-CH_2NH_2$	H	0
6.	$-CH_3$	$-NH_2$	$-CH_2CONH_2$	H	0
7.	$-OH$	$-OH$	$-CH_2OH$	CH_3	0

roots also differs from that relative to microorganisms of the same type (pyrimidine plus thiazole), *e.g.*, *Phycomyces*, by being much less pronounced. It is, however, somewhat similar to the situation found in flagellates by LWOFF *et al.* (see p. 121), where the vinyl group $CH=CH_2$, in position 5 of thiazole does not destroy the activity.

The specificity of action of nicotinic acid on pea roots (*Pisum*) has likewise been studied. Twenty-three substances, all more or less apparently related to nicotinic acid, were tested by BONNER. The only ones which were active were those which yield nicotinic acid upon simple hydrolysis. The active compounds include: nicotinic acid, nicotinamide, the methyl-, ethyl-, propyl-, and butyl-nicotinates, and nicotinuric acid.

Pea roots are unable to hydrolyze nicotino-3-nitrile and to oxydize the methyl group of β-picoline necessary for the formation of nicotinic acid. In general these results are comparable with those obtained with *Staphylococcus aureus* and *Shigella dyssenteriae* as well as with dogs. (See p. 143 for the specificity of action of nicotinic acid on microorganisms). A single exception exists in the case of β-picoline, which possesses canine anti-blacktongue activity.

REFERENCES:-

1922: W. J. ROBBINS. — Cultivation of excised root tips and stem tips under sterile conditions. Bot. Gaz. 73:376.

——— ——— ——— . — Effect of autolyzed-yeast and peptone on growth of excised root tips in the dark. Bot. Gaz. 74:59.

1923: W. J. ROBBINS and W. E. MANEVAL. — Further experiments on growth of excised root tips under sterile conditions. Bot. Gaz. 76:274.

1933: P. R. WHITE. — Plant tissue cultures. Results of preliminary experiments on the culturing of isolated stem-tips of *Stellaria media*. Protoplasma 19:97.

1934: P. R. WHITE. — Potentially unlimited growth of excised tomato root tips. Plant Physiol. 9:585.

1935: R. J. GAUTHERET. — Recherches sur la culture des tissus végétaux. Essais de culture de quelques tissus méristématiques. Paris, Le François, 279 p.

1936: F. KÖGL und A. J. HAAGEN SMIT. — Biotin und Aneurin als Phytohormone. Zeitschr. f. physiol. Chem. 243:209.

1936: S. VON HAUSEN. — The rôle of vitamin C in the growth of higher plants. Ann. Acad. Sc. Fennicae, Series A, 46, 134 p.

1937*a*: P. R. WHITE. — Separation from yeast of materials essential for growth of excised tomato roots. Plant Physiol. 12:777.

1937*b*: ——— ——— . — Amino acids in the nutrition of excised tomato roots. Plant Physiol. 12:793.

1937*c*: ——— ——— . — Vitamin B_1 in the nutrition of excised tomato roots. Plant Physiol. 12:803.

1937: W. J. ROBBINS and M. A. BARTLEY. — Thiazole and the growth of excised tomato roots. Proc. Nat. Acad. Sc. Washington 23:385.

——— ——— ——— ——— ——— . — Vitamin B_1 and the growth of excised tomato roots. Science 85:246.

1937: J. BONNER and F. ADDICOTT. — Cultivation in vitro of excised pea vitro. Preliminary experiments on the rôle of accessory substances. Proc. Nat. Acad. Sc. Washington 23:453.

1937: J. BONNER and F. ADDICOTT. — Cultivation in vitro of excised pea roots. Bot. Gaz. 99:144.

1938: F. T. ADDICOTT and J. BONNER. — Nicotinic acid and the growth of isolated pea roots. Science 88:577.

1938: J. BONNER and D. BONNER. — Ascorbic acid and the growth of plant embryos. Proc. Nat. Acad. Sc. Washington 24:70.

1938: J. BONNER. — Nicotinic acid and the growth of isolated pea embryos. Plant Physiology 13:865.

1938*a*: W. J. ROBBINS and M. A. BARTLEY-SCHMIDT. — Vitamin B_6, a growth substance for excised tomato roots. Proc. Nat. Acad. Sc. Washington 25:1.

1938*b*: ——— ——— ——— ——— . — Growth of excised roots of the tomato. Bot. Gaz. 99:671.

1938: J. BONNER. — Thiamin (vitamin B_1) and the growth of roots: the relation of chemical structure to physiological activity. Amer. J. Bot. 25:543.

1938: J. BONNER and E. R. BUCHMAN. — Syntheses carried out *in vivo* by isolated pea roots, I. Proc. Nat. Ac. Sc. Washington 24:431.

1939: W. RYTZ jr. — Beitrag zum Aneurinstoffwechsel bei höheren Pflanzen. Ber. schweiz. bot. Ges. 49:339.

1939: J. BONNER and P. S. DEVIRIAN. — Growth factor requirements of four species of isolated roots. Amer. J. Bot. 26:661.

1939: F. T. ADDICOTT and P. S. DEVIRIAN. — A second growth factor for excised pea roots: nicotinic acid. Amer. J. Bot. 26:667.

1939: F. T. ADDICOTT. — Vitamin B_1 in relation to meristematic activity of isolated pea roots. Bot. Gaz. 100:836.

1939: D. M. BONNER, A. J. HAAGEN SMIT & F. C. WENT. — Leaf growth

1939: D. M. BONNER, A. J. HAAGEN SMIT and F. W. WENT. — Leaf growth hormones, I. A bio-assay and source for leaf growth factors. Bot. Gaz. 101:128.

1940: E. BURLET. — Über die pflanzlichen Organkulture und ihre Anwendung bei physiologischen Untersuchungen. Ber. Schweiz. Bot. Ges. 50:519.

1940: J. BONNER. — On the growth factor requirements of isolated roots. Amer. J. Bot. 27:692.

1940: P. R. WHITE. — Vitamin B_6, nicotinic acid, pyridine, glycine and thiamin in the nutrition of excised tomato roots. Amer. J. Bot. 27:811.

Chapter VII.

VITAMINS AND THEIR ACTION ON PLANTS SYNTHESIZING THEM. II. TISSUE CULTURES, CUTTINGS, FORMATION OF ORGANS, ETC.

Root tissue cultures. — In the preceding chapter the vitamin requirements of certain organs were compared and an analysis of the behavior of various plant tissues seems appropriate now. In this connection special attention must be given to plant tissue cultures which have, for the most part, defied the efforts of numerous investigators. Whereas green tissues are difficult to culture, undifferentiated meristematic tissues are very satisfactory for this purpose due to their great ability to proliferate under normal conditions.

The pioneering work of HABERLANDT (1902) on tissue cultures deserves mention here since it established once and for all the importance of this field. Because his experiments were concerned with traumatism and the subsequent production of active substances of a hormonal nature, this particular aspect of the problem can be omitted from the following discussion.

Notable progress in the technique of tissue culture was made through the work of GAUTHERET and WHITE. According to the recommendations of these authors, cultures of embryonic organs, *e.g.*, roots and embryos, will be distinguished clearly from those of meristematic tissues (particularly the cambium). The latter represent tissue cultures in the real meaning of the term and as such they are comparable to animal tissue cultures. When properly used the term "tissue culture" implies that differentiation of cells does not accompany growth[1]. The general procedure of tissue culture is as follows: a fragment of tissue is inoculated and given some time to grow; then the newly formed tissue is separated from the fragment of inoculum and both are transferred to, and cultivated in the same medium or a fresh one. If desirable, several transfers may be made. In the case of actively growing tissues NOBECOURT observed that the volume increased one hundred times during one

[1]More recent investigations by WHITE have forced him to revise somewhat his concept concerning tissue cultures. He observed that individual cells possess a much greater capacity for differentiation than once supposed. He summarized his results as follows: "Cultures of callus from a hybrid *Nicotiana* which, on a semi-solid nutrient, have been maintained in an undifferentiated state through many passages can be induced to form leafy branches of a high degree of differentiation by immersing them in a liquid nutrient" (WHITE, 1939, 1941, 1942).

passage. Using the hypothetical case of a culture whose volume increases twenty times per passage, a stock culture could be grown until it had increased its volume twenty times, then could be cut into 20 pieces and these in turn could be cultured and subcultured until 20 passages had been made. By this time the volume of tissue would be equal to 20^7 or 1,280,000,000 times the original volume.

Frequently the course of development is retarded or modified by the differentiation of parenchyma tissue (see GAUTHERET 1935).

Excellent results were obtained by GAUTHERET (1935) with cells isolated from the root cap of lupine and particularly with the cambium of *Alnus, Populus, Salix, Acer,* and *Ulmus.* Knop's solution + glucose were used. GAUTHERET found that mannitol could not be utilized by the cultures. He was able to maintain the cultures for a long period of time without any differentiation taking place. Using the cambium of *Salix capraea,* GAUTHERET (1935) was able to study a constellation of growth substances and found that heteroauxin has a beneficial action when used in small dosages (up to 10^{-10}) but that it is toxic when used in larger dosages. Knop's solution with 2 per cent glucose and 1.5 per cent agar was used as a medium. The chlorhydrate of cysteine has a beneficial action, the optimal concentration being 10^{-6}. However, the essential factor is vitamin B_1. With this vitamin, cultures could be maintained for a period of five months either by transferring the newly formed tissue separated from the fragment of inoculum, or by transferring the fragment itself.

Numerous observations of this sort have since been confirmed on carrot tissues. Fragments of root tissue are easily isolated and cultured. NOBÉCOURT (1939*a*) obtained excellent results and maintained cultures for a considerable period of time. He also found that growth factors are required. GAUTHERET, also working with carrot tissues, found heteroauxin beneficial and thiamin, as well as the chlorhydrate of cysteine, indispensable. The ratios, final weight/initial weight, indicating the growth of the culture are:

mineral solution alone	2.0
— — + glucose	6.5
— — — + 0.001% cysteine	8.6
— — — + 0.000,001% heteroauxin	8.4
— — — + 0.0001% thiamin	10.9

This represents another case in which thiamin appears to be essential although it was supplied in large dosage. This vitamin is the one most frequently involved in the loss of capacity for synthesis. This loss, however, in the case of carrots must not be complete because NOBÉCOURT (1939*b*) observed that a fragment of carrot tissue, after having undergone five passages in fourteen months on a *thiamin-free* synthetic medium, is capable of initiating growth in *Phycomyces* and in *Phytophthora.* Thus these tissue cultures have been able to synthesize at least part of the necessary thiamin. GAUTHERET's experiments, in which thiamin was used in high dosage (0.0001%), raise the question of whether the action was strictly of vitamin nature.

Stem tissue cultures. — The first experiments with stem cultures of *Zea* and *Pisum* by ROBBINS (1922) were unsuccessful. The method employed was the same as that used in the culture of growing points of stems. In 1933 WHITE took up the problem and, working with small fragments of cambial tissue isolated from the stem of *Stellaria media,* he succeeded in obtaining growth and differentiation as evidenced by the production of rudimentary leaves and floral parts. Stem tissue culture should be investigated further since it has not received nearly as much attention as the culture of excised roots and of cambial tissues of roots.

Formation of organs. — Cuttings. — The production of roots and other organs on cuttings is largely dependent on hormones of the auxin or heteroauxin type. A specific root-producing substance was demonstrated by R. BOUILLENNE and WENT, Jr. (1933) and called *rhizocaline* (with *Acalypha* in particular). The general characteristics of this substance were not determined, although it was observed that they differed from those of thiamin (thiamin being represented by an extract from rice polishings). The tendency has been to consider "rhizocaline" as a substance belonging to the auxin group, since heteroauxin has a marked effect upon root formation. Recently BOUILLENNE (1938) resumed work on this problem with hypocotyls of *Impatiens Balsamina* and demonstrated that "rhizocaline" is not the same as auxin but he failed to establish its identity. He called attention to the very interesting fact that the addition of thiamin to a basal medium of saccharose and amino acids increased the number of roots formed. However, in his experiments its action was not very noticeable since the basal medium alone produced similar results. The action of thiamin as an accessory factor for the formation of roots on cuttings still remains to be studied. Biotin, however, has been shown by WENT and THIMANN (1937) to increase root formation on cuttings of *Pisum* (etiolated) by one hundred per cent when used in a concentration of 5000 US per cc. along with auxin and saccharose. These experiments clearly demonstrate the hormonal phenomenon as distinguished from the vitamin phenomenon defined on p. 38. Hormones (auxin and substances with similar activity) are concerned with the formation of organs, whereas vitamins (thiamin, biotin, etc.) take part in assimilation and nutrition by increasing the amount of protoplasm and without affecting the structure of the plant.

LAZAR (1935, 1936) suggested that carotin plays an active rôle in root formation on cuttings of *Impatiens Balsamina.* However, these results have not been confirmed.

Germination of seeds and the culture of whole plants. — From the earliest stages germination is accompanied by highly complex phenomena regarding digestion and translocation, according to the evidence obtained by SCHANDER with rice seeds. It can be assumed that these phenomena are influenced by active substances, particularly vitamin B_1 contained in the pericarp. Since thiamin is readily synthesized by the mature plants and the supply of it in the cotyledons should suffice during the early stages of germination, there seems to be little reason to investigate its action on seed germination and on the growth of whole plants. BONNER and GREENE (1938), however, have demonstrated that the addition of thiamin to seedlings definitely favors the production of dry matter and elongation of shoots. With plants growing in sand they obtained the following ratios (length of shoot with vitamin/length of shoot of control) for certain plants: *Aleurites Fordii* (rooted cuttings), 1.86; *Bougainvillea glabra* (seedlings), 1.3; *Arbutus unedo* (seedlings), 1.3; *Eucalyptus ficifolia* (seedlings), 1.5. In *Ceratonia siliqua* growing in sand the elongation of the shoot is increased almost 100 per cent (Fig. 5). With plants grow-

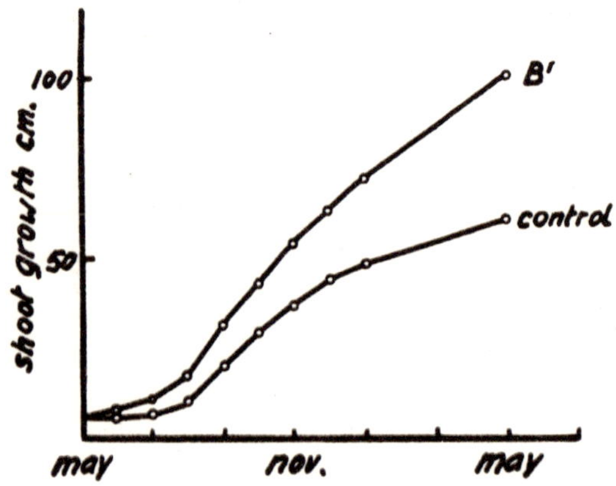

FIG. 5. — Growth of seedlings of *Ceratonia siliqua* in sand culture, with and without vitamin B_1 (from J. BONNER and GREENE, 1939).

ing in soil the effect is even more pronounced as shown by the following ratios: *Arbutus unedo,* 2.3; *Prunus ilicifolia,* 1.6; *Bryophyllum* sp., 2.5. The production of dry matter is also significantly increased by thiamin: *Xanthium pennsylvanicum,* 160 per cent (short photoperiod); *Brassica alba,* 453 per cent and *B. nigra,* 372 per cent (short photoperiod); and *Cosmos,* 103 per cent (long photoperiod). The optimal concentration of vitamin B_1 for the promotion of dry weight deposition appears to be 0.01 mg. per liter of nutrient solution (sand cultures), but a detectable effect was exerted by nutrient solution containing only 0.0001 mg. or 0.1γ of vitamin B_1 per liter.

It can be shown by means of the *Phycomyces* test that the vitamin added to the soil or sand accumulates in the leaves. It seems that the response of the plant to the addition of thiamin is more pronounced when the quantity normally contained in the leaves is small (less than 6.0 mg. per kg. dry weight). In tomatoes and peas, which normally have a high thiamin content (18.0 and 13.0 mg. per kg. dry weight, respectively), there is no visible reaction to thiamin added to the soil.

Other investigators, encouraged by the striking results of BONNER and GREENE have conducted similar experiments but have been unable to obtain any increase in growth by applying thiamin to the nutrient medium. ARNON (1940) investigated the effect of thiamin on several species of plants studied by BONNER and GREENE (tomato, cocklebur, cosmos, mustard, lettuce) and found, under the particular set of conditions, no beneficial effects whatever. His plants were grown in "carefully controlled nutrient solutions" and were supplied with thiamin at the rate of 0.01 and 0.05 mg. per liter. He concluded that, "results do not support the view that intact plants grown from seed can benefit from additions of thiamin to an otherwise favorable nutrient medium". Similar observations were made by MINNUM (1941*a* and *b*) with radish and cauliflower, GORHAM (1941) with *Lemna*, HAMNER (1940) with cabbage, cocklebur, cosmos, dahlia, mustard, radish and zinnias, and SWARTZ (1941) with chrysanthemum, marigold and cosmos.

Several other vitamins have also been studied with regard to their action on intact plants. According to DENNISON (1940) riboflavin has a beneficial effect on egg plant and tobacco growing in sand culture but MINNUM (1941*b*) was unable to detect any action of this vitamin on radish or cauliflower cultured in a similar manner. The latter also found pyridoxine and pantothenic acid ineffective. The same was established for nicotinic acid (SWARTZ, 1941). HAVAS recorded that ascorbic acid was beneficial to the germination of wheat seeds. It also favors the growth of tobacco plants according to DENNISON (1940).

For a long time the favorable effect of various substances on the development of embryos has been suspected. Xanthine, hypoxanthine, guanine, and nucleic acid are now known as beneficial to the embryo of wheat. Pyrimidine derivatives seem to be among the active principles contained in bacterial infusions of peat studied by BOTTOMLEY in his early work on auximones.

General observations such as those of DAGYS (1935, 1936) are also of value. He showed that the young tissues of certain trees, and also the sap produced in the spring by birch are rich in growth factors of the bios type. They can be assayed by the yeast test and thus their production and activity can be followed throughout the course of the year. In the spring when trees emerge from the dormant condition and growth is active these growth factors certainly play an important part in the metabolism of the plant.

We should like to emphasize the fact that, although the leaves are autotrophic concerning growth factors, the embryos, young roots, and cuttings are partially heterotrophic with respect to certain of these factors. This heterotrophism affects particularly the young embryonic organs which may later become autotrophic. There are indications that fully developed plants may be deficient in their ability to synthesize certain vitamins and as a consequence respond to an exogenous supply. It is not certain that a favorable response of a plant to a vitamin is always the result of a loss in its capacity for synthesis. When a vitamin is active only if supplied in large quantities it is very likely that the action is non-specific and not essential.

Another striking observation is that in a number of cases several factors are involved. The importance of thiamin is clearly demonstrated by the fact that it is always found as an essential growth factor. There is unquestionably a certain unity among the various losses of capacity for synthesis which affect the various organs of different plants often widely separated systematically. In order to arrive at these conclusions it was necessary to isolate the various organs and tissues and study them separately. Consequently the results obtained from such cultures are somewhat artificial. Nevertheless they give precise information concerning the chemical correlations between various organs, *e.g.*, leaf—root.

Pollen grains and pollen tubes. — Pollen grains must be considered as heterotrophic because they develop at the expense of the tissue of the style. Nevertheless, many pollen grains are rich in growth factors. They are particularly rich in thiamin or its components since a dialyzate from pollen has a strong growth promoting action on *Phycomyces* (SCHOPFER 1934). Furthermore, BRANDSCHEIDT (1930) observed that the pollen grains of *Tulipa* and *Cornus*, placed in distilled water, definitely favored the germination of spores of fungi and the growth of the thallus (*Phycomyces, Mucor strictus, Rhizopus, Cunninghamella*).

BRINK (1924), studying the culture of pollen grains in artificial media, found that yeast extract considerably favored the growth of pollen tubes and that the active principle in the extract was thermostabile. It is not necessarily of an organic nature since boron in very weak dosage is definitely known to favor germination. However, it is very likely that thiamin may turn out to be the active principle in yeast extract. Recently it has been noted that germinating pollen grains exercise a mutual stimulation upon one another and that extracts of embryonic tissues have a strong action on pollen which is apparently of hormonal nature. Nevertheless, vitamins also play a part. Vitamin B_1 increases the percentage of germination of pollen of some varieties of *Carica quercifolia* (DANDLIKER, COOPER and TRAUB, 1938; and COOPER, 1939). Natural riboflavin is the most active substance on pollen, but, since it is contaminated by boron, the activity must be attributed to the latter. Vitamin C also has marked activity. It is natural that the germination and growth of the pollen tube should depend upon the factors present in the tissues of the style and the ovary. The principal vitamins, acting as essential regulators in the higher plants are thiamin, nicotinic acid, pyridoxine, and ascorbic acid. Nothing is known concerning the actions of vitamins K and E. The latter is known to have an inhibiting effect on *Melandrium album* (SCHOPFER and BLUMER).

REFERENCES:-

1902: G. Haberlandt. — Kulturversuche mit isolierten Pflanzenzellen. Sitzungsber. Akad. Wiss. Wien, Math. Naturwiss. Kl. 111:69.

1924: R. A. Brink. — The physiology of pollen. Amer. J. Bot. 11:218, 283, 351, 417.

1930: P. Brandscheidt. — Zur Physiologie der Pollenkeimung und ihrer experimentellen Beeinflussung. Planta 11:368.

1933: P. R. White. — The SH radical and some other sources of sulfur as affecting growth of isolated root tips of wheat seedlings. Protoplasma 19:132.

1933: R. Bouillenne et F. W. Went. — Recherches expérimentales sur la néoformation des racines dans les plantules et les boutures des plantes supérieures (substances formatrices de racines). Ann. Jardin bot. Buitenzorg 43, 175 p.

1934: W. H. Schopfer. — Sur l'existence dans les pollinies d'Orchidées d'un facteur de croissance de microorganisme. C. r. Soc. phys. hist. nat. Genève 51:29.

1935: O. Lazar. — L'influence du carotène sur la néoformation des racines chez *Impatiens balsamina* L. C. r. Soc. biol. Paris, 120:799 and 1374; 121:886.

1935: J. Dagys. — Wuchsstoffe der Mikroorganismen in embryonalen Geweben und im Blutungssafte. Protoplasma 24:14.

1936: J. Dagys. — Die Hefewuchsstoffe in Knospen und Blättern. Protoplasma, 26:20.

1937: F. W. Went and K. V. Thimann. — Phytohormones. New York, The Macmillan Company, 294 p.

1937: R. Gautheret. — Nouvelles recherches sur la culture du tissu cambial. C. r. Acad. Sc. Paris, 205:572.

1938: J. Bonner and J. Greene. — Vitamin B_1 and the growth of green plants. Bot. Gaz. 100:226.

1938: W. B. Dandliker, W. C. Cooper and H. Traub. — Vitamin B_1 and germination of pollen. Science 88:622.

1938: R. et M. Bouillenne. — Contribution à l'étude des facteurs de la néoformation et de la croissance des racines. Rhizocaline, hétéroauxines, acides aminés, vitamine B_1. Bull. Soc. roy. Belgique 71:43.

1939: W. C. Cooper. — Vitamins and the germination of pollen grains and fungus spores. Bot. Gaz. 100:844.

1939*a*: P. Nobécourt. — Sur la pérennité et l'augmentation de volume des cultures de tissus végétaux. C. r. Soc. biol. Paris 130:1270.

1939*b*: —— ——. — Synthèse de la vitamine B_1 dans les cultures de tissus végétaux. C. r. Soc. biol. Paris 133:530.

1939: J. Bonner and J. Greene. — Further experiments on the relation of vitamin B_1 to the growth of green plants. Bot. Gaz. 101:491.

1939: P. R. White. — Controlled differentiation in a plant tissue culture. Bull. Torrey Bot. Club 66:507-513.

1940: R. Savelli. — Sur une action de nature hormonale exercée par le tissu embryonnaire de *Vicia faba*. C. r. Acad. Sc. Paris 210:705.

1940: D. I. Arnon. — Vitamin B_1 in relation to the growth of green plants. Science 92:264-266.

1940: R. Dennison. — Growth response of plants to riboflavin and ascorbic acid. Science 92:17.

1940: C. L. Hamner. — Effects of vitamin B_1 upon the development of some flowering plants. Bot. Gaz. 102:156-168.

1941: Daphne B. Swartz. — Effect of various growth-promoting substances upon several species of plants. Bot. Gaz. 103:366-373.

1941: P. R. Gorham. — Measurement of the response of *Lemna* to growth-promoting substances. Amer. Jour. Bot. 28:98-101.

1941*a*: E. C. Minnum. — Effect of vitamin B_1 on the yield of several vegetable crop plants. Amer. Soc. Hort. Sci. Proc., 38:475-476.

1941*b*: —— ——. — Effect of vitamins on growth of radish and cauliflower. Bot. Gaz. 103:397-400.

1941: P. R. White. — Plant tissue cultures. Biol. Rev. 16:34.

1942: —— ——. — "Vegetable dynamicks" and plant tissue cultures. Plant Physiol. 17:153.

Chapter VIII.

THE BIOSYNTHESIS OF VITAMINS.

Despite its great importance, the biosynthesis of the principal vitamins is poorly understood.

Vitamin A and carotinoids. — Although the relationships between the carotinoids and plastids (chromoplasts) are well known since the work of GUILLIERMOND and his school, our knowledge of vitamin A is quite limited. Recently, JOYET-LAVERGNE (1936, 1937) applied the Carr-Price reaction in a microchemical study of plant and animal cells (fish liver, and the Sporozoa, *Steinina ovalis* and *Gregarina polymorpha*). In this work the provitamin and vitamin A were determined jointly, since they can not be distinguished by the microchemical application of this reaction. Plants widely separated taxonomically gave a positive reaction: *Iris germanica, Crinum Powelli, Zea Mays, Acer* sp., *Linaria cymbalaria, Elodea canadensis, Phaseolus* sp., algae such as *Zygnema pectinatum,* higher fungi such as *Saccharomyces Ludwigi, Nadsonia fulvescens, Spermophthora Gossypii,* and lower fungi such as *Pythium deBaryanum.* The blue tint produced by the carotinoid test is localized for the most part in the chondriome and the nucleolus, appearing in the plastids or, more specifically, in the mitochondrial region of the plastids. This might tempt one to assume that vitamin A is a constituent of the chondriome and the nucleolus. But since the reaction is not specific the best that can be done is to assume that the color produced in the animal cell is due to vitamin A, whereas that produced in the plant cell is due to one or several carotinoids formed there. This suggests that the transformation of carotene into vitamin A occurs in the chondriome, a view entirely in agreement with the available knowledge of the origin of carotinoids in plant cells. These yellow pigments arise in the chromoplasts, which in turn are formed either from leucoplasts or from chloroplasts following the degeneration of chlorophyll. Since the plastids are formed from the chondriome, and since the latter is known to play an important rôle in syntheses, the logical seat of vitamin A production is the chondriome. Although the presence of preformed vitamin A in plants is not precluded, no actual study of this problem has been made. An investigation of the action of various plant extracts on carotene *in vitro* might yield information concerning the existence of a carotinase in plants.

Even if the chondriome can be demonstrated as the seat of synthesis, carotinoids in other parts of the cell must be accounted for, *e.g.*, those adsorbed on *chondrioconts* or found in liquid or solid state in chloroplasts or leucoplasts, in emulsoid or solid condition in the cytoplasm, or even dissolved in the vacuole.

The synthesis of carotene is for the most part carried on by plant cells, though carotene may occur endogenously in animal cells. The latter are characterized by their ability to transform carotene into vitamin A by enzymatic action (hydrolysis). Conclusive proof for this statement requires that this transformation be carried out *in vitro*. However, the existence of a carotinase acting *in vitro* has not yet been demonstrated with certainty. Recently WILLSTAEDT showed that a culture of animal tissue (fibroblast) in a medium lacking vitamin A could be improved by the addition of pure carotene but only in the presence of a small fragment of liver tissue (the liver is the organ in which this transformation occurs in the living animal). This demonstration is not conclusive proof of carotinase activity because the favorable action of the fragment of liver might have been due to other substances supplied.

The conditions associated with the appearance of carotinoids are well known in various groups of plants.

Fungi. — In *Mucor hiemalis* the production of carotinoids, dissolved in the fats of the hyphae, is largely dependent upon the composition of the medium. The optimal production occurs only when the carbon-nitrogen ratio is correct (the nitrogen being furnished in the form of asparagine) (SCHOPFER, 1928). In *Phycomyces Blakesleeanus* the carotene occurs probably in the β-form or a closely related carotinoid as indicated by the marked vitamin A activity of extracts from this fungus (SCHOPFER and JUNG, 1935). The carotene content of the thallus has been assayed spectroscopically and biologically. It rises as the supply of asparagine in the medium is increased, reaching 0.09 to 0.1 per cent when the concentration of asparagine is advanced to 0.4 per cent. When the cultures are kept in the dark or in red light they remain white, but in blue or violet light they produce carotinoids abundantly. The latter wave lengths are strongly absorbed by carotene and apparently are the most favorable to its synthesis in this organism.

Rhodotorula rubra, which likewise is rich in carotinoids, requires an independent source of carbon for their synthesis. Glycerol, in particular, permits abundant pigmentation[1] (FROMAGEOT and TCHANG, 1938).

Bacteria. — *Mycobacterium Phlei* has served to demonstrate the conditions required for carotinoid formation (assayed colorimetrically) (INGRAHAM and STEENBOCK, 1935). Glycerol again is the most active agent for the production of carotinoids, although

[1]In the presence of glycerol, the pigmentation is maintained as with lactic acid + glucose. In the presence of glucose alone as the source of carbon, the pigmentation disappears rapidly. This destruction is certainly dependent upon the presence of catalase which is found in abundance wherever there are large quantities of carotinoids. If catalase becomes deficient, H_2O_2 destroys the carotinoid pigments.

it does not greatly increase the number of cells (Fig. 6). The following alcohols, methyl, ethyl, n-propyl, n-butyl, isobutyl and n-amyl, also favor the formation of pigments. Isopropyl alcohol has a particularly marked effect, comparable with that of glycerol or of ethylene glycol. With the latter the pigmentation reaches 3280 Y units as compared with 50 Y units produced by the control. On the other hand, β-ionone and concentrates of vitamin A are ineffective. The composition of the medium influences not only the quantity of pigments produced but also the relative proportions of the various carotinoids. In *Mycobacterium Phlei* the following have been identified spectroscopically, α- and β-carotenes, esters of lutein, of

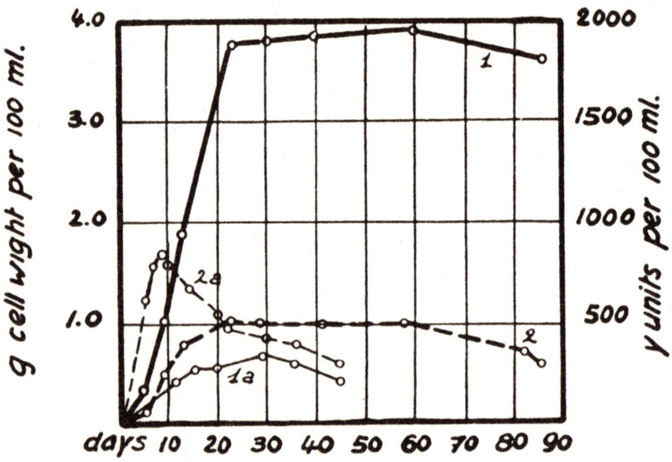

FIG 6. — Synthesis of carotinoids by *Mycobacterium Phlei* (from INGRAHAM and STEENBOCK, 1935). 1. total pigments; 2, weight of cells with 5% glycerol. 1a, total pigments; 2a, weight of cells with 10% glucose.

zeaxanthin, of azafrin, and of cryptoxanthin. In the case of this bacterium, light is not required for the biosynthesis of carotinoids.

Algae: In the algae, carotin formation is largely dependent on an unbalanced condition of the medium (F. CHODAT, 1938). This lack of balance arises from the impoverishment of the medium in nitrogen. The absolute concentration of glucose and light is unimportant (F. CHODAT and F. MEIER). In the green alga, *Dictyococcus cinnabarinus,* the production of pigments can be followed throughout the course of its culture. In a medium favorable to carotin production the accumulation of carotinoids in cultures continues for several months. The esterified xanthophylls accumulate very rapidly from the third to the sixth week and, during that time, they are twice as abundant as carotenes. During this early period the xanthophylls diminish progressively as they are being esterified (F. CHODAT and F. WENZINGER, WENZINGER, 1940).

The flagellates behave differently (A. and M. LWOFF, 1930). The carotene is localized in the eye spot, a structure found even in forms lacking chlorophyll such as *Polytoma uvella. Hematococcus pluvialis,* growing on a medium composed of mineral salts and asparagine or peptone, is stimulated by the addition of sodium acetate to produce carotinoids in such quantities that they mask the chloroplasts. This synthesis is entirely independent of the concentration of the medium, the *p*H (between 6.3 and 8.6), and light.

The first impression gained from studying the above groups of organisms (*Mucoraceae,* red yeasts, bacteria, and *Chlorophyceae*) is that they differ widely with respect to the biosynthesis of carotin. The higher plants present additional differences.

Emphasis should be placed on the fact that the carotinoids found in microorganisms have vitamin A activity and that they function as the provitamin. Vitamin A activity has been observed in: *Chlorella* (COWARD), *Nitschia closterium* (JAMESON, DRUMMOND and COWARD, and AHMAD), in phytoplankton in general (DRUMMOND and GUNTHER), *Chlorococcus* (GUNDERSON and SKINNER), *Trentepohlia aurea* (LEDERER), in lichens (ELLIS, PALMER, and BARNUM), in *Rhodotorula Sanniei* and *Phycomyces Blakesleeanus* (SCHOPFER and JUNG, 1935).

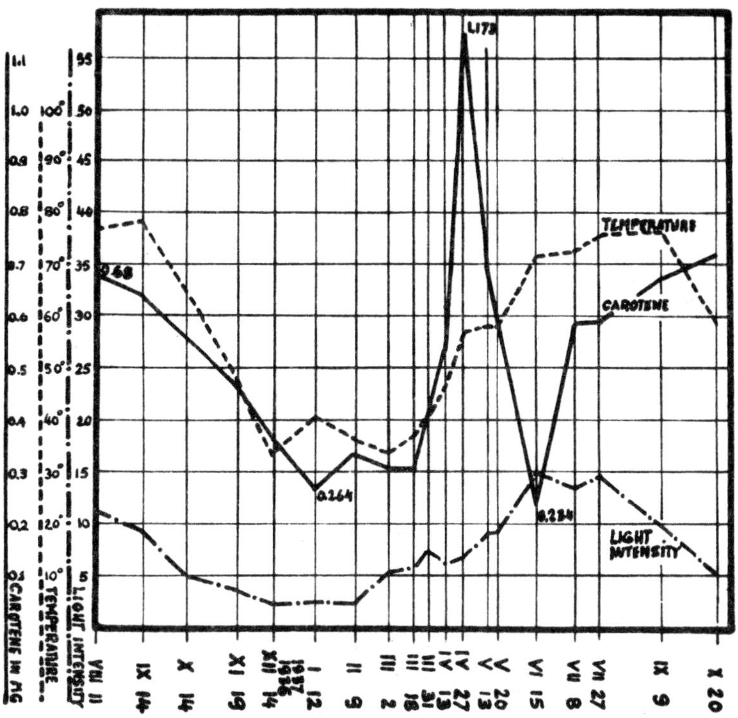

FIG. 7. — Carotene content of Dutch Sweet Clover in relation to temperature and light intensity (from BECK and REDMAN, 1940). The Roman numerals on the abscissa designate the months of the year.

Flowering plants. — In higher plants the biosynthesis and content of carotinoid are dependent upon various factors, some of which are now known for *Pisum* (Torstoi variety) through the work of VIRTANEN and his school. In general, all conditions which favor the development of the plant also increase the percentage of carotinoids. The bacteria of the soil do not exert a direct effect, but act indirectly by favoring the fixation of nitrogen in the soil (VIRTANEN and collaborators, 1933).

When the plants are cultivated in a synthetic medium in quartz sand, KNO_3 is a more favorable nitrogen source and more effective than $(NH_4)_2SO_4$ in the production of carotene. The carotene content is highest at the time of flowering (0.2 per cent of the dry wt.) (*Pisum, Phleum pratense, Dactylis glomerata*). Subsequently it diminishes rapidly (Fig. 7). In following the variation in carotene content in Dutch Sweet Clover throughout the year, the maximum occurs during April regardless of temperature and light intensity. However, any relation which might exist between those environmental factors and carotene content is indirect (BECK and REDMAN, 1940). The maximal carotene content coincides with the beginning of intensive growth (Fig. 7).

The probable mechanism involved in the biosynthesis of carotene is best studied in fruits with high carotinoid content. In tomatoes, DUGGAR (1913) showed that light is not required for reddening of the fruit (formation of lycopin), but that a temperature of less than 30°C. and the presence of oxygen are essential. An atmosphere of nitrogen or hydrogen causes the fruit to decompose. This temperature relationship may possibly be associated with the inhibition of an enzyme. The results concerning temperature and light have been confirmed.

The fruit and calyx of *Physalis Alkekengi* are deep red in color and contain physalin (*dipalmitic ester of zeaxanthin,* $C_{40}H_{56}O_2$) (KUHN and WIEGAND, 1929). At first the calyx is green and contains chlorophyll, lutein, and carotene, but as it turns yellow the carotene content increases and the xanthophylls decrease. At maturity, physalin is abundant and the free xanthophylls which remain are represented particularly by zeaxanthin. These chemical changes might lead to the assumption that *phytol* which is liberated upon the decomposition of chlorophyll is made available for the synthesis of carotenes. Upon oxidation lycopin forms cyclic compounds and can give rise to oxides of carotinoids, which after esterification may lead to the formation of physalin. Esterification immobilizes the pigment in the form of reserves. However, the amount of phytol present in the calyx and the berries of *Physalis* is not large enough to explain the formation of 0.6-1.0 per cent zeaxanthin (computed on the basis of dry weight). Hence, it is necessary to conclude that the phytol exists as a reserve or is formed *de novo*.

According to a hypothesis advanced by KARRER and his students, a dehydrogenation of phytol occurs, followed by a condensation, which leads to the formation of lycopin and then carotene (see KARRER and WALKER 1934, and WALKER 1935). Phytol is therefore fundamental to the synthesis of chlorophyll and carotinoids, part of it being used in the formation of each of these pigments. The production of phytol for use in carotene formation probably is not delayed till the time of fruiting as might be suspected, but it continues throughout the life of the plant and thus provides a reserve of phytol for the rapid formation of carotene during the ripening period.

WENZINGER (1940) applied this hypothesis for the derivation of carotene from phytol to the autotrophic green alga, *Dictyococcus cinnabarinus*. It has also been advocated by LUBIMENKO. According to the proposed scheme, isoprene compounds undergo a condensation into protophytol. The latter may possibly have a unique orientation toward the carotinoids. In the case of the green plants the reaction may proceed in either of two directions: 1) protophytol → carotinoids, or 2) protophytol → phytol → chlorophyll. The direction in which the reaction goes is probably dependent on the nutrition of the plant. This problem can be studied more advantageously in algae than in higher plants.

According to F. CHODAT, carotinogenesis is of two different types, one of spontaneous nature, *i.e.*, independent of the medium, the other induced by the medium. In the latter case, the composition of the medium is the controlling factor in the production of carotinoids. This action can be explained as follows: if the alga receives a normal supply of nitrogen, the carbon will be used up indiscriminately for the synthesis of proteins and lipides. If the nitrogen content of the medium is increased, protein synthesis is accelerated until it uses up all the available carbon leaving none for the synthesis of lipides and the accompanying synthesis of carotinoids. If the nitrogen supply is decreased the synthesis of lipides and carotinoids will be increased. This hypothesis is very attractive but unfortunately lacks experimental support. Moreover, the relation existing between the production of lipides and that of carotinoids is not understood.

Functions of carotinoids. — There is still very little known concerning the function of these substances. It has been proposed that they intervene ecologically in various ways: for protection, particularly in the case of amylase, against ultra-violet rays, as a sensitizer or transformer of energy, etc. The most interesting function is the absorption of blue and violet light by carotene, producing phototropisms in *Phycomyces* (BÜNNING) and the *Avena* coleoptile (JOHNSTON). The fundamental rôle of carotene is undoubtedly due to its great capacity to absorb oxygen. Accordingly it should function as an oxygen carrier and an antioxidant. It probably acts in conjunction with an enzymatic system (see p. 187), although no factual data are as yet on hand. If more information were available concerning the function of carotene it would be easier to determine the mechanism involved in its biosynthesis.

The general conclusion which can be drawn concerning the mechanism of carotene synthesis in all organisms containing chlorophyll is that phytol acts as a precursor. In organisms lacking chlorophyll, the mechanism of synthesis, although simple, is still obscure. In non-green plants each group seems to require a specific carbon source for carotinogenesis: an excess of glucose for the

Mucoraceae, glycerol for *Mycobacterium Phlei* and *Rhodotorula Sanniei,* and sodium acetate for *Hematococcus pluvialis.* The carbon source is not the sole activator since asparagine, which supplies both carbon and nitrogen, favors the formation of carotinoids in the alga *Dictyococcus* and the fungus *Phycomyces.* The only explanation for the production of carotinoids is that given by the author in 1928, *viz.,* since an excess of carbon favors the synthesis of lipides, the synthesis of carotinoids is thought to be comparable to it although the mechanism involved is still unknown. Regardless of what happens in the synthesis of lipides, the explanation of the biosynthesis of carotinoids in green plants must take into account the effect of the carbon-nitrogen ratio of the medium. The importance of this ratio has been demonstrated in the case of the *Mucoraceae* (SCHOPFER, 1928).

Vitamin E (α-tocopherol). — This fat soluble vitamin is also synthesized from phytol (KARRER and DEMOLE, 1938). Despite the high α-tocopherol content of certain plant organs, wheat germ (embryo) in particular, we know almost nothing concerning its biosynthesis. It is localized for the most part in the chloroplasts, where it reaches a concentration of 0.08 per cent as compared with only 0.002 per cent (on dry wt. basis) in the cytoplasm (DAM, GLAVIND and PRANGE). In the wheat germ, which is high in α-tocopherol but low in chlorophyll, the synthesis of this vitamin is probably dependent upon other cytoplasmic constituents. The presence of the vitamin in the chloroplasts may be merely the result of a previous synthesis during the formation of the green plastids, *e.g.,* in the mitochondria (chondriome). The whole problem is open for investigation.

Vitamin K. — The biosynthesis of this fat soluble vitamin has been studied much more extensively than that of α-tocopherol (DAM, GLAVIND and NIELSEN, 1940). Vitamin K is of particular interest to botanists since, like chlorophyll it contains phytol (see p. 45). This vitamin is mostly confined to the chloroplasts, where it represents 0.006 per cent of the green weight; the cytoplasm contains only 0.0001 per cent (cabbage). DAM and his collaborators have studied the action of light on the biosynthesis of vitamin K in the embryos of *Picea canadensis.* They showed that plants grown in light produce more than twice as much vitamin K as those grown in the dark:

	VITAMIN K, γ	NO. OF PLANTS
seeds	0.0004	
plants in the light	0.039	5300
plants in the dark	0.018	3400

The vitamin K content of the seeds is so small that the increase observed was undoubtedly due to a subsequent synthesis. The surprising fact is that vitamin K can be synthesized in the dark. *Pisum,* on the other hand, is unable to synthesize this vitamin in the dark. The fact that seedlings of *Picea canadensis* form chlorophyll and vitamin K even in darkness raises the question of a possible direct relationship between chlorophyll and vitamin K. A comparison of the plants grown in the dark with those grown in light shows readily that the chlorophyll content varies directly with the vitamin K content. These data suggest a relationship between the two similar to that assumed for carotene and chlorophyll, *i.e.*, their synthesis is dependent on the presence of free phytol. But this view is not supported by the fact that the disintegration of chlorophyll in autumn is not accompanied by an increase in the vitamin K content. On the other hand, investigations with variegated leaves show that vitamin K is found in the areas lacking chlorophyll. It is possible, however, that this vitamin originated in adjoining green regions. Nevertheless, it is certain that the phytol group can be synthesized in the absence of light. For the present it can be assumed that the cells which contain many chloroplasts also contain large amounts of available phytol in the form of reserves for later use elsewhere in the plant. It is quite possible that the chloroplasts represent the main seat of vitamin K formation, although there is no indication that the synthesis of vitamin K is directly dependent upon the activity of chlorophyll. Two syntheses from phytol may occur: phytol → vitamin K, and phytol → chlorophyll. These syntheses may go on simultaneously, but that of vitamin K may also take place independently, *viz.*, from phytol translocated from the chloroplasts. Likewise vitamin E which also contains phytol, must be formed in the dark from phytol translocated to the embryo.

Furthermore, it is necessary to consider the relationship between vitamins K and E. Their distribution is by no means identical. Yet the eventual transformation of the one into the other, either in the chloroplasts or in places where both are found in abundance, is not precluded.

The analogues of vitamin K occurring in plants are not well known. Those which are inactive on test animals may be active on plants.

In the case of the fat soluble vitamins, phytol plays an important rôle since it represents an essential constituent, or a starting point for the three vitamins:

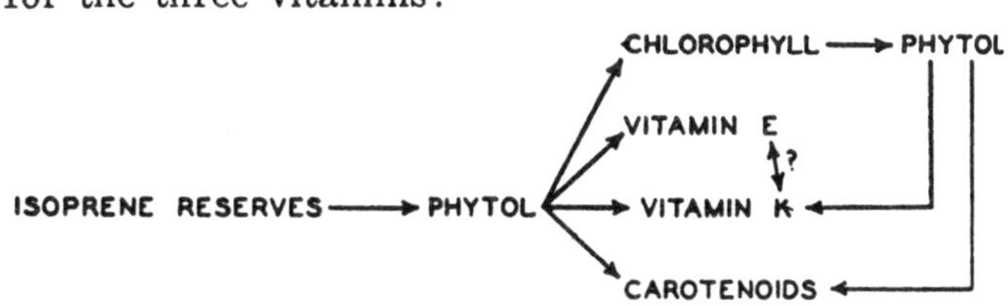

Vitamin B_1. — Nothing is known concerning the cytology of this water soluble vitamin. It has been shown that either thiamin or its product of dehydrogenation, thiochrome, is able to penetrate into the cytoplasm and the vacuole of *Allium* (upper epidermis of the bulb scales). No information is available regarding the location of the vitamin in the cell. The colorimetric tests used for the determination of thiamin *in vitro* are not applicable to a microchemical study.

As far as the water soluble vitamins, thiamin, riboflavin, and pyridoxine, are concerned, it can be said that the active form of any of these is represented by a coenzyme bound to a proteinaceous substance. While the active form is found only in the cytoplasm, the free form can accumulate in the vacuole where it normally crystallizes at times, *e.g.*, riboflavin (see p. 148). Obviously the combined form (vitamin + carrier) is difficult to test chemically since the vitamin proper (the active group giving the color reactions), represents only a small part.

By comparison the metabolism of thiamin in the plant is much better understood. This vitamin is abundant in seeds (cotyledons), decreases somewhat during the course of growth, is formed again in the new seed and accumulates there (HLAVATY). W. RYTZ jr. (1939) was able to follow the production of this vitamin in pea plants by means of the *Phycomyces* test. It is possible, however, that this test does not detect substances formed during the course of development which are analogous to thiamin and have a similar vitamin action. J. BONNER and BUCHMAN (1938) have been able to show that peas require and are able to utilize a thiamin active on *Phycomyces*, but that it manufactures analogous substances which are *inactive* on *Phycomyces*. This observation, like that of RYTZ, brings up the problem of the multiplicity of thiamins. Various homologues and analogues are known to have marked physiological activity despite the fact that their chemical structure differs more or less from that of normal thiamin. A thiamin in which the CH_3 group in position 2 of the pyrimidine portion is replaced by a C_2H_5 exhibits a greater activity than the typical thiamin on microorganisms, particularly *Phycomyces*.

BONNER and GREENE (1938) working with pea seedlings, 4 cm. in length (epicotyl), divided them into two lots, one of which was placed in light, the other in the dark. After eight days, they assayed the thiamin in the terminal bud, in the leaves, and in the root (*Phycomyces* test) and found that the roots of the plants grown in light contained two and a half times more thiamin than those grown in the dark, whereas the leaves and terminal bud contained three times that amount. The young tissues are therefore very active in synthesizing thiamin. Similar observations have also been made by RYTZ jr.

ONDRATSCHEK (1940) working with the green alga *Hematococcus pluvialis* also found that light favors the synthesis of thiamin. This alga, grown in the dark on a mineral medium without sugar, requires principally thiamin as a growth factor. When grown in light with sodium acetate and asparagine, it requires chiefly ascorbic acid; and although thiamin is beneficial it is not essential for growth due to the fact that it is synthesized in insufficient quantity. When the alga is grown in the dark with acetate and asparagine, thiamin becomes the essential factor because it cannot be synthesized.

The thiamin synthesized in the leaves must be translocated to the roots which are more or less heterotrophic for this factor. Tomato roots are able to synthesize pyrimidine and are therefore dependent on the leaves for only thiazole, whereas pea roots are dependent on the leaves for both pyrimidine and thiazole.

The study of the vitamin requirements of embryos and roots makes it possible to obtain detailed information concerning the biosynthesis of thiamin. Pea roots, since they do not require thiamin but can utilize its two components, obviously have the ability to unite the two components into a thiamin molecule. This step in the biosynthesis of thiamin is similar to that of its chemical synthesis *in vitro*.

It is possible to trace the biosynthesis of thiamin even further. Instead of starting with pyrimidine and thiazole as precursors, BONNER and BUCHMAN (1938) used the precursors of thiazole, thioformamide and chloracetopropyl alcohol. These precursors represent the constituents from which thiazole can be synthesized *in vitro*. When these two constituents of thiazole are supplied along with pyrimidine to pea roots, growth is as good as when thiamin is supplied. Thus the root is able: 1) to synthesize thiazole from its precursors, and 2) to synthesize thiamin from its components. However, the pea root is evidently unable to synthesize either the precursors of thiazole or the components of thiamin since these have to be supplied. The synthesis of thiazole can also be accomplished with acetopropyl alcohol *in vivo*, but not *in vitro*.

The chemical synthesis of thiamin consists principally of three operations: 1) the synthesis of pyrimidine, 2) the synthesis of thiazole, and 3) the union of the two components into a molecule of thiamin hydrobromide and the transformation of the latter into the hydrochloride. Pea roots are able to carry out two of these operations: the synthesis of thiazole from its components, and the union of thiazole and pyrimidine. It is not known whether pea roots are able to utilize the precursors of pyrimidine. Pea roots do not necessarily have to receive the whole thiamin molecule from the leaves, since pyrimidine and the precursors of thiazole are able to serve the same purpose. To bring about the union of components of thiazole and thiamin, specific enzymes undoubtedly are required, although they have never actually been demonstrated.

The metabolism of thiamin and analogous substances in peas presents additional points which must be clarified. It has already been stated that pea roots are able to utilize substitution products of thiamin which are inactive on *Phycomyces*. The extracts of roots grown with these products are inactive on *Phycomyces*. Thus it is impossible to assume that these substitution products are transformed into true thiamin. It is necessary to consider the analogues of thiamin as different from the latter.

The fate of thiamin during the growth of green plants is not know in detail. However, the vitamin is known to be esterified with pyrophosphoric acid to form cocarboxylase. Thiamin is active only by virtue of the cocarboxylase which it helps produce, but its ultimate fate is unknown.

The fate of thiamin in *Phycomyces* is better known. After it has been absorbed by a mature culture, the thiazole component is destroyed. This probably is the first stage in the thiamin cycle (see p. 117). It is unknown whether the same changes occur in higher plants (J. BONNER and BUCHMAN, 1938).

Thiamin is probably the best known vitamin found in plants. It represents the first example of vitamin correlations between the various organs of the plant (leaf and root). In addition, it illustrates the production, evolution, transformations and accumulation of a vitamin throughout the entire course of development.

REFERENCES:-

1913: R. J. DUGGAR. — Lycopersicin, the red pigment of the tomato and the effect of conditions upon its development. Washington Univer. Stud. 1:22.

1928: W. H. SCHOPFER. — Recherches sur la sexualité des champignons. Le problème de la biochimie comparée du sexe. Bull. Soc. Bot. Genève 20:149.

1929: R. KUHN und W. WIEGAND. — Der Farbstoff der Judenkirsche (*Physalis Alkekengi* und *Physalis Franchetti*). Helv. chim. Acta 12:499.

1930: M. LWOFF et A. LWOFF. — Détermination expérimentale de la synthèse massive de pigment caroténoide par le Flagellé *Hematococcus pluvialis* Flot. C. r. Soc. Biol. Paris 105:454.

1933: A. VIRTANEN, S. VON HAUSEN und S. SAASTAMOINEN. — Untersuchung über die Vitaminbildung in Pflanzen, I. Biochem. Zeitschr. 267:179.

1934: P. KARRER und O. WALKER. — Untersuchungen über die herbstlichen Färbungen der Blätter. Helv. chim. Acta 17:43.

1934: E. LEDERER. — Les caroténoïdes des plantes. Actualités scientifiques et industrielles, Hermann & Cie, Paris.

1935: O. WALKER. — Überblick über die Chemie der Carotenoide und ihre Verbreitung in der Natur. Dissert., Zürich.

1935: W. H. SCHOPFER et A. JUNG. — Recherches sur l'activité vitaminique A du thalle d'une Mucorinée. C. r. Soc. Biol. Paris 120:1093.

1935: W. H. SCHOPFER. — Sur l'identification d'un caroténoïde de champignon. C. r. Soc. Biol. Paris 118:3.

1935: M. A. INGRAHAM and H. STEENBOCK. — The relation of microorganisms to carotenoids and vitamin A, II. The production of carotenoids by *Mycobacterium Phlei.* Biochem. J. 29:2553.

1936: Ph. JOYET-LAVERGNE. — Contributions cytophysiologiques à l'étude du rôle de la vitamine A. Bull. Soc. Chim. Biol. Paris 18:1041.

1936: W. H. SCHOPFER et V. KOCHER. — Sur la cristallisation du carotène de *Phycomyces.* Actes Soc. helv. Sc. nat., p. 320.

1937: PH. JOYET-LAVERGNE. — La vitamine a dans la cellule. Protoplasma 28:131.

1938: C. FROMAGEOT et J. L. TCHANG. — Sur la synthèse des pigments caroténoïdes de *Rhodotorula Sanniei.* Arch. f. Mikrobiol. 9:424.

1938: J. BONNER and J. GREENE. — Vitamin B_1 and the growth of green plants. Botan. Gaz. 100:226.

1938: P. KARRER et V. DEMOLE. — Synthèse et titration biologique de la vitamine E. Schweiz. Med. Wochenschr. 68:954.

1938: J. BONNER and E. R. BUCHMAN. — Syntheses carried out *in vivo* by isolated Pea roots, I. Proc. Nat. Acad. Sc. Washington 24:431.

1938: F. CHODAT. — Etudes sur la genèse des caroténoïdes. Arch. Sc. Phys. et Nat., Genève, 20:96.

1939: W. RYTZ jr. — Beitrag zum Aneurinstoffwechsel bei höheren Pflanzen. Ber. schweiz. bot. Ges. 49:339.

1940: F. WENZINGER. — Evolution des pigments caroténoïdes chez une algue verte. Bull. Soc. bot. Genève 30:129.

1940: H. DAM, J. GLAVIND und N. NIELSEN. — Weitere Untersuchungen über die Bildung und Bedeutung des Vitamin K im Pflanzenorganismus. Zeitschr. physiol. Chemie 265:80.

1940: K. ONDRATSCHEK. — Experimentelle Untersuchungen über den Einfluss von Wirkstoffen auf die Vermehrung einiger mixotropher Algen. Arch. f. Mikrobiol. 11:89.

1940: W. A. BECK and R. REDMAN. — Seasonal variations in the production of plant pigments. Plant Physiology 15:81.

Chapter IX.

THE BIOSYNTHESIS OF VITAMINS (Contd.)

Vitamin C. — This vitamin is closely related to the sugars and, according to the observations of GIROUD and his collaborators, is formed mainly in green tissues. However, vitamin C occurs also in non-green tissues of higher plants and a very similar substance is found in fungi. Despite these facts GIROUD's observations resulted in numerous attempts to localize the seat of vitamin C synthesis in the chloroplasts.

Cytology. — The Molisch reaction (1921) which consists of a reduction of $AgNO_3$ by the plant cell, occurs only in the chloroplasts. On the basis of this early evidence, GIROUD and his students (1934*a*) attributed the reducing action in the chloroplasts to ascorbic acid and perfected the test for ascorbic acid by acidulating the silver nitrate. This improved test, when applied to the fruits of *Capsicum annuum,* brings about a pronounced blackening of the plastids. Not all tissues containing chlorophyll react in the same manner. In *Gladiolus,* which is known to be rich in ascorbic acid, PEKAREK observed a blackening of the reticulum of the chloroplast, while the grana of chlorophyll remained unstained. A blackening of the chloroplasts of *Polygonatum multiflorum* was seen by WEBER in all cases except one in which the opposite reaction took place, *i.e.,* the cytoplasm became blackened but not the plastids.

GIROUD's test lacks specificity and, in fact, no satisfactory test for the study of the histochemistry of ascorbic acid is available at the present time.

An explanation of the Molisch reaction has been proposed by GAUTHERET (1935) and accepted by MIRIMANOFF (1938*b*). Despite GIROUD's statement that the reduction of silver nitrate can be obtained in the dark, GAUTHERET advanced the very plausible hypothesis that the reduction is in reality a photolysis of $AgNO_3$, in which the chlorophyll acts as an optical sensitizer irrespective of ascorbic acid, and that the chloroplasts represent merely the medium in which this reaction takes place. This hypothesis refutes all other hypotheses according to which chlorophyll and ascorbic acid are directly responsible for the reduction of silver nitrate.

In this discussion it is necessary to distinguish two phenomena and, accordingly, two categories of arguments: 1) the coexistence of ascorbic acid and chlorophyll and the production of the former dependent on the latter, and 2) the reduction of $AgNO_3$ by ascorbic acid in the chlorophyll containing regions. The merits of the various arguments will be apparent from the following discussion of the mechanism of biosynthesis of ascorbic acid.

One fact is certain — with few exceptions, seeds are devoid of vitamin C. They must, however, contain a precursor of ascorbic acid since the vitamin is formed upon germination. For a long time vitamin charts have shown that barley seeds have no antiscorbutic value, whereas young seedlings are clearly active. According to HARRIS and RAY (1933) the increase in ascorbic acid in *Pisum* during the course of development is as follows:

	MG.
seeds before germination	0
seeds soaked for 24 hours (not germinated)	0.08
seeds germinated for 24 hours	0.69
seeds germinated for 72 hours	0.82
seeds germinated for 96 hours	0.86

(assay by the method of TILLMANS, reduction of 2,6-dichlorophenolindophenol).

For certain seeds (*Oleaceae*) the maximum content is attained at the end of a 24 hour germination period.

In the later development of the plant ascorbic acid is formed continually in its green organs as well as in those lacking chlorophyll (GIROUD *et al.* 1934*b*, RANDOIN *et al.* 1935). The work of REID (1937) shows that the accumulation of ascorbic acid is independent of chlorophyll and that the highest concentration of this vitamin is found in actively growing organs, particularly in the regions of elongation, leaf buds and floral buds. The corolla lacks chlorophyll but contains more ascorbic acid than the green calyx. According to the work of GLICK (1937) on barley seedlings the vitamin C content of the coleoptile decreases regularly, whereas that of the enclosed first leaf increases from the 5th day at which time this structure is green but not yet very active in photosynthesis.

One point on which all observations agree is that leaves have a higher ascorbic acid content than other parts of the plant. The maximal content in the leaf is reached just before flowering. Afterward, the amount decreases rapidly.

Although there seems to exist a certain relationship between ascorbic acid and chlorophyll, it is apparently not causal in character but merely the result of their parallel behavior. DIRSCHENDORFER (1937) states that fruits with well developed chlorenchyma (rose, paprika) contain more ascorbic acid than those with poorly developed chlorenchyma (apple, pear). These differences may be caused by the previous activity of chlorophyll, or by other phenomena linked with chlorophyll activity but they do not prove that there is any direct effect of photosynthesis on the synthesis of ascorbic acid.

The interpretations of the results are contradictory. This fact demonstrates that the relationship between photosynthesis and ascorbic acid cannot be determined by a superficial study of biosynthesis. Rather it is necessary to study experimentally the probable mechanism of the biosynthesis of vitamin C starting with the possible precursors.

The nature of the precursors has been determined experimentally through the culture of excised embryos, *i.e.*, after the removal of their cotyledons. We recall from the work of VON HAUSEN (1936) that these cultures require ascorbic acid during the early stages of development, but that later they become able to synthesize this growth factor. Since little, if any, photosynthesis occurs in these cultures, the synthesis of this vitamin must take place at the expense of a carbon source of the mineral medium, which may well be a sugar. Actually, RAY (1934) demonstrated that in sterile synthetic media, mannose can be considered as the precursor since it permits an abundant formation of ascorbic acid by excised embryos.

The synthesis of ascorbic acid is not confined to plants. Extracts of animal organs such as the spleen, liver, and cardiac muscles are likewise capable of synthesizing ascorbic acid *in vitro* from mannose. The mechanism of the transformation is enzymatic in nature, the enzyme being a dehydrogenase which can be extracted from the organs in question (GUHA and GHOSH, 1934 and 1935). A similar enzyme, capable of synthesizing ascorbic acid from mannose, has been extracted from germinating seeds of *Phaseolus mungo*. The mechanism is therefore well established. The synthesis of ascorbic acid by extracts from living organisms is essentially similar to that of the chemical laboratory in that sugars serve as the starting material. In the latter xylose or sorbose is the sugar employed. The typical vitamin C is *l*-xylo-ascorbic acid; the analogues obtained with other sugars are of lesser activity. If the ascorbic acid produced in these experiments on biological synthesis is actually the same as that obtained from xylose by ordinary chemical synthesis, the organism obviously must convert mannose into xylose as one of the steps in this biosynthesis.

These strictly chemical reports enable us to discuss critically the relations between photosynthesis and the biosynthesis of ascorbic acid which some workers have attempted to establish in a superficial manner. If a sugar must serve as the basis for the synthesis of ascorbic acid *in vitro* and *in vivo*, it is only natural that the biosynthesis of ascorbic acid occurs where the sugar is formed in the greatest quantity, *i.e.*, in the chloroplasts. This explains why the green tissues are usually the richest in ascorbic acid. However, if only sugar is required for the biosynthesis of ascorbic acid, the presence of chlorophyll is not essential. Young seedlings which have not yet turned green and are not able to carry on an appreciable amount of photosynthesis, are able nevertheless to synthesize ascorbic acid from a precursor of carbohydrate nature furnished by the reserves of the seed. Later on, when the plant is completely autotrophic, it is only natural that ascorbic acid should be more abundant where the sugar is synthesized. Nothing is known, however, concerning the manner in which this enzymatic synthesis is connected with photosynthetic activity. Nor is it known at what time or level of photosynthesis the synthesis of ascorbic acid begins.

Certain experimental results support the view that the synthesis of ascorbic acid is linked with photosynthesis. K. and M. WEISSENBÖCK (1940) observed that seedlings obtained from intact seeds (oily seeds: *Helianthus, Cucurbita, Raphanus, Papaver;* or starchy seeds: *Zea, Avena, Triticum*) and cultivated in light without CO_2 maintain their ascorbic acid content at the same level as those supplied with CO_2. On the other hand, plants grown from seeds deprived of their reserves and cultivated without CO_2 (no photosynthesis) undergo a detectable decrease in their ascorbic acid content. But not all species behave alike. Seedlings of *Vicia faba* deprived of their cotyledons do not suffer a great decrease in ascorbic acid content, indicating a greater independence of ascorbic acid synthesis and photosynthesis.

F. WEBER (1940) found that rhizomes of *Stachys,* which had become green under the action of light, contained more ascorbic acid than the controls.

Consequently an indirect connection between ascorbic acid synthesis and photosynthesis can be assumed in view of the fact that ascorbic acid is produced at the expense of sugar. These relations between sugar content and ascorbic acid have been confirmed by MOLDTMANN (1939) with various Monocotyledons. In general, factors which favor photosynthesis also favor ascorbic acid synthesis.

Dependence of the biosynthesis of ascorbic acid on external factors. — By virtue of the principle just pointed out, certain external factors can be shown to exert a distinct action. Light, for example, is an important factor (SUGAWARA, REID, 1937) but its effectiveness varies with the wave length. The effect of light, in order of decreasing activity, is as follows: white, red, orange, green, blue. Red light, which favors photosynthesis, is likewise the most effective for the production of ascorbic acid in etiolated seedlings of *Zea Mays, Pisum sativum, Brassica pekinensis, Phaseolus vulgare, Raphanus sativus, Hordeum vulgare* and *Avena sativa* (SUGAWARA, 1939).

Other factors not directly associated with photosynthesis are involved, *e.g.*, the composition of the mineral medium. Manganese, added to a liquid synthetic medium, increases the production of ascorbic acid as much as 75 per cent in *Avena, Triticum, Hordeum, Phaseolus mungo,* and *Cicer arietinum.* This metal is definitely known to act as a catalyzer of the enzymatic reaction (probably as a coenzyme) in the synthesis of vitamin C (RUDRA, 1939).

VON HAUSEN (1936) studied the influence of KCl, $Ca_3(PO_4)_2$, and $Ca(NO_3)_2$ on pea plants grown in sand culture. The greatest amount of vitamin C (measured by the Tillmans test) is produced when these salts are properly balanced. With the first two salts the quantity of ascorbic acid diminishes in direct proportion to the decrease in dry weight as the dosage of salt increases. In the case of $Ca(NO_3)_2$ the maximum vitamin C content and dry weight are attained with 1000 mg. per liter of solution; with higher dosages, the dry weight decreases rapidly, but the vitamin C content continues to increase (Fig. 8). These experiments indicate that the external medium affects the biosynthesis of ascorbic acid either directly or indirectly because of its action on the carbon and nitrogen metabolism.

As might be expected, temperature also plays a rôle since the biosynthesis of vitamin C involves an enzymatic reaction requiring an optimal temperature. Plants of *Bryophyllum calycinum*, grown in the dark at 20°C. for 48 hours undergo a decrease in ascorbic acid content; at 7° and also at 37°C. the content increases. Plants grown in the dark are affected directly by temperature, *i.e.*, temperature controls the enzymatic process which brings about the

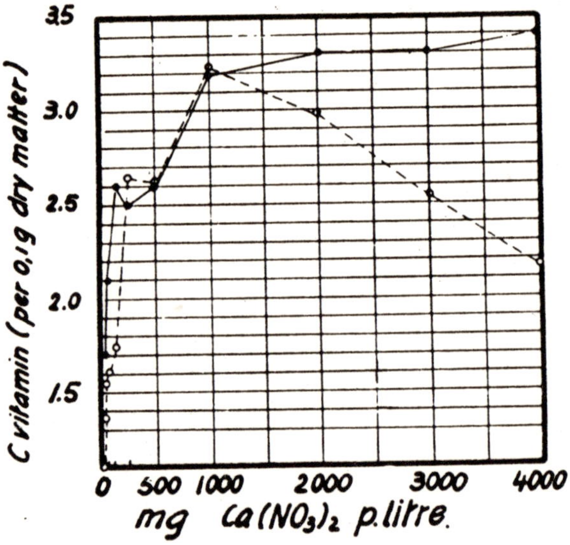

FIG. 8. — Influence of various concentrations of $Ca(NO_3)_2$ on the biosynthesis of vitamin C in *Pisum* (from v. HAUSEN, 1936). Vitamin C is expressed in ml. of color indicator (dichlorophenol-indolphenol) used in titrating 0.1 g. of dry matter.

synthesis of ascorbic acid. Plants grown in light are affected directly and indirectly, *e.g.*, temperature acts on one step in photosynthesis, and thus on the production of sugar which in turn affects the production of ascorbic acid.

Relation between the syntheses of carotinoids and ascorbic acid. — It is unusual that the fruits richest in carotinoids are frequently the richest in ascorbic acid (GIROUD and collaborators, see GIROUD, 1938). This relation has been found also in animals (adrenal glands for example). Is it purely a matter of chance? It is difficult to establish a direct chemical relationship or to attribute this parallelism to pure chance. As is known, ascorbic acid is easily destroyed by oxidation; and carotene acts as an anti-oxidizing agent. This suggests the possibility that carotene protects the ascorbic acid against oxidation (GIROUD), thus permitting indirectly an accumulation of vitamin C. This relation has been confirmed and recognized by various authors. F. WEBER made a real contribution towards the solution of the problem by showing that seedlings of various species grown at a low temperature are inhibited in their production of chlorophyll and that the carotinoids are abundant and the ascorbic acid content is also high, in fact, higher than in seedlings kept either at ordinary temperature and in light or at low temperature and in the dark. These experiments again demonstrate that the synthesis of ascorbic acid is not directly connected with photosynthesis, provided the precursor, sugar, is present. They also show that the coexistence of carotinoids and ascorbic acid is not accidental (MOLDTMANN, 1939; NEUBAUER, 1939; WEBER, 1940).

Let us now return to the Molisch reaction. After demonstrating that photosynthesis is not the direct cause of ascorbic acid formation, are we still in a position to regard vitamin C as responsible for the reduction of silver nitrate? It is difficult to prove this view especially if we recall some of the numerous exceptions: the staining of the stroma in place of the chlorophyll, cytoplasm stained black rather than the plastid, etc. On the basis of these facts, the logical assumption is that the reduction of silver can be brought about either by the photolytic action of light or by reducing sugars. This interpretation would also explain the existence of ascorbic acid in non-green tissues (ERTL, 1939).

Furthermore, according to PEKAREK's photomicrographs, the reduction of the silver reagent occurs in the colorless stroma of the plastid. Sugar is most likely to be the reducing agent, although the direct action of ascorbic acid can not be excluded, since it is water soluble and may be found in the stroma of the plastids.

According to certain investigations, ascorbic acid occurs largely in a combined form. This view is controversial and in general has not been accepted.

It is also necessary to consider the possibility that ascorbic acid may be connected with photosynthesis and the production of sugars in an entirely different manner. After prolonged exposure to heat formaldehyde leads to the production of carbohydrates. WEST and NEY observed that ascorbic acid markedly catalyzes this reaction. It may eventually be shown that the synthesis of sugar is regulated by ascorbic acid which in turn is produced from sugar.

Finally, it may be asked whether ascorbic acid must always be considered as a true vitamin, in view of the fact that the dosage required in many cases is rather large, that it is widely distributed, and that it is present in enormous quantities in certain tissues (*Gladiolus*). It is found in the secretion of the tentacles of *Drosera* (WEBER; MIRIMANOFF, 1939). The rôle of ascorbic acid is possibly that of a nutrient and carbon reserve in addition to its part played in the phenomena of oxidation-reduction. The last mentioned rôle represents its real vitamin function (see p. 193).

Other vitamins. — We know practically nothing concerning the conditions required for the biosynthesis of the B_2 vitamins (riboflavin, pyridoxine, and nicotinic acid). The synthesis of riboflavin in *Aspergillus niger* is dependent on the composition of the medium. A deficiency of magnesium in the medium favors the production of this vitamin (LAVOLLAY and LABORAY).

The biosynthesis of pantothenic acid is somewhat better known. This growth factor of yeast and antipellagra vitamin for the chick, has been shown by R. J. WILLIAMS and his school to be composed of beta-alanine and a lactone of α-γ-dihydroxy-β, β-dimethylbutyric acid. An organism's capacity for synthesis may be confined to one of the components. For example, organisms are known to produce the lactone, but not the beta-alanine. Upon the addition of the latter these organisms are able to synthesize pantothenic acid.

The biosynthesis of vitamins represents the most interesting aspect of the whole problem. This field is destined to furnish many important discoveries.

REFERENCES:-

1921: H. MOLISCH. — Beitrag zur Mikrochemie der Pflanzen. Zur Silberreduktion der Chlorophyllkörner. Ber. d. d. bot. Ges. 39:136.

1933: L. J. HARRIS and S. N. RAY. — Specificity of hexuronic acid as antiscorbutic factor. Biochem. J. 27:580.

1934: S. N. RAY. — On the nature of the precursor of the vitamin C in the vegetable kingdom, I. Vit. C in the growing pea seedling. Biochem. J. 28:996.

1934*a*: A. GIROUD, C. P. LEBLOND et R. RATSIMAMANGA. — Signification de la réduction des sels d'argent au niveau des plastes chlorophylliens. C. r. Soc. Biol. Paris 117:614.

1934*b*: A. GIROUD, R. RATSIMAMANGA et C. P. LEBLOND. — Parallélisme entre la vitamine C et la chlorophylle. C. r. Soc. Biol. Paris 117:612.

1934: B. C. GUHA and A. R. GHOSH. — Synthesis of ascorbic acid (vitamin C) by means of tissues *in vitro.* Nature 134:739.

1935: B. C. GUHA and A. R. GHOSH. — Biological synthesis of ascorbic acid. Nature 135:871.

1935: L. RANDOIN, A. GIROUD et C. P. LEBLOND. — La teneur en acide ascorbique des tissus chlorophylliens et achlorophylliens. Bull. Soc. chim. biol. Paris 17:1649.

1935: R. J. GAUTHERET. — Recherches sur la formation de la chlorophylle dans les racines et la réduction des sels d'argent par les chloroplastes. Thèse Sciences, Paris.

1935: S. VON HAUSEN. — The rôle of vitamin C in the growth of higher plants. Ann. Acad. Sc. Fennicae, Series A, 46, 134 p.

1936: A. GIROUD, R. RATSIMAMANGA, C. P. LEBLOND, H. CHALOPIN et M. RABINOWICZ. — Relation entre l'acide ascorbique et les caroténoïdes. Bull. Soc. chim. biol. Paris 18:573.

1937: O. DIRSCHENDORFER. — Über den histochemischen Nachweis von Vitamin C (*l*-Ascorbinsäure) in Pflanzen. Protoplasma 28:516.

1937: D. GLICK. — The quantitative distribution of ascorbic acid in the developing barley embryo. C. r. Trav. Lab. Carlsberg 21:203.

1937: M. E. REID. — Localisation of ascorbic acid in the cowpea plant at different periods of development. Amer. J. Bot. 24:445.

1938: A. GIROUD. — L'acide ascorbique dans la cellule et les tissus. Protoplasma Monographien, Bd. 16, Berlin, Gebr. Bornträger, 187 p.

1938*a*: A. MIRIMANOFF. — Le dosage et la localisation de la vitamine C dans les tissus végétaux. Rev. Cytologie et Cytophysiologie végét. 3:119.

1938*b*: A. MIRIMANOFF. — A propos de la réaction de Molisch. Rev. gén. Bot. 50:333.

1938: M. E. REID. — The effect of light on the accumulation of ascorbic acid in young cowpea plants. Amer. J. Bot. 25:701.

1938: J. WOLF. — Über die Schwankungen des Gehaltes an Ascorbinsäure in abgeschnittenen Blättern von *Bryophyllum calycinum* bei verschiedenen Temperaturen. Planta 28:725.

1939: M. NEUBAUER. — Das Vitamin C in der Pflanze. Protoplasma 33:345.

1939: H. G. MOLDTMANN. — Untersuchungen über den Ascorbinsäuregehalt der Pflanzen in seiner Abhängigkeit von inneren und äusseren Faktoren. Planta 30:297.

1939: TOMOTO SUGAWARA. — Studies on the formation of ascorbic acid (vitamin C) in plants, II. The influence of radiation of different wavelengths on the ascorbic acid contents in etiolated seedlings. Jap. J. Bot. 10:325.

1939: M. N. RUDRA. — Die Rolle des Mangans bei der biologischen Ascorbinsäure-Synthese. Biochem. Zeitschr. 301:238.

1939: A. MIRIMANOFF. — Remarques sur la sécrétion des tentacules de *Drosera*. Protoplasma 33:211.

1939: O. ERTL. — Über die Silbernitratreduktion der Plastiden. Protoplasma 33:275.

1940: K. und M. WEISSENBÖCK. — Vitamin C-Gehalt im Licht CO_2-frei gezogener Pflanzen. Protoplasma 34:585.

1940: F. WEBER. — Vitamin C-gehalt ergrünter Rhizom-Knollen von *Stachys*. Protoplasma 34:153.

1942: C. D. DARLINGTON. — Vitamin C and chromosome number in *Rosa*. Nature 150:404.

1942: E. J. MORGAN. — Parsley as a rich source of vitamin C. Nature 150:92.

1942: M. PYKE, R. MELVILLE and H. SARSON. — Vitamin C in walnuts. Nature 150:267.

1942: M. E. REID. — Effect of variations in light intensity, length of photo-period, and availability of nitrogen upon accumulation of ascorbic acid in cowpea plants. Bull. Torrey Bot. Club 69:204.

1942: C. G. VINSON and F. B. CROSS. — Vitamin C content of persimmon leaves and fruits. Science 96:430.

Part 2.

VITAMINS IN RELATION TO PLANTS UNABLE TO SYNTHESIZE THEM. — GROWTH FACTORS OF MICROORGANISMS.

In the preceding chapters attention has been called to the fact that green plants which are completely autotrophic may be, as far as isolated organs and tissues are concerned, partially heterotrophic for vitamins. If the plant is intact this partial heterotrophism of certain organs or tissues is not apparent, and in most cases, the plant is able to grow because one of the organs acts as a source of supply. There are, however, certain plants, mostly lower ones, which are completely heterotrophic for one or several growth factors. In this respect they react like animals, and the factor or factors which they are unable to synthesize must be supplied by the culture medium. As far as vitamins are concerned these are the most interesting plants of all since they readily demonstrate the necessity of vitamins in plant nutrition. From the historical and chronological point of view, these organisms should have been studied first, because they are responsible for introducing vitamins into plant physiology. The more logical sequence, and the one followed here, considers the green plants first since they demonstrate more readily the principle that the loss of capacity for synthesis is responsible for the need of an exogenous source of vitamins. Recognizing this principle we may now study the microorganisms which are heterotrophic with respect to vitamins.

Chapter X.

THIAMIN AND ITS COMPONENTS.

The complete thiamin molecule. — Thiamin (vitamin B_1) seems to play a more important rôle than the other vitamins, but this is largely due to the fact that it has been studied more than the others.

In certain cases, thiamin is the only growth factor required for vegetative development, *i.e.*, the microorganism has suffered only one loss in its capacity to synthesize vitamins. The best known case is that of the phycomycete *Phycomyces Blakesleeanus*. This fungus, which is so common in all laboratories, grows very well on media prepared from various natural products such as malt extract, yeast extract, etc. It will not grow on a synthetic medium composed of glucose 3 per cent, asparagine 0.1 per cent, $MgSO_4$ 0.05 per cent, and KH_2PO_4 0.15 per cent. The author observed that zygote formation occurred only if the carbon source was represented by Kahlbaum's brand of maltose (SCHOPFER, 1931-32). It was demonstrated that this sugar contained an impurity of vitamin nature. Despite extensive studies no conditions were found under which *Phycomyces* could produce zygotes without the growth factor occurring in maltose as an impurity. Vegetative development also required the same growth factor. The characteristics of this unidentified substance indicated that it belonged to the group of B vitamins and it was found to be almost, if not completely, replaceable by pure crystalline vitamin B_1 (SCHOPFER, 1934). In all cases preparations of the vitamin from natural materials (JANSEN, WINDAUS, VAN VEEN, PETERS, and R. R. WILLIAMS) had the same activity as the synthetic vitamin (IG Farbenindustrie, Hoffman-La Roche, etc.). It was certain that the activity was not that of an impurity on the vitamin crystals.

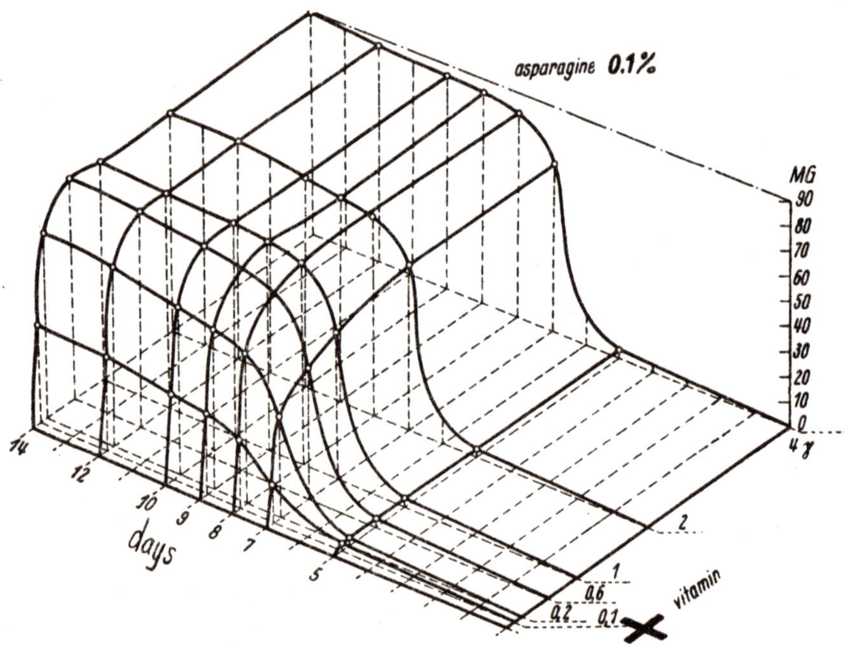

FIG. 9. — Growth of *Phycomyces Blakesleeanus* on a synthetic medium as a function of vitamin B_1 dosage and time. The optimal dosage is indicated by the curve obtained with 0.6γ of thiamin/25 cc. of medium. Asparagine concentration, 0.1 per cent (from SCHOPFER, 1937a).

The action of vitamin B_1 is quantitative. Within certain limits an increase in the dosage causes a corresponding increase in growth but when the optimum is reached further additions are without effect (see Fig. 9). The optimal dosage, as shown in the following table, is 0.5γ per 25 cc. of medium containing 0.1 per cent asparagine.

TABLE VI. — Action of thiamin on *Phycomyces Blakesleeanus* (from SCHOPFER, Erg. Biol., 1939, Springer, Berlin).

dosage of vitamin in γ per 25 cc. of medium:	0	0.01	0.02	0.06	0.1	0.2	0.4	0.8	1.6
dry wt. of culture in mg.:	2-5	8.5	13.1	25.9	36.7	56.7	75.9	84.4	83.4

The three dimensional graph clearly shows that additions of thiamin exceeding 0.5 to 0.6γ have little effect (Fig. 9).

The action of the vitamin is more or less dependent upon the composition of the medium. By increasing the asparagine content the optimal dosage of vitamin is raised and the production of dry matter is also increased. By decreasing the asparagine content the optimal dosage of vitamin is reduced and the maximal weight is also decreased. Apparently a relationship exists between the asparagine content of the medium and the optimal dosage of active vitamin. Furthermore, a rather constant relationship exists between the optimal dosage of vitamin (which varies with the asparagine content of the medium) and the amount of dry matter produced.

TABLE VII. — Action of thiamin on *Phycomyces Blakesleeanus*. Various relationships (modified from SCHOPFER, Arch. f. Mikrobiol., 1935).

asparagine per cent	asparagine mg. in 25 cc. medium	optimal dosage of vitamin, γ/25 cc. medium	dry weight of crop in mg.	ratio of vitamin/dry wt. of crop
0.01	2.5	0.05	10	0.005
0.05	12.5	0.2	40	0.0051
0.1	25	0.6	90	0.0068
0.4	100	1	190	0.0052

The ratio in the last column indicates the quantity of vitamin required to produce 1 mg. of dry matter under optimal conditions. This figure is always approximately 0.005γ. The close agreement between the optimal dosage of vitamin and the asparagine content of the medium is unexplainable. Thiamin cannot be shown to participate in nitrogen metabolism. This subject merits further investigation.

The ratio, weight of thallus/optimal dosage of vitamin (the inverse of the preceding ratio), is about 100,000. When the vitamin is supplied in amounts which are not optimal, the ratio varies with the dosage. With suboptimal dosages, the ratio is very high, 500,000 to 600,000. These results have been confirmed by ROBBINS.

The mycelium of *Phycomyces* has a great affinity for thiamin, being able to extract it from an adsorbate placed in the culture medium. This extraction is effected at ordinary room temperature at which no elution of the vitamin into the medium occurs.

That the need for thiamin is due to a loss in capacity for synthesis can be demonstrated experimentally with *Phycomyces*.

The spores from a normal culture placed in a vitamin-free medium give rise to a short germ tube but no further growth occurs. The cultures are suffering from an avitaminosis. An extract from these cultures does not activate a new culture of the fungus. On the other hand, a concentrated extract prepared from a great number of spores (several million) readily activates a new culture. Obviously the spores contain a small quantity of vitamin which was furnished by the thallus producing them. The amount is rather small, being barely enough for germination. No further growth takes place because the capacity to synthesize the vitamin has been lost.

Other members of the *Mucoraceae* are also dependent on thiamin, *e.g.*, *Mucor Ramannianus, Absidia ramosa,* and *Parasitella simplex.* These same genera also contain species which are autotrophic for vitamins, *e.g.*, *Mucor hiemalis, Absidia repens,* and *A. orchidis.*

One highly significant fact which has been demonstrated is that extracts of species which are autotrophic for thiamin (*Mucor hiemalis, Mucor mucedo, Absidia repens, A. glauca, A. orchidis, Sporodinia grandis*) are able to activate growth in species which are heterotrophic for thiamin, when the latter (particularly *Phycomyces*) are cultivated on vitamin-free synthetic media.

According to ONDRATSCHEK (1940*b*) the same is true for algae (*Hormidium*). Extracts of species autotrophic for thiamin can serve as a vitamin source for species heterotrophic for this vitamin. That the need of a vitamin is due to a loss of capacity for synthesis is now a well established principle of general applicability.

Certain species of the genus *Rhizopus* (*R. Oryzae, chinensis, tritici, tonkinensis, nodosus, suinus, japonicus*) are autotrophic for thiamin. They present an unusual phenomenon, *viz.*, their growth is inhibited by excessive dosages of thiamin (SCHOPFER, 1935, confirmed by ROBBINS and KAVANAGH). This phenomenon apparently depends upon the culture conditions, since JANKE did not observe it when *Rhizopus* was cultured on glass beads placed in the medium. The mechanism of this inhibition is beginning to receive consideration (SCHOPFER, 1942). This period of inhibition, which up till now was the only one demonstrated, is preceded by a short period of acceleration when thiamin functions as a growth factor. This period, observed only in young cultures, is prolonged by lowering the temperature (range studied, 18-36°C.). These phenomena are functions of the temperature, age of culture, and composition of the medium. JANKE's inability to confirm the author's observations of 1935 is explained by the fact that the use of ammonium tartrate (used by JANKE) instead of asparagine prevents or retards the passage from the accelerating to the inhibiting stage.

Multiplicity of factors for Phycomyces. — Thiamin is almost as effective as natural products, *e.g.*, yeast extract, for the vegetative development of this fungus. As early as 1935 the author demonstrated a highly thermostabile factor (much more stabile than thiamin) which he designated as factor M. It is constituted partially by the combination pyrimidine plus thiazole. These components of thiamin are able to replace the latter quantitatively; they are also highly thermostabile.

When considering vegetative development it is necessary to take into account one essential condition - the nitrogen content of the medium. As the nitrogen content (asparagine) of the medium is increased, a limited dosage of vitamin (suboptimal) exerts a very marked effect. A medium containing 0.05 per cent asparagine (0.0125 g. per 25 cc. of medium) and 2γ of thiamin produces 38 mg. of dry matter. If an additional 0.0125 g. of asparagine is added to the medium and the thiamin is left at 2γ, the dry matter produced is 90 mg. Asparagine appears to function as a co-factor.

A beneficial effect can be obtained by adding an extract of a natural product (fruit, potato, etc.) to the medium containing thiamin. Inasmuch as no consideration is given to the nitrogen added by this extract it is difficult to determine whether the active agent is a new growth factor. It is entirely possible that there are other factors involved (ROBBINS, 1939*b*) although there is no actual proof of additional growth factors of vitamin nature for *Phycomyces*. It is also necessary to take into account the mineral elements (metals) with catalytic action since their effect is not to be overlooked.

Recently UMRATH claimed that *Phycomyces* requires one of the B substances of NIELSEN when growing on a medium in which ammonium sulphate serves as the nitrogen source. His conclusion that the active substance is formed from ammonium tartrate heated for 45 minutes at 120°C. is not justifiable. Since ammonium sulphate is a poor source of nitrogen for *Phycomyces* the action of the ammonium tartrate is undoubtedly that of a source of assimilable nitrogen rather than growth factors (SCHOPFER). If the medium contains a source of nitrogen which is difficult to assimilate, any extract supplying assimilable nitrogen will have a beneficial effect providing thiamin is present.

The action of the vitamin is correlated with the chemical composition of the medium. LEONIAN and LILLY, studying the growth of *Phycomyces* under new cultural conditions, demonstrated the rôle of mineral elements (metals) which ROBBINS had already observed. They also studied the action of a certain number of substances as nitrogen sources, and in the main, confirmed the author's observations concerning the nitrogen nutrition of *Phycomyces, viz.*, that ammonium sulphate and arginine are poor sources of nitrogen. They were the first to observe that certain organic acids (succinic and fumaric acid) increase considerably the production of dry matter when used with poor sources of nitrogen (arginine, and ammonium nitrate). Agar, by virtue of the mineral matter which it contains, is also beneficial. Their cultures in all cases were supplied with excessive dosages of thiamin. The general conclusion to be drawn from these and other experiments concerning nitrogen nutrition is that thiamin increases the production of dry matter in proportion to the supply of assimilable nitrogen. Although these observations of LEONIAN and LILLY are of interest they confuse rather than clarify the problem. It still remains to be determined whether the action of organic acids is due to the acids themselves or to metallic impurities associated with them. From all indications the action observed is that of nutrients (carbon compounds or minerals) rather than that of a new organic growth factor.

For the development of zygotes, the author observed already in 1934 that thiamin supplied in quantities sufficient for vegetative development was unable to produce gametes and normal zygotes. ROBBINS (1939*a*, *b*) confirmed this observation and found that certain substances are present in agar which favor the formation of zygotes in the presence of thiamin. This substance (or substances) is contained in other natural products (potato, brown sugar, oat meal and corn meal). They can be extracted in part by dilute methyl alcohol and more completely by aqueous pyridine. The unidentified substance (or substances), referred to as factor Z, is produced by *Phycomyces* (ROBBINS, 1940) but not in amounts sufficient for its maximal development. This factor was found (ROBBINS and HAMNER, 1940) to be multiple with at least two substances composing it, one fraction (Z_1) adsorbed by charcoal and the other (Z_2) not adsorbed. A method for estimating these two factors in solutions of unknown composition was described by ROBBINS (1941) and was used in assaying extracts from agar, potatoes and peptone. Although these substances appear to act as true growth factors there is no positive proof that they are such. Attempts to identify them with known factors were unsuccessful. Factor Z_1 seems to belong in the vitamin B complex.* It is probably not identical with biotin, pantothenic acid, riboflavin, pyridoxine, thiamin, glutamine, or para-amino benzoic acid, and Z_2 is not identical with glutamine or para-amino benzoic acid. Both fractions are clearly beneficial not only to gametic reproduction but also to spore germination and early mycelial growth. Aside from these two unidentified factors, *Phycomyces* requires only thiamin (pyrimidine and thiazole). Factor M which is beneficial to this fungus is constituted partly by the combination pyrimidine + thiazole. Asparagine may be considered as a co-factor. The actions of other substances on this fungus must, for the present, be interpreted as those of ordinary nutrients.

Ustilago violacea, a basidiomycete parasitizing the anthers of *Melandrium album,* has also been studied extensively in regard to growth factors. The growth of this organism is dependent upon thiamin. A very weak dosage is sufficient for optimal development. The optimal dosage is dependent upon the amount of inoculum used. This again illustrates the phenomenon responsible for the Liebig-Pasteur controversy. When the amount of inoculum is small, the optimal dosage may be considered as 0.01 to 0.02γ/25 cc. of medium (growth measured by means of a nephelometer) (SCHOPFER and BLUMER, 1938). When the amount of inoculum is large only a small dosage of thiamin is required for normal development of the culture (0.005γ per 25 cc. of medium). The large inoculum has carried along growth factors which it contained. It must be recalled, however, that with thiamin alone

*Factor Z_1 is hypoxanthine (ROBBINS and KAVANAGH, 1942).

the growth, although excellent, is inferior to that obtained with a medium prepared from natural products. Apparently other factors are required but they have not yet been identified.

The genus *Ustilago* exhibits the same phenomenon observed in *Mucor, viz.*, some species are heterotrophic for thiamin while others are autotrophic for this factor. *U. violacea* and *U. scabiosae* are heterotrophic for thiamin, whereas the majority of species (*U. tritici, U. avenae, U. levis, U. bromivora,* etc.) are autotrophic for this factor. Obviously there is no relationship between parasitism and the need for thiamin. The extracts of species autotrophic for thiamin are able to promote the growth of heterotrophic species by supplying growth factors. These extracts are likewise effective on *Phycomyces.*

The number of organisms which are known to require thiamin is large and is continually increasing. They represent all of the principal groups: bacteria, fungi, algae, rhizopods, and ciliates.

Bacteria. — A few of the bacteria are heterotrophic for thiamin: *Propionibacterium pentosaceum* (TATUM, WOOD and PETERSON, 1936), *Staphylococcus aureus* (KNIGHT, 1937), *Rhizobium* (NILSSON *et al.*). On the other hand, many species are autotrophic for this vitamin: *B. coli, B. smegmatis, B. timothy, B. Moelleri, B. adherens, B. subtilis, B. mycoïdes* (SÜNDERLIN and WERKMAN), *B. vulgatus, B. mesentericus, B. lactis aerogenes, Vibrio alcaligenes* (SCHIEBLICH). It is frequently observed that bacterial contaminants are able to promote growth of cultures of a fungus heterotrophic for thiamin grown on a vitamin-free medium.

The following fungi require thiamin: Phycomycetes; Oomycetes: *Pythium Butleri* (ROBBINS and KAVANAGH, 1938*a*), *Phytophthora cinnamomi, P. parasitica* and other species (ROBBINS, 1938*a*); Zygomycetes: *Phycomyces Blakesleeanus, Mucor Ramannianus, Absidia ramosa,* and *Parasitella simplex* (SCHOPFER).

Ascomycetes, Hemiascomycetes: *Saccharomyces cerevisiae, Nematospora Gossypii* (KÖGL and FRIES, 1937); Hypocreales: *Nectria coccinea* (FRIES, 1938), *Sphaerulina trifolii* (ROBBINS and KAVANAGH, 1938), *Valsa pini, Helvella infula, Lophodermium pinastri, et al.* (FRIES, 1938).

Basidiomycetes, Ustilaginales: *Ustilago violacea* and *U. scabiosae* (SCHOPFER and BLUMER, 1938), *Tilletia tritici* (DEFAGO, 1940), *Tilletia horrida* (SCHOPFER and BLUMER, 1939); Polyporales: *Polyporus adustus, P. abietinus, P. benzoïnus, P. fomentarius et al.* (FRIES, 1938), *Polyporus Spraguei, Stereum frustulosum, Fomes igniarius* (NOECKER, 1938); Agaricales: *Tricoloma nudum* (FRIES), *Schizophyllum commune* (ROBBINS and KAVANAGH, 1938*a*); Cantharellales: *Hydnum erinaceus* (NOECKER); Dacromycetales: *Dacromyces stillatus* (FRIES).

Fungi Imperfecti: *Dematium nigrum, Rhodotorula rubra* (SCHOPFER, 1938), *Rhodotorula Sanniei* (FROMAGEOT), *Torula Laurentii* and *T. fermentati* (ROBBINS and KAVANAGH, 1938*b*).

In addition to the thiamin requiring fungi listed above several others* have been found by LEONIAN and GREENE (1938).

Among the algae the following require thiamin: Flagellates, Phytomastigina (plant organisms): *Euglena gracilis* and *E. viridis* (ONDRATSCHEK, 1940*a*), *Polytoma ocellatum, P. caudatum, Chilomonas paramaecium, Polytomella caeca* (LWOFF and DUSI, 1937); Zoomastigina: *Strigomonas oncopelti, S. fasciculata, S. culicidarum* (MARG. LWOFF); *Chlorophyceae, Volvocales: Chlamydomonas orbicularis* and other species of *Chlamydomonas* (ONDRATSCHEK, 1940*a*), *Chlorogonium tetragamum* (ONDRATSCHEK, 1940*a*), *Hematococcus pluvialis* (ONDRATSCHEK, 1940*a*); *Ulotrichales: Hormidium Barlowi, Uronema gigas* (ONDRATSCHEK, 1940*a*). As might be expected, the thiamin requirements are much more marked in the colorless members of the Flagellatae and *Chlorophyceae* than in the green forms (autotrophs).

The rhizopod *Acanthamoeba castellani* (LWOFF) and the ciliate *Glaucoma piriformis* require thiamin (A. and MARG. LWOFF, 1937).

It is beyond the scope of this book to enumerate all the organisms requiring thiamin as a growth factor. The examples given above clearly demonstrate the fundamental importance of this vitamin in living organisms (see SCHOPFER, 1939).

In a great number of cases thiamin is only one of several essential factors, and its action is maximal only when the other factors are present. With *Staphylococcus aureus* the other factors are nicotinic acid and biotin. With certain races of *Saccharomyces cerevisiae* the other factors are bios I (inositol), bios II*a* (beta-alanine), bios II*b* (biotin), and possibly pyridoxine. The combination bios + thiamin is required by several fungi (see FRIES, 1938).

The flagellates and *Chlorophyceae* studied by ONDRATSCHEK in the presence of ascorbic acid require either thiamin or ascorbic acid, the former being essential in the colorless forms, the latter essential in the green forms.

Strigomonas fasciculata and *S. culicidarum* require hemin in addition to thiamin (M. LWOFF).

Organisms requiring constellations of growth factors (two or more) exclusive of thiamin will be studied in later chapters.

In general it may be said that in each group of organisms, the majority are autotrophic for thiamin and that only a few are heterotrophic for this factor.

The components of thiamin as growth factors. — Up till this point, our discussion concerning thiamin implied that the activity was due to the whole molecule. However, in the case of ***Staphylococcus aureus***, it is known that thiamin, once considered as essential for growth, can be replaced by its two components, pyrimidine and thiazole in equimolar concentrations. This organism has the

---

*ROBBINS and KAVANAGH (1942) list many additional fungi requiring thiamin.

ability to unite the two components into a whole molecule (KNIGHT, 1937). The same discovery was made with *Phycomyces Blakesleeanus* by SCHOPFER and JUNG (1937) and by ROBBINS and KAVANAGH (1937) and confirmed by BONNER and ERICKSON (1938).

In the case of *Phycomyces* it can be demonstrated that the growth curve obtained with thiamin corresponds exactly to that obtained with equimolar concentrations of pyrimidine and thiazole. When the two components are supplied in concentrations which are not equimolar, the response of the organism is strange and inexplainable and not as would be predicted on the basis of the law of limiting factors, according to which the component which is found in the smallest dosage should play the rôle of the limiting factor. A suboptimal amount of pyrimidine and a supraoptimal amount of thiazole give results almost as good as the mixture of the two components in equimolar and optimal dosages. The reciprocal, a suboptimal dosage of thiazole and a supraoptimal dosage of pyrimidine, gives similar results.

0.4γ pyrimidine + 0.4γ thiazole: 92 mg. dry wt.
0.2γ pyrimidine + 0.2γ thiazole: 65 mg. dry wt.
0.2γ pyrimidine + 0.8γ thiazole: 80 mg. dry wt.
0.8γ pyrimidine + 0.2γ thiazole: 85 mg. dry wt.

The unbalanced combinations do not give exactly the same results as those from balanced optimal combinations but the difference is very small (SCHOPFER, 1939). These results of the author differ somewhat from those of J. BONNER and BUCHMAN (1939) who find that proportionately larger amounts of thiazole than pyrimidine are required by *Phycomyces.* They attribute their results to the fact that thiazole is readily destroyed whereas pyrimidine remains intact. Observations that equimolar concentrations of the components of thiamin are no more effective than unbalanced combinations were first made with *Ustilago scabiosae* (BLUMER and SCHOPFER). This fungus is like *Phycomyces* in that it requires both pyrimidine and thiazole and exhibits the same behavior toward unbalanced combinations of these thiamin components. To speak of a transformation of one of the constituents into the other would be presumptuous. The author is of the opinion that the process is probably much more complicated and can not be explained at the present. This strange behavior has been observed in other organisms, *e.g.*, *Torula Laurentii* (ROBBINS and KAVANAGH, 1938), and the flagellates *Polytomella caeca* and *Chilomonas paramaecium* (LWOFF and DUSI).

Contrary to expectation, it was demonstrated (SCHOPFER, 1937*b*) that an organism apparently dependent on thiamin, *Rhodotorula rubra* (Baarn strain) actually requires only pyrimidine, and that another thiamin requiring organism, *Mucor Ramannianus,* is able to grow when supplied with only the thiazole component (MÜLLER and SCHOPFER, 1937). Shortly thereafter other organisms were found which behave similarly. Organisms requiring only pyrimidine are: *Schizophyllum commune, Pythium Butleri*

(ROBBINS and KAVANAGH, 1938*a*), *Dematium nigrum* and *Parasitella simplex* (SCHOPFER, 1938). Organisms requiring only thiazole are the flagellates *Polytoma ocellatum* and *P. caudatum* (LWOFF and DUSI). The only microorganism in the plant kingdom known to require only thiazole as a growth factor is *Mucor Ramannianus.* These reports, for the most part, have been confirmed and it is now possible to establish definite categories in accordance with the needs of the organisms for the one or the other component.

These observations were supplemented by the following facts: a ciliate, *Glaucoma piriformis,* and also the parasitic flagellates, *Strigomonas oncopelti, S. fasciculata, S. culicidarum* are able to thrive only in the presence of the whole molecule of thiamin; the components of the vitamin are ineffective (MARG. LWOFF, 1938). The same is true for several species of *Phytophthora*: *P. parasitica, P. palmivora, P. Boehmeriae, P. capsici, P. Drechsleri, P. cryptogea, P. cinnamomi* (ROBBINS, 1938*a*).

We are now able to establish the following groups (SCHOPFER, *cf.* ROBBINS, 1938*b*).

1) Organisms autotrophic for thiamin: *Absidia repens.*
2) Organisms requiring only pyrimidine: *Rhodotorula rubra* type.
3) Organisms requiring only thiazole: *Mucor Ramannianus* type.
4) Organisms requiring both pyrimidine and thiazole: *Phycomyces* type.
5) Organisms requiring the entire molecule of thiamin: *Glaucoma-Phytophthora* type.
6) A separate group of organisms which are partially inhibited by excess thiamin: *Rhizopus nigricans* and other *Rhizopus* species.

Several of these groups are represented by higher plants and animals. We have already seen that the tomato root belongs to the *M. Ramannianus* type, and the pea root to the *Phycomyces* type. Of the animals, the pigeon, which is sometimes used as a test object for thiamin, is actually able to utilize the combination pyrimidine + thiazole (ROBBINS, HOGAN and RICHARDSON, and confirmed by E. and R. ABDERHALDEN). Apparently these various types occur throughout both the plant and animal kingdoms. These categories were established in the field of microbiology on the basis of observations on bacteria and fungi, particularly *Phycomyces.* In which of the above groups does Man belong? That has not yet been determined and it is still uncertain that the vitamin action is always dependent on the entire molecule of thiamin.

It is necessary, furthermore, to determine whether the establishment of these categories is strictly dependent on the composition of the medium. The exact quantitative replacement of thiamin by the combination pyrimidine + thiazole or by one of the components alone is possible only with carefully controlled cultural conditions.

Economic coefficient of growth factors. — We have considered the possibility of establishing an economic coefficient of the action of a vitamin represented by the ratio, dry weight of the organism/weight of vitamin required. We have also called attention to the fact that the ratio varies in *Phycomyces* with the age of the culture and with the time at which the determination is made (see p. 102). If it is determined (early) in young cultures, as was done by FRIES, the figures are very large. They indicate the necessity of the vitamin in the early stages of growth when the capacities for primary syntheses are not influenced by secondary factors. The author determines this coefficient later when the organism has attained its full development under a given set of conditions. These conditions involve the composition, volume, and depth of the medium and also the dosage of vitamin. The dosage must be optimal if the coefficient is to be of any value. If the vitamin is in excess, the results obtained have little value. Under the conditions prescribed by the author the following ratios have been found:

TABLE VIII. Action of thiamin on various microorganisms. Economic coefficient (from SCHOPFER and BLUMER, 1938).

T = thiamin P = pyrimidine	Optimal dosage of vitamin per 100 cc. of media	Economic coefficient	
Bacteria:			
Staphylococcus aureus (T)	0.3γ	—	(KNIGHT)
Propionibacterium pentosaceum (T)	0.5γ	—	(TATUM, WOOD and PETERSON)
Phycomycetes:			
Phycomyces Blakesleeanus (T)	2.0γ	200,000	(SCHOPFER)
Ascomycetes:			
Rhodotorula rubra (P)	0.8γ	200,000	(SCHOPFER)
Basidiomycetes:			
Polyporus adustus (T)	0.166γ	380,000	(KÖGL and FRIES)
Ustilago violacea (T)	0.060γ	2,500,000	(SCHOPFER and BLUMER)
Fungi Imperfecti:			
Dematium nigrum (P)	0.08γ	1,000,000	(SCHOPFER)
Algae:			
Chilomonas and other colorless flagellates (T)	1.0γ	—	(ONDRATSCHEK)
Hematococcus pluvialis (T)	100.0γ	—	(ONDRATSCHEK)

Apparently these ratios vary with the organism, the most sensitive to thiamin being *Ustilago violacea*, the least sensitive *Hematococcus pluvialis*.

The reciprocal calculation (optimal dosage of vitamin/dry weight of crop) indicates the amount of vitamin required under a given set of conditions, to yield 1 mg. of dry matter:

Phycomyces Blakesleeanus	0.5 γ/100 mg.=0.005 γ for 1 mg.
Rhodotorula rubra	0.2 γ/40 mg.=0.005 γ for 1 mg.
Dematium nigrum	0.02γ/20 mg.=0.001 γ for 1 mg.
Ustilago violacea	0.01γ/25 mg.=0.0004γ for 1 mg.

The economic coefficient has, in reality, no great significance. The dry weight of the crop largely determines the ratio. The author recognizes the fact that the production of dry matter and the accumulation of reserves depend on numerous factors which have no connection with the dosage of vitamin B_1. Nevertheless, these figures do indicate the great sensitivity of the organisms and the delicacy of their reactions.

Furthermore, attention should be called to the fact that thiamin and pyrimidine furnish ratios which have a definite relationship.

After having established categories based on the needs of thiamin and its components, the problem which now confronts us is the cause of these differences.

REFERENCES:-

1931/32: W. H. SCHOPFER. — Recherches expérimentales sur la formation des zygotes chez *Phycomyces blakesleeanus.* Influence des substances vitaminiques. Bull. Soc. bot. suisse, 1931, 40:87; 1932, 41:73.

1934: W. H. SCHOPFER. — Les vitamines cristallisées B comme hormones de croissance chez un microorganisme. Arch. f. Mikrobiol. 5:513.

1935: W. H. SCHOPFER. — Etude sur les facteurs de croissance. Action de la vitamine cristallisée B_1 sur *Rhizopus* et d'autres Mucorinées. Zeitschr. f. Vitaminforschung 4:187.

1936: E. TATUM, H. G. WOOD and W. H. PETERSON. — Vitamin B_1, a growth stimulant for propionic acid bacteria. Biochem. J. 30:1898.

1937: W. J. ROBBINS and F. KAVANAGH. — Intermediates of vitamin B_1 and growth of *Phycomyces.* Proc. Nat. Acad. Sci. U.S.A. 23:499.

1937: W. F. MÜLLER et W. H. SCHOPFER. — L'action de l'aneurine et de tinic acid and Vitamin B_1. Biochem. J. 31:731.

1937: F. KÖGL und N. FRIES. — Über den Einfluss von Biotin, Aneurin und Mesoinosit auf das Wachstum verschiedener Pilzarten. Zeitschr. physiol. Chem. 249:93.

1937: W. F. MÜLLER et W. H. SCHOPFER. — L'action de l'aneurine et de ses constituants sur *Mucor Ramannianus* Möll. C. r. Ac. Sc. Paris 205:687,

1937: A. et M. LWOFF. — L'aneurine, facteur de croissance pour le Cilié *Glaucoma piriformis.* C. r. Soc. Biol. Paris 126:644.

1937: A. LWOFF et H. DUSI. — Le thiazol facteur de croissance pour *Polytoma ocellatum.* Importance des constituants de l'aneurine pour les Flagellés leucophytes. C. r. Acad. Sc. Paris 205:882.

1937*a*: W. H. SCHOPFER. — Recherches sur le métabolisme de l'azote d'un microorganisme acellulaire (*Phycomyces blakesleeanus*). Le rôle des facteurs de croissance. Protoplasma 28:381.

1937*b*: W. H. SCHOPFER. — L'action des constituants de l'aneurine sur les levures. (*Rhodotorula rubra et R. flava*). C. r. Acad. Sc. Paris 205:445.

1937: W. H. SCHOPFER et A. JUNG. — L'action des produits de désintégration de l'aneurine sur *Phycomyces.* Le second facteur de croissance des Mucorinées. C. r. Acad. Sc. Paris 204:1500.

1938*a*: W. J. ROBBINS and F. KAVANAGH. — Vitamin B_1 and the growth of certain fungi. Amer. J. Bot. 25:229.

1938*b*: W. J. ROBBINS and F. KAVANAGH. — Intermediates of vitamin B_1 and the growth of *Torula*. Plant Physiology 13:611.

1938*a*: W. J. ROBBINS. — Thiamin and the growth of species of *Phytophthora*. Bull. Torrey Bot. Club 65:267.

1938*b*: W. J. ROBBINS. — Organisms requiring vitamin B_1. Proc. Nat. Acad. Sc. U.S.A. 24:53.

1938: M. LWOFF. — L'aneurine, facteur de croissance pour les *Strigomonas* (Flagellées Trypanosomides). C. r. Soc. Biol. Paris 128:241.

1938: L. H. LEONIAN and V. GREENE LILLY. — Studies on the nutrition of Fungi, I. Thiamin, its constituents, and the source of nitrogen. Phytopathology 28:531.

1938: N. L. NOECKER. — Vitamin B_1 in the nutrition of four species of wood-destroying fungi. Amer. J. Bot. 25:345.

1938: N. FRIES. — Über die Bedeutung von Wuchsstoffen für das Wachstum verschiedener Pilze. Symbolae botanicae Upsalienses III, 2.

1938: W. H. SCHOPFER. — La pyrimidine facteur de croissance de microorganisme (*Rhodotorula*, Mucorinées, *Dematium*). Protoplasma 31:105.

1938: W. H. SCHOPFER und S. BLUMER unter Mitwirkung von V. KOCHER. — Untersuchungen über die Biologie von *Ustilago violacea*, II. Wirkung des Aneurins und anderer Wuchsstoffe vitaminischer Natur. Arch. f. Mikrobiol. 9:305.

1938: J. BONNER and J. ERICKSON. — The *Phycomyces* assay for thiamin (vitamin B_1); the methods and its chemical specificity. Amer. J. Bot. 25:685.

1939: W. H. SCHOPFER et S. BLUMER. — Recherches sur la répartition de l'héterotrophie par rapport à l'aneurine chez les champignons. Arch. f. Mikrobiol. 11:205.

1939: J. BONNER and E. R. BUCHMAN. — The synthesis and destruction of vitamin B_1 by *Phycomyces*. Proc. Nat. Acad. Sci. U.S.A. 25:164-171.

1939*a*: W. J. ROBBINS. — Growth substances in agar. Amer. Jour. Bot. 26:772.

1939*b*: W. J. ROBBINS. — Growth substances and gametic reproduction by *Phycomyces*. Bot. Gaz. 101:428.

1939: W. H. SCHOPFER. — Vitamine und Wachstumsfaktoren bei den Mikroorganismen, mit besonderer Berücksichtigung des Aneurins. Erg. d. Biologie 16:1-172.

1940*a*: K. ONDRATSCHEK. — Experimentelle Untersuchungen über den Zusammenhang zwischen Vitaminbedürfnis und Verlust des Synthesevermögens einiger mixotropher Algen. Arch. f. Mikrobiol. 11:226.

1940*b*: K. ONDRATSCHEK. — Experimentelle Untersuchungen über den Einfluss von Wirkstoffen auf die Vermehrung einiger mixotropher Algen. Arch. f. Mikrobiol. 11:89.

1940: G. DEFAGO. — Effets de l'aneurine, de ses composants et de l'hétéroauxine sur la croissance de trois parasites du blé. Phytopath. Zeitschrift 13:293.

1940: W. J. ROBBINS. — The effect of extracts of *Phycomyces* upon its development. Amer. Jour. Bot. 27:559.

1940: W. J. ROBBINS and K. C. HAMNER. — Effect of potato extracts on growth of *Phycomyces*. Bot. Gaz. 101:912.

1941: W. J. ROBBINS. — Further observations on factor Z. Bot. Gaz. 102:520.

1942: W. J. ROBBINS and V. KAVANAGH. — Vitamin deficiencies of the filamentous fungi. Bot. Rev. 8:411.

1942: W. J. ROBBINS and F. KAVANAGH. — Hypoxanthine, a growth substance for *Phycomyces*. Proc. Nat. Acad. Sci. U.S.A. 28:65.

1942: W. H. SCHOPFER. — Recherches sur le besoin en facteurs de croissance et le pouvoir de synthèse de *Rhizopus suinus*. C. r. Soc. Phys. Hist. Nat. Genève 59:101.

Chapter XI.

THIAMIN AND ITS COMPONENTS (Contd.)

Analysis of the capacity for synthesis relative to thiamin. — The fact that some organisms require only one of the components of thiamin while others require the whole molecule, poses the question: what causes these differences? The only tenable hypothesis is that the loss of capacity for synthesis does not affect all organisms alike. The fact that *Rhodotorula* requires only the pyrimidine component simply means that this red yeast has the ability to synthesize the thiazole component. Conversely, organisms requiring only thiazole are able to synthesize pyrimidine. Another hypothesis is that the complete molecule of thiamin is ultimately formed in the organism (MÜLLER and SCHOPFER, 1937). These hypotheses have been tested experimentally by the author through the use of thiamin assays of *Phycomyces* cultures supplied with pyrimidine and thiazole, and also cultures of *Rhodotorula rubra* and *Mucor Ramannianus* supplied with pyrimidine and thiazole, respectively. The ordinary methods of analysis such as the thiochrome test gave negative results. The animal test employed by the author in coöperation with A. JUNG gave no better results. The reason for our failure to obtain a positive reaction with these tests may be that the amount of thiamin is too small, but it is more probable, as BONNER has shown, that the thiamin is destroyed during the course of development.

Our efforts to obtain a more sensitive assay for thiamin led to the use of the *Phytophthora* and *Phycomyces* tests. The latter fungus, when cultured on a thiamin-free synthetic medium, responded markedly to extracts obtained from cultures of: *Phycomyces* on pyrimidine and thiazole, *Rhodotorula* on pyrimidine, and *Mucor Ramannianus* on thiazole, thus proving that the extracts of all three fungi contain the two constituents, pyrimidine and thiazole. Similar experiments carried out with the *Phytophthora* test (SCHOPFER, 1940) likewise resulted in growth, proving that the complete molecule of thiamin has been reconstructed. If *Rhodotorula* requires only pyrimidine it is because thiazole is normally synthesized. The converse is true for *Mucor Ramannianus.*

A still more exact demonstration can be made with an experiment on the artificial symbiosis of *Mucor Ramannianus* and *Rhodotorula rubra.* When each of these is inoculated separately on a vitamin-free synthetic medium neither is able to develop. But when they are inoculated together, they grow luxuriantly and the two partners establish an artificial symbiosis. This proves that *Rhodotorula* synthesizes thiazole which it supplies to *Mucor Ram-*

annianus and, conversely, that *Mucor Ramannianus* synthesizes pyrimidine which it furnishes to *Rhodotorula.* This experiment on symbiosis is rather delicate to perform and is successful only under properly controlled conditions (MÜLLER and SCHOPFER, 1937; SCHOPFER, 1938).

ROBBINS and KAVANAGH (1938*a*) have conducted similar experiments with various species of *Torula* and have arrived at the same conclusions. This aspect of the problem can therefore be considered as closed, at least for the present.

In general it can be said that when a substance is not required it is because it is actually synthesized by the organism in question.

It may be recalled that roots of peas and tomatoes are partially heterotrophic for thiamin. Actually, the two components, pyrimidine and thiazole suffice for the growth of pea roots (p. 89), and thiazole alone suffices for tomato roots. The same type of partial heterotrophism has also been found in animals. Pigeons, which up till now seemed to require the entire molecule of vitamin B_1 are actually able to utilize the two components, pyrimidine + thiazole (ROBBINS, HOGAN and RICHARDSON 1937 and confirmed by E. ABDERHALDEN, 1938).

On the basis of requirements of various organisms relative to thiamin and its components it is possible to construct a composite table showing how this type of heterotrophism transects the various phyla of organisms.

TABLE IX. — DEGREES OF THIAMIN HETEROTROPHISM IN VARIOUS ORGANISMS (from SCHOPFER, 1939).

synthesized:	thiamin or P + T	pyrimidine	thiazole	——	——
required:	——	thiazole	pyrimidine	pyrimidine + thiazole	thiamin
	Absidia orchidis higher green plants (intact)	*Mucor Ramannianus* tomato roots	*Rhodotorula rubra*	*Phycomyces* *Staphylococcus* pea roots pigeon	*Glaucoma* *Phytophthora* *Strigomonas* Rat

From left to right there is a progressive increase in heterotrophism which cuts across all of the major groups of organisms, the same type occurring in the flagellates, fungi, higher green plants and animals. This shows that the origin of heterotrophism, relative to thiamin, is dependent on a property found in all living beings. It is *polyphyletic* and may occur in any organism irrespective of its systematic position.

In addition to the various well defined types, intermediates occur which can be demonstrated among microorganisms. For example, *Parasitella simplex* is heterotrophic for thiamin and when supplied with the entire molecule the fungus grows luxuriantly. When supplied with pyrimidine it also grows well under a given set of cultural conditions. However, the production of dry matter is generally reduced by about one fourth. The behavior of this organism can be explained by assuming that it possesses most likely the ability to synthesize thiazole, but not in sufficient quantity. Despite additions of excessive dosages of pyrimidine, growth is not improved. Apparently the limiting factor in this case is the capacity to synthesize the necessary thiazole.

The capacity for synthesis may vary during the course of a series of experiments. An organism requiring pyrimidine + thiazole and possessing the full ability to condense them into thiamin (*Ustilago scabiosae*), may lose this ability in part. As a result its growth with P + T is no longer as good as with the entire molecule of thiamin. This case represents a progression toward heterotrophism. On the other hand an organism may progress toward autotrophism. For example, *Rhodotorula rubra* (Lister strain No. 4585), cultivated in a vitamin-free synthetic medium, grows very poorly because it has a very limited capacity for the synthesis of thiamin (LWOFF). When cultivated by LWOFF for 14 months on a vitamin-free synthetic medium, it improved in its growth three or four fold during this period. Obviously, it increased its capacity for synthesis. However, in this case the capacity for synthesis was not completely lost and by a process of "training" or adaptation, the very weak capacity existing at the beginning was augmented. A distinction must be made between a capacity for synthesis which is *weak* and one which is *absent* and regarded as genetically lost. From experimental evidence it appears that an organism may progress to complete heterotrophism but may not return to autotrophism, in other words, the complete loss of capacity for synthesis is definitive and irreversible.

How can one be certain that a capacity for synthesis is lost *definitively?* If an organism becomes unable to grow in a given medium and if transfers are no longer possible, the loss was probably of this nature. However, other cultural conditions are possible and until all of these have been exhausted it is unsafe to assume that the capacity of synthesis is permanently lost.

Numerous experiments demonstrate that the capacity for synthesis is dependent upon external factors, both chemical and physical. *Pythium Butleri* requires pyrimidine when cultivated on a medium containing, among other things, 16.4 g. mineral salts per liter; but if the concentration of mineral salts is reduced to 1.64 g. the organism is able to develop equally well without pyrimidine. Apparently the strong concentration of mineral salts inhibits the synthesis of pyrimidine but not that of thiazole. A decrease in the concentration again permits a partial synthesis of pyrimidine (ROBBINS and KAVANAGH, 1938*b*). *Rhodotorula rubra* requires pyrimidine when cultured on glucose but when cultured on glycerine it seems to be able to synthesize this growth factor (FROMAGEOT and TCHANG). A quantitative and qualitative modification of the medium brings about a change in the capacity for synthesis. The physical factors, temperature for example, likewise affect this capacity. ROBBINS observed that *Phycomyces* is able to synthesize factor Z (required for zygote formation) if the temperature is held below 20°C. but if the temperature is raised to 25°C., it is no longer able to do so.

These examples serve to demonstrate that the capacity for synthesis is only relative in its intensity and that caution must be exercised before concluding that it has been lost completely.

A growth factor cannot be defined on the basis of whether its synthesis is entirely lacking or only apparently so, as LWOFF has suggested. In both cases the function is the same and the substance is required for maximal development.

That the ability to synthesize is relative was also observed by FRIES (1938). To express this in the case of thiamin, he set up the following categories:

1) organisms not requiring thiamin (*Aspergillus niger*).
2) organisms in which the need for thiamin is facultative, *i.e.*, thiamin is beneficial but not essential for growth (*Lenzites sepiaria*).
3) organisms in which the need for thiamin may be obligate or facultative, depending upon the cultural conditions.
4) organisms in which the need for thiamin is obligate (*Polyporus adustus*).

This classification agrees fully with the views of the author.

Biosynthesis of thiamin in microorganisms. — Sufficient information is now available to disclose the manner in which this vitamin is synthesized and also the mechanism of its metabolism. Its function as a fragment of an enzyme is well established. Since the work of LOHMANN and SCHUSTER it is known that this vitamin is none other than the active group in cocarboxylase which functions as the coenzyme in carboxylase and is responsible for the decarboxylation of pyruvic acid (see p. 190). These investigations carried on with yeast represent the starting point for the modern work with this vitamin. A similar demonstration was made by HILLS (1938) with *Staphylococcus aureus*. The coenzyme function of thiamin has been further confirmed by HAAG (1940) with *Phycomyces* grown on a medium based on glucose. If thiamin is lacking, microorganisms, like animals, accumulate pyruvic acid.

The coenzyme cocarboxylase is a pyrophosphate of thiamin.

The metabolism of thiamin in *Phycomyces* may be represented as follows:

1) absorption of pyrimidine and thiazole.
2) condensation of the two components into a molecule of thiamin.
3) esterification of thiamin into the pyrophosphate of thiamin.

Little is known concerning the ultimate fate of thiamin. The author has observed that thalli of *Phycomyces* which are old and fully developed no longer give a positive reaction for thiamin and lack growth factor activity. The thiamin has therefore disappeared. BONNER and BUCHMAN (1938) working with *Phycomyces* observed a liberation of pyrimidine and a destruction of the thiazole nucleus probably by opening the ring at the second position.

The same authors working with *Pisum* demonstrated an earlier step in the biosynthesis of thiamin. They showed that the root of *Pisum* is able to condense the precursors of thiazole (acetopropyl alcohol and thioformamide) into a molecule of thiazole. The same experiment attempted with *Phycomyces* yielded negative results.

With *Pisum* the following steps in the metabolism of thiamin are known:

1) condensation of acetopropyl alcohol and thioformamide into thiazole.
2) condensation of thiazole and pyrimidine into thiamin.
3) esterification of thiamin into cocarboxylase.

An organism completely autotrophic for thiamin must therefore be able to effect the synthesis of pyrimidine, thiazole, thiamin (by condensation) and cocarboxylase. An organism is rendered dependent upon an exogenous supply if it loses its ability to accomplish any one of the following steps: synthesis of the precursors of pyrimidine, condensation of these precursors into a molecule of pyrimidine, synthesis of the precursors of thiazole or their condensation into thiazole, condensation of pyrimidine and thiazole into thiamin, and the esterification of thiamin to cocarboxylase. Accordingly, two phenomena are involved, *viz.*, the synthesis of the components and their condensation.

These concepts are based on information concerning the synthesis of this vitamin *in vitro* (R. R. WILLIAMS and CLINE, 1937; TODD and BERGEL, 1937; ANDERSAG and WESTPHAL, 1937; GREWE, 1937). It has not been proven that the steps in the biosynthesis are the same as those in the synthesis *in vitro*, but from what is known the two methods of synthesis appear to be similar. One of the usual procedures in the chemical synthesis consists of the following steps: 1) condensation of acetamidine and the ester of formylsuccinic acid into pyrimidine. 2) condensation of methyl-α-chlor-γ-acetopropylketone and thioformamide into thiazole. 3) condensation of pyrimidine and thiazole into thiamin. The steps in biosynthesis as found by animal studies are very similar to those of synthesis *in vitro*. Experiments with the precursors of pyrimidine have not yet been carried out.

These condensations forming thiazole and thiamin very probably are effected by means of enzymes. The same is true for the splitting of the thiamin molecule into its components. In 1937 the author advanced the hypothesis of the existence of an *aneurinase* (thiaminase of BONNER and BUCHMAN) with reversible action. BONNER and BUCHMAN have also suggested the possible existence of a thiazolase. The existence of these enzymes has not yet been demonstrated experimentally. All tests carried out *in vitro* have given unsatisfactory results.

Specificity of action of thiamin and its components. — A large number of substitution products of thiamin and its components is available and thus a detailed study of the specificity of action becomes possible. Since thiamin functions as a fragment of a coenzyme, a very marked specificity of action might be expected *a priori*.

A critical and detailed study of the results obtained with several organisms has been published (SCHOPFER, 1939).

The specificity of action has been studied in detail for the following: *Phycomyces* (SCHOPFER, SINCLAIR, ROBBINS and KAVANAGH, BONNER and ERICKSON), flagellates (LWOFF and DUSI, A. and M. LWOFF), protozoans (*cf.* SCHOPFER, 1939), *Staphylococcus aureus* (KNIGHT and MCILWAIN), *Ustilago violacea* (SCHOPFER and BLUMER), *Dematium nigrum* and *Rhodotorula rubra* (SCHOPFER, 1938), *Schizophyllum commune,* haplont and diplont (SCHOPFER and BLUMER), *Pisum* root (BONNER and BUCHMAN, 1938), and the pigeon (SCHULTZ, 1940). Most of our information on this aspect of vitaminology was secured by experiments with microorganisms. Studies with animals were begun only recently.

Substitution in the thiamin molecule. — In order to help the reader understand the following discussion, the structural formula of thiamin is herewith presented:

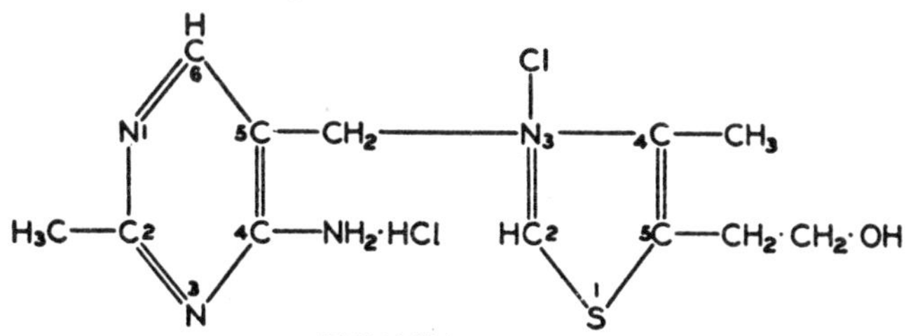

THIAMIN

Substitution products of thiamin in which the NH_2 group in the fourth position of the pyrimidine nucleus has been substituted are inactive on *all* organisms studied. This group is therefore of fundamental importance; it is known to be responsible for the decarboxylation of pyruvic acid.

Normal thiamin is, however, not the most active compound. A thiamin having in the second position of the pyrimidine nucleus a C_2H_5 in place of CH_3 has a greater activity than normal thiamin on *Phycomyces,* and also on animals. Their relative activity is expressed by the ratio of ethyl thiamin to methyl thiamin having the same physiological activity. This ratio for *Phycomyces* was found to be 0.83:1.0 (SCHOPFER). The ratio for the pigeon is 0.85:1.0 (SCHULTZ, 1940).

The importance of the NH_2 group is demonstrated by the fact that thiochrome is known to be inactive on all organisms requiring thiamin or pyrimidine + thiazole which have been studied.

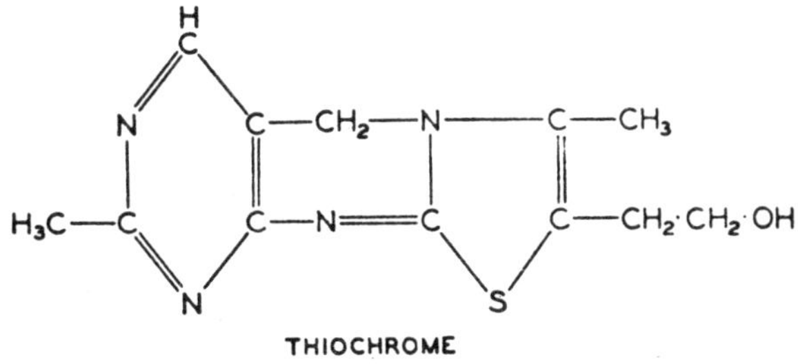

THIOCHROME

When the beta-hydroxyethyl group in the fifth position of thiazole is replaced by a hydroxypropyl the compound is inactivated for *Phycomyces* but remains active for *Glaucoma piriformis, Strigomonas,* and *Staphylococcus aureus.* A substitution of an H in place of the CH_3 group in the second position of thiazole destroys the activity for *Staphylococcus aureus.*

The pyrophosphoric ester of thiamin with the pyrophosphoric group in the fifth position of thiazole (cocarboxylase) is active on *Phycomyces.*

The isovitamin B_1 having the CH_3 group in the sixth position and an H in the second position of pyrimidine is inactive on *Glaucoma piriformis* but active on *Staphylococcus aureus.*

The substitution of an atom of Cl in the second position of the pyrimidine nucleus in place of the CH_3 group does not reduce the activity for *Staphylococcus aureus.*

Analogues of thiamin have been employed in which the pyrimidine nucleus has been replaced by an entirely different group, benzyl or 4,5-methyl-imidazol but have been found inactive.

Organisms differ in their response to substitutions except for the substitution of NH_2 in the fourth position of the pyrimidine nucleus which inactivates the molecule for all organisms studied.

Substitutions in the intermediates of thiamin (pyrimidine and thiazole). — It is impossible to treat this subject in full detail. Additional information may be obtained by consulting the references at the end of the chapter.

If an organism which is dependent on pyrimidine + thiazole is supplied with an inactive analogue of thiamin in which one nucleus possesses an active structure while the other nucleus is inactive, a positive action can be obtained by adding an active form of the nucleus which is inactive in the thiamin analogue. For example, thiochrome (thiazole inactive, pyrimidine active) when used in admixture with the proper thiazole is active in the same sense as the vitamin on *Phycomyces* and *Staphylococcus.* Chlorothiamin (pyrimidine inactive with Cl in the second position, thiazole active) in admixture with the proper pyrimidine is active on *Staphylococcus.* The organism is able to split the methylene bond connecting the pyrimidine with the thiazole nucleus ($C\text{-}CH_2\text{-}N$). The active component of the analogue condenses with the active component which was added thereby producing natural thiamin.

The most specific test object among the P + T organisms is *Phycomyces,* which has been studied extensively. The only thiazoles, apart from normal thiazole, which are active on this fungus are those which have the $CH_2.CH_2OH$ in the fifth position replaced by the groups CH_2CH_2Cl or $CH_2CH_2O.CO.C_2H_5$. *Staphylococcus* and the flagellates have a less pronounced specificity and respond to compounds which are completely inactive on *Phycomyces.* The second position of thiazole is of prime importance for all organisms except the flagellates studied by LWOFF and DUSI. If the H in this position is substituted by any other group the activity of thiazole is nullified for all organisms except the flagellates, which tolerate a CH_3 in this position.

In general the flagellates and the root of *Pisum* are the least specific of all the organisms in their reactions to substitutions in the pyrimidine and thiazole nuclei. There is a marked similarity between the two (see p. 70).

For pyrimidine, as already observed while considering the entire molecule of thiamin, a C_2H_5 group in the second position in place of the CH_3 group does not lower the activity. The only pyrimidines which are active for *Phycomyces* are those with the groups CH_2NHCHS, $CH_2OC_2H_5$, or CH_2Br in the fifth position. The specificities of *Staphylococcus*, flagellates, and *Ustilago violacea* are very similar to those of *Phycomyces*. In the case of *Rhodotorula rubra* and *Dematium* the specificity is less pronounced. These two organisms require only the pyrimidine component of thiamin.

In general it may be said that the degree of specificity depends on the concentration of the active substance. Analogues of thiamin or its components which are inactive when used in the dosage ordinarily employed for natural thiamin (or its components) become active when supplied in rather large dosage. This is true particularly in the case of *Rhodotorula rubra*, and has been confirmed by various authors. This phenomenon will be discussed later on.

The specificity of action of thiamin on animals is treated in great detail by SCHULTZ (1940) who investigated the action of 39 homologues and analogues of thiamin on pigeons and found 22 compounds active. His work dealt only with the complete molecule of thiamin in which the substitutions were made in the pyrimidine and thiazole nuclei. By employing, in some instances, very large dosages of these substances, SCHULTZ obtained an antineuritic activity. Some examples are presented in the following table.

TABLE X. — SPECIFICITY OF ACTION OF THIAMIN ON PIGEONS (from SCHULTZ, 1940).

No.	pyrimidine		thiazole		activity	
	2	4	5	4	A	B
1.	C_2H_5	NH_2	CH_2CH_2OH	CH_3	2.1	0.84
2.	CH_3	NH_2	CH_2CH_2OH	CH_3	2.5	1.0
3.	$(CH_3)_2CH$	NH_2	CH_2CH_2OH	CH_3	8.3	3.3
4.	CH_3	NH_2	CH_2CH_2OH	C_2H_7	91.	36.
5.	CH_3	NH_2	CH_2CH_2OH	H	200.	80.
6.	CH_3	$NH.CH_3$	CH_2CH_2OH	CH_3	536.	214.
7.	CH_3	NH_2	$C_2H_4.O.C_2H_5$	CH_3	1000.	400.
8.	H	$NH_2$6-CH_3	CH_2CH_2OH	CH_3	11800.	4500.

A = number of gammas necessary to produce vitamin activity of one international unit.

B = activity of the vitamin based on natural thiamin with an activity of 1.0 (compound No. 2); figures obtained by dividing the figures under A by 2.5.

It is surprising to find that products devoid of activity when supplied in the usual dosage (optimal dosage of natural thiamin) become active when supplied in concentrations as much as 4500 times stronger than usual. A substitution for the NH_2 group in the fourth position of the pyrimidine (No. 6) diminishes the vitamin activity on pigeons 214 times but does not completely destroy it. A transfer of the CH_3 group from the second position of the pyrimidine to the sixth position (isovitamin) (No. 8) causes a decrease in activity to 1/4500 of that of the natural vitamin. On the other hand a C_2H_5 group in the second position of pyrimidine in place of the CH_3 group (No. 1) *increases* the activity. The same observation was made by ROBBINS and KAVANAGH (1938) and by the author with *Phycomyces* (SCHOPFER, 1941*a*). Except for the ethyl-thiamin (No. 1) and the isovitamin (No. 8) the above compounds have not been tested on microorganisms or plants, hence no comparisons can yet be made.

The phenomenon observed by SCHULTZ, *viz.*, that the specificity of a vitamin decreases as the dosage is increased, conforms with the findings of the author.

It is necessary, however, to consider carefully what is meant by specificity. The author considers the action of thiamin on *Phycomyces* as specific because of the fact that a substitution destroys or markedly reduces the activity of *the usual dosage.* When activity is obtained with a much larger dosage, it is not known what takes place. The logical assumption is that the substitution product is transformed into the usual (active) form as a result of various reactions - dehydrogenation, hydrogenation, amination and deamination. In this case the apparent lack of specificity is simply proof that the organism can readily bring about the transformation indicated above. Inasmuch as we do not understand exactly what takes place the concept of specificity remains superficial. The most noteworthy result of SCHULTZ's investigations is that the substitutions in the fourth position of pyrimidine and in the second position of thiazole (which are so important for the enzymatic functions of thiamin) do not completely destroy the vitamin activity.

These studies clearly indicate the multiplicity of the B_1 vitamins. To say that the substituted product is transformed into normal thiamin does not solve the problem. There are cases in which it is necessary to seek another explanation. Hence BONNER and BUCHMAN presented the following results: three thiazoles are able to support the growth of pea roots; they differ from normal thiazole in that they have $CHOH.CH_3$ or $CH_2CHOH.CH_3$ in place of CH_2CH_2OH in the fifth position, or have a CH_3 group in the second position. These thiazoles cannot be utilized by *Phycomyces.* It cannot be assumed that these thiazoles, which are inactive for *Phycomyces,* have been transformed by *Pisum* into true thiamin. Evidently substances are formed which have a different structure but a similar physiological function. RYTZ Jr. has found that in a mature pea plant the vitamin which is active on *Phycomyces* is no longer present although it was abundant during the young stages. It cannot be assumed that the vitamin is no longer required. As the plant matured it must have formed other factors analogous with the vitamin but inactive on *Phycomyces.*

Thiamin, in the broad sense of the term, is a physiological entity comprising several substances differing widely in their chemical constitution but possessing similar physiological action.

Heterovitamins of B_1 have recently been discovered and synthesized by BAUMGARTEN and DORNOW (1940) and DORNOW (1940). The analogy existing between the pyridine group and thiazole has been known (HANTZSCH, OCHIAI and NAGASAWA) for some time. By replacing the thiazole with a pyridine joined to the pyrimidine by the same methylene bond as in normal thiamin, BAUMGARTEN and DORNOW obtained two heterovitamins:

(I) 2-METHYL-3-HYDROXYETHYL-N-(2-METHYL-4-AMINO-PYRIMIDYL-(5))-METHYL-PYRIDINE BROMIDE HYDROBROMIDE

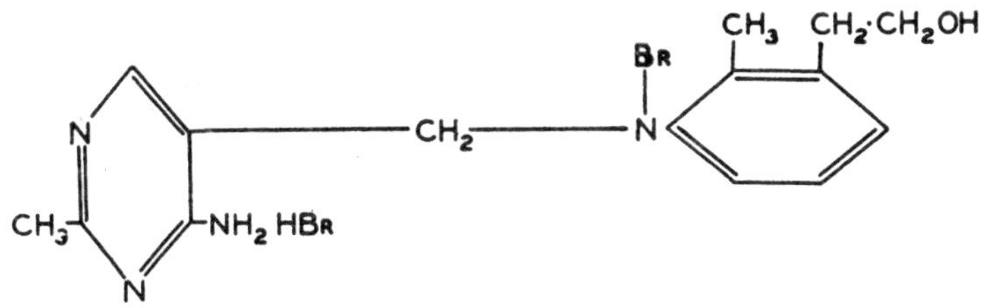

(II) 3-HYDROXYETHYL-N-(2-METHYL-4-AMINO-PYRIMIDYL-(5))METHYL-PYRIDINE BROMIDE HYDROBROMIDE

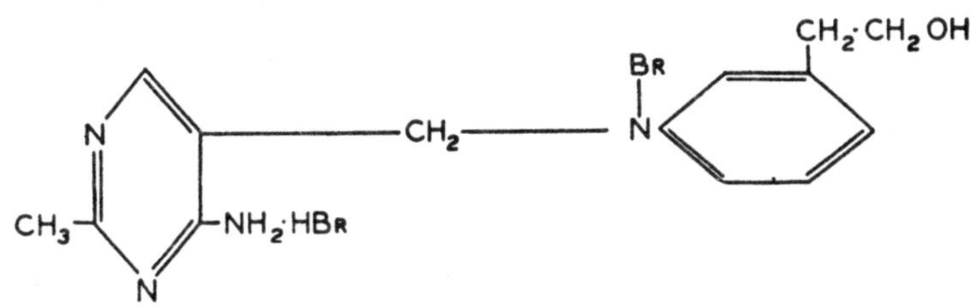

The first of these possesses about 1/26 and the second 1/240 of the activity of thiamin and represents a greater activity than certain analogues tested by SCHULTZ.

The author has found that these heterovitamins have a weak activity on *Phycomyces* and *Ustilago violacea* (SCHOPFER 1941*b*). Substances even more distantly related to the components of thiamin have been found to be active when supplied in high dosage. Accordingly, ROBBINS working with *Phycomyces* demonstrated that the combination pyrimidine + thiazole (1 mμ mole of each) can be replaced by the combination pyrimidine (1 mμ mole) + β(4-methyl thiazolyl-5)-alanine (1000 mμ moles), the latter replacing the thiazole. When only 1 mμ mole of the β(4-methyl thiazolyl-5)-alanine was employed no effect was observed. Nothing is known concerning the mechanism of this replacement. Probably a direct or indirect action of the β(4-methyl thiazolyl-5)-alanine is responsible for the synthesis of thiazole.

It is thus necessary to await the demonstration of other substances having the same physiological properties as thiamin. These may completely change the complexion of the problem.

The extensive investigations regarding the specificity of thiamin have already contributed substantially to our knowledge of growth factors and vitamins in plants.

REFERENCES:-

1937: A. R. TODD and F. BERGEL. — Aneurin, Part VII. A synthesis of aneurin. J. Amer. Chem. Soc. 59:364.

1937: H. ANDERSAG und K. WESTPHAL. — Über die Synthese des antineuritischen Vitamins. Ber. d. d. chem. Ges. 70:2035.

1937: R. GREWE. — Das Aneurin. Erg. Physiol. 39:252.

1937: W. F. MÜLLER et W. H. SCHOPFER. — L'action de l'aneurine et de ses constituants sur *Mucor Ramannianus.* C. r. Acad. Sc. Paris 205:687.

1937: W. J. ROBBINS, A. G. HOGAN and L. R. RICHARDSON. — Pyrimidine and thiazole intermediates as substitutes for vitamin B_1. Proc. Nat. Acad. Sc. U.S.A. 23:388.

1937: R. R. WILLIAMS and J. K. CLINE. — Synthetic vitamin B_1. J. Amer. Chem. Soc. 59:216.

1938: G. M. HILLS. — Aneurin (Vitamin B_1) and pyruvate metabolism by *Staphylococcus aureus.* Biochem. J. 32:388.

1938: J. BONNER and E. R. BUCHMAN. — Synthesis carried out *in vivo* by isolated pea roots, I. Proc. Nat. Acad. Sc. U.S.A. 24:431.

1938: W. H. SCHOPFER. — Symbiose et facteurs de croissance. C. r. 1er Congrès Intern. des Microbiologistes de Langue française, Paris, p. 28.

1938*a*: W. J. ROBBINS and F. KAVANAGH. — Intermediates of vitamin B_1 and the growth of *Torula.* Plant Physiol. 13:611.

1938*b*: W. J. ROBBINS and F. KAVANAGH. — Thiamin and the growth of *Pythium Butleri.* Bull. Torrey Botan. Club 65:453.

1938*c*: W. J. ROBBINS and F. KAVANAGH. — Evidence for a second thiamin. Proc. Nat. Acad. Sci. U.S.A. 24:229.

1938: E. und R. ABDERHALDEN. — Die Heilung der B_1-Avitaminose bei der Taube durch Zufuhr der Thiazol- und Pyrimidinkomponente des Aneurins. Pflügers Arch. 240:746.

1938: W. H. SCHOPFER. — La pyrimidine (2-methyl-4-amino-5-aminomethylpyrimidine) facteur de croissance de microorganismes (*Rhodotorula,* Mucorinées, *Dematium*). Protoplasma 31:105.

1938: N. FRIES. — Über die Bedeutung von Wuchsstoffen für das Wachstum verschiedener Pilze. Symbolae botanicae Upsalienses III, Bd. 2:1-188.

1939: W. H. SCHOPFER. — Vitamine und Wachstumsfaktoren bei den Mikroorganismen . . . Erg. Biol. 16:1-172.

1940: A. DORNOW. — Über ein neues Hetero-Vitamin B_1, das 3-Oxyäthyl-N- [(2-methyl-4-amino-pyrimidyl-(5))-methyl] -pyridinumbromid. Ber. d. d. chem. Ges. 73:156.

1940: P. BAUMGARTEN und A. DORNOW. — Über das 2-Methyl-3-oxyäthyl-N- [(2-methyl-4-amino-pyrimidyl-(5))-methyl] -pyridinumbromid, ein Heterovitamin B_1. Ber. d. d. chem. Ges. 73:44.

1940: W. H. SCHOPFER. — Le problème du métabolisme de l'aneurine chez les plantes. Officina Wander, Berne, p. 323.

1940: F. SCHULTZ. — Über die antineuritische Wirksamkeit von Vitamin B_1-Homologen und Analogen. Zeitschr. physiol. Chem. 265:113.

1940: E. HAAG. — Sur le mécanisme du test Phycomyces de l'aneurine. C. r. Soc. Phys. Hist. nat. Genève 57:136.

1941: W. F. MÜLLER. — Zur Wirkstoffphysiologie des Bodenpilzes *Mucor Ramannianus.* Ber. schweiz. bot. Ges. 51:165.

1941*a*: W. H. SCHOPFER. — Le problème de la spécificité d'action des vitamines. Etude de quelques analogues de l'aneurine (vitamine B_1). C. r. Soc. Phys. Hist. nat. Genève 58:58.

1941*b*: W. H. SCHOPFER. — Les hétérovitamines B_1 et leur action sur les microorganismes. C. r. Soc. Phys. Hist. nat. Genève 58:64.

Chapter XII.

YEAST AND BIOS.

The yeasts present an extremely difficult problem with respect to growth factors, since they involve highly complex constellations of factors, some of which have not yet been identified. They represent the first in which vitamins were found.

It should be recalled that two distinct phenomena occur in yeasts: 1) growth which is conditioned by cell division and is measured either by determining the weight of the living material produced or by counting the number of cells; 2) fermentation which is measured by determining the CO_2 evolved. The two phenomena may be parallel but this is not obligatory, *i.e.*, a culture may increase in weight in the absence of fermentation and vice versa. Each of these phenomena possesses its own specific activators. Growth involves an activation of cell division and an increase in mass, whereas fermentation involves an activation of an enzymatic process concerned with the decomposition of sugar.

Again it is necessary to go back to PASTEUR who first succeeded in culturing yeast on a synthetic medium. His controversy with LIEBIG is classical: PASTEUR (1871) reported that he was able to cultivate his organism by inoculating a pin head, "tête d'épingle", of yeast. LIEBIG (1871) tried to repeat the experiment of PASTEUR but was unable to do so. According to LIEBIG, no growth took place when he used an inoculum the size of a pin head. Because of the death of LIEBIG no agreement between these two investigators could be reached, and the problem remained unsolved. The credit for clarifying the situation goes to WILDIERS (1901). He was able to show that the size of the "pin head" was responsible for the discord. To LIEBIG this size of inoculum must have been much smaller than to PASTEUR. WILDIERS was led to conclude that the cultures developed for PASTEUR because, in using a sizeable mass of inoculum, he introduced certain unknown substances produced by the cells themselves. LIEBIG on the contrary in using a smaller inoculum had not introduced a sufficient amount of this substance because the mass of inoculum was too small. WILDIERS called this strange unknown substance "bios". This marked the real beginning of the science of growth factors in plants, although it will be recalled that PASTEUR in 1860 had observed that the addition of natural products to the culture medium was highly beneficial to the growth of microorganisms.

The history of bios from 1901 to 1925 is indeed most interesting. The existence of such a substance was denied by one group but vigorously supported by another. The latter group studied it extensively trying especially to connect it with known substances. Under the influence of R. J. WILLIAMS the belief prevailed for a while that bios, which was then thought to be a single substance, was identical with the group of B vitamins (particularly B_1). Later, this view had to be renounced when certain differences in their chemical properties were discovered. For example, SOUZA and MACCOLLUM found that the B vitamin found in wheat germ when treated with hot alkali (caustic soda) suffered a destruction of its antineuritic activity without losing any of its bios activity. Bios was shown to have nothing in common with metallic catalyzers and to be different from the cozymase of HARDEN and YOUNG, and hence not concerned specifically with fermentation.

It became necessary to abandon the use of yeast as a simple test for thiamin although this technique was partly sound. Since the growth of yeast is favored by thiamin it can be used as a test object for this vitamin providing other factors are present.

Repeatedly it was thought that a crystalline substance with the properties of bios had been obtained, but such was never proved.

Two fundamental works gave the bios problem renewed impetus. Miss COPPING (1929) showed that various races of yeast respond differently, *viz.*, the wild yeasts thrive in vitamin-free synthetic solutions and therefore are autotrophic for growth factors, whereas the tame yeasts, those which have been cultured for years, have become more or less heterotrophic and require the bios substances. The second piece of work, that of E. J. FULMER, W. W. DUECKER and V. E. NELSON (1923), showed that bios was not a single factor but rather a complex of factors differing in their properties. At this time the American and Canadian school (LASH MILLER and his collaborators, 1930) began to dominate the field and to overshadow completely the Louvain school (IDE, 1921) to which WILDIERS belongs.

The methods employed for separating the bios fractions are as follows: bios I can be separated from bios II by precipitating the former by alcoholic baryta or basic lead acetate, the latter remaining in the filtrate.

Bios I. — Bios I occurs in abundance in tea leaves, and it was from this source that Miss E. EASTCOTT (1928) isolated this fraction in the form of *i*-inositol. The bios II which she used was only a concentrate. Separately the two bios fractions are only slightly active. They must be used together if they are to assume their complete activity. The data presented by Miss EASTCOTT in 1928 clearly show the manner in which one fraction acts as the limiting factor on the other.

The bios II fraction can be purified and analyzed. It has been shown to consist of two constituents, one adsorbed by animal charcoal (bios II*b*), and the other not adsorbed (bios II*a*), according to the nomenclature used by LASH MILLER (1933), EASTCOTT and SPARLING (1932), LASH MILLER, EASTCOTT and MACHONACHIE (1933). The bios II*b* of LASH MILLER is designated as Bios II by

KÖGL, and bios II*a* (MILLER) as bios III (KÖGL). It is on the basis of this method of separation of the two bios II components that all the subsequent investigations have been carried out. Both of these components have been chemically isolated.

Bios II. — Bios II*a* is not adsorbed by animal charcoal and remains in the filtrate. The study of this filtrate led to the discovery of several substances. At first its action was thought to be due to hydroxyaminobutyric acid, which was isolated from tomato juice and found to be active on yeast by Miss STANTIAL and SAUNDERSON. However, this acid is inactive in its synthetic form. Hence its action must have been produced by an impurity.* During the same year, LASH MILLER (1936) and R. J. WILLIAMS and E. ROHRMAN 1936 presented evidence showing that certain amino acids have bios activity. MILLER showed that leucine had a pronounced effect, when used in the presence of concentrates of bios II*b*. WILLIAMS and ROHRMAN demonstrated the importance of β-alanine which is extremely active when used in the presence of other bios materials (races of yeast employed: Wildiers', Gebrüder Mayer's, Lash Miller's, Kögl's Race M). The activity of β-alanine is greatest in the presence of aspartic acid. β-alanine was also used by LASH MILLER. With his yeasts the best results were obtained in the presence of a concentrate of bios II*b* together with *l*-leucine and β-alanine. In general the amino acids are active only when supplied in large dosages, hence it is not certain that they act as true growth factors (see p. 132, action of amino acids). The filtrate remaining after the extract has been treated with animal charcoal undoubtedly contains other active factors which will be discussed later. This fraction (bios II*a* of LASH MILLER) corresponds to bios III of KÖGL (see bios III, p. 128).

Bios II*b* is the fraction adsorbed by animal charcoal and corresponds to Bios II of KÖGL. Investigations of this substance have given excellent results. The essential constituent of this fraction is biotin which acts in conjunction with pantothenic acid (see p. 129). Biotin although present in yeast was not extracted from it, but instead from egg yolk (KÖGL and TÖNNIS, 1936). The chemical relation between certain lipoids (lecithin) and bios had already been pointed out in some of the earliest publications on this subject (IDE, DEVLOO). The isolation of biotin was first accomplished by KÖGL and TÖNNIS. They started with 250 kilograms of egg yolk and, by a series of masterly operations, succeeded in isolating a small quantity of crystalline material. A highly active substance was obtained by the following procedure: precipitation by acetone, by alcohol, by basic and neutral lead acetate, an adsorption of the precipitate by animal charcoal, precipitation by phosphotungstic acid, sublimation, and esterification, a precipitation with phosphotungstic acid, and finally distillation

---

*The α-aminobyturic acid, β- and γ hydroxybutyric acids and also α-amino-β hydroxy-isobutyric acid have no action whatever.

in vacuum. The 250 kg. of egg yolk yielded only 1.1 mg. of crystalline biotin (melting point 148°C.), with an activity of 25,000,000,000 US/gram, each unit (US) representing 1/25,000γ. The *Saccharomyces* unit (US) is the quantity of biotin which in 5 hours, under the very precise conditions employed by KÖGL, brings about a 100 per cent increase in the number of cells. This unit appears small but it corresponds to 120,000,000,000 molecules.

Biotin is one of the most active substances known, its action being still visible in a dilution of 1:400,000,000,000. The methyl ester of biotin, according to KÖGL and DU VIGNEAUD, has the empirical formula $C_{11}H_{18}O_3N_2S$. Like thiamin, biotin contains sulphur. The possible relationship between thiamin and biotin, from a chemical point of view, has been postulated (GREWE, SCHOPFER, 1934-1939), but has not been definitely established. The identity of biotin with vitamin H has been demonstrated (GYÖRGY *et al.*, 1940). The chemistry of this vitamin has been advanced by HOFFMAN, MELVILLE, and DU VIGNEAUD (1941), by KÖGL and MAN, and by KÖGL and PONS (1941). Its structural formula has been established by DU VIGNEAUD (1942).

If the standard growth curve obtained by means of yeast extract is compared with that obtained by means of the combination, bios I (*i*-inositol) plus bios II (biotin), a considerable difference is found. An additional factor (or factors) is present in the filtrate from the animal charcoal. The addition of this filtrate to bios I and II permits a growth comparable with that obtained by furnishing the crude extract of yeast. This observation led KÖGL to conclude the existence of a third substance, bios III (bios II*a* of LASH MILLER) (VAN HASSELT, 1935), the composition of which has not yet been fully determined.

Bios III. — It is strange but not altogether unexpected that the action of bios III can be completely replaced by thiamin. It should be recalled that this was the first factor found essential for the growth of yeast (R. J. WILLIAMS). Its action was confirmed by NIELSEN and HARTELIUS (1938*a*). There are, however, some inconsistencies which have not yet been solved. Thiamin is adsorbed by animal charcoal whereas bios III is not. Furthermore, thiamin is much less thermostabile than the crude bios III of the filtrate. This might indicate that the activity of bios III is due to the components of thiamin since both of these are much more thermostabile than thiamin itself.

The results obtained by various investigators (KÖGL, NIELSEN, DAGYS, ALMOSLECHNER, K. RIPPEL, *et al.*) are difficult to compare since each employed different methods to measure the activity of these substances, *e.g.*, the weight of the yeast formed, the number of cells produced from one cell, the number of cells formed per unit volume of medium, the nephelometric measurement (absorption of light measured by means of a nephelometer involving the use of a photoelectric cell). Still other methods are based on the bios produced by a culture during the course of its development, and measure the rate of bios production or the amount produced during the entire period of growth. The various methods involve different phenomena with mechanisms which differ. K. RIPPEL has made a critical examination of this problem which should be helpful in clearing up the difficulties.

Pantothenic acid is probably associated with the bios II*a* fraction. It has been investigated extensively by R. J. WILLIAMS and his collaborators and deserves particular attention. Already in 1933, R. J. WILLIAMS, C. M. LYMAN, G. H. GOODYEAR, J. H. TRUSDAIL and D. HOLIDAY (1933) discovered a growth factor for yeast, of universal occurrence in both plant and animal tissues, which they called pantothenic acid. It was an hydroxyaliphatic acid of relatively high molecular weight, for which WILLIAMS and MOSHER determined the dissociation constant by fractional electrolysis. In 1938, WILLIAMS and collaborators isolated from liver a highly active preparation for which, in 1939, the formula $(C_8H_{14}O_5N)_2$ was given.

One of the components of pantothenic acid was found to be beta-alanine, for which WILLIAMS and ROHRMAN (1936) had already discovered bios activity. The other component was shown to be a lactone and was identified as the lactone of α-γ-dihydroxy-β, β-dimethylbutyric acid (WILLIAMS and collaborators, 1940*a*). Pantothenic acid (+, optically active) is therefore considered as: α, γ-dihydroxy-β, β-dimethylbutryl-β-alanide with the formula:

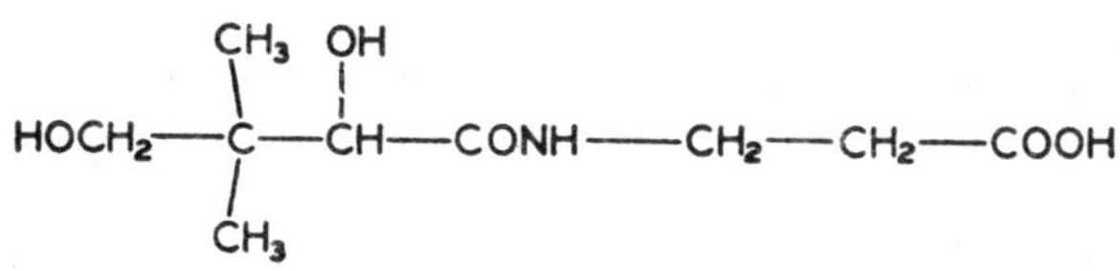

This structure has been confirmed by partial and total synthesis. A partial synthesis was accomplished by treating beta-alanine with the impure natural lactone of the concentrate. The synthetic lactone (form —, optically inactive) is identical with the lactone obtained by the hydrolysis of natural pantothenic acid. Pantothenic acid (+) was synthesized not only from the natural lactone but also from the synthetic lactone (form —). The latter synthesis yielded a product having the same physiological activity as the pantothenic acid resynthesized from the natural form.

Finally, the racemic and (—) forms of pantothenic acid were synthesized from the racemic lactone and its (+) form. The activities of the three forms are:

pantothenic acid (+)	100%
pantothenic acid, racemic mixture	50%
pantothenic acid (—)	0

This remarkable series of investigations, pursued with tenacity since 1933, has achieved remarkable results. As a consequence thereof, pantothenic acid must be added to the list of substances indispensable for certain yeasts.

The position of pantothenic acid in the usual classification of bios (I, II*a*, II*b*) is still uncertain. At first it seemed that pantothenic acid could be placed in the bios II*b* fraction along with biotin, but this idea had to be abandoned when it was shown by R. J. WILLIAMS, R. E. EAKIN, and E. E. SNELL (1940*b*) that, with various races of yeast, the two substances have an additive action.

Pyridoxine (adermin, vitamin B_6). — In addition to the factors already discussed, the recently discovered vitamin B_6 has been shown by MÖLLER to be necessary for lactic acid bacteria, and likewise as a growth factor for yeast (R. E. EAKIN and R. J. WILLIAMS, 1939).

Thus we have at least six chemically identified factors with bios activity which according to the American investigations promote the growth of yeast almost but not quite as well as the standard (extract of liver).

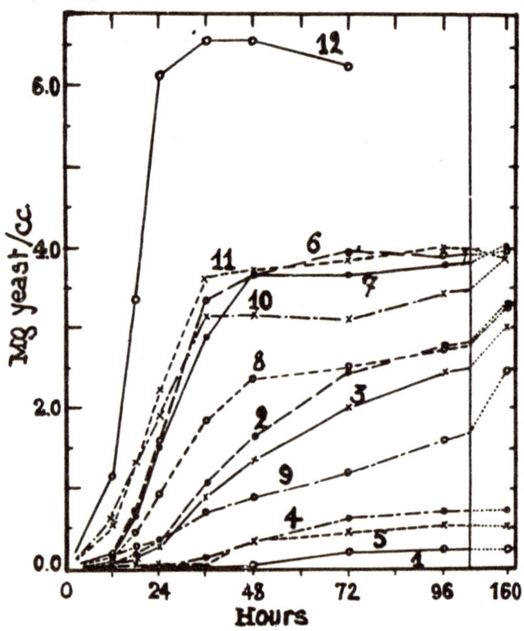

FIG. 10. — Growth of yeast, *Saccharomyces cerevisiae* (Gebrüder Mayer race), in the presence of various combinations of factors (from R. J. WILLIAMS, R. E. EAKIN, and E. E. SNELL, 1940*b*).

Figure 10 shows the interrelationship of action among the factors.

The basal medium was enriched with the following supplements:

	(γ/1 cc. of medium)
inositol	5.0
thiamin	0.04
biotin	0.0001
pantothenic acid	0.006
β-alanine	1.0
pyridoxine	0.04

These supplements were used in various combinations. Their action is shown by the growth curves:

Curve 1: basal medium (alone)
2: " + pantothenic acid
3: " + β-alanine
4: " + biotin
5: " + inositol, thiamin, biotin, pyridoxine
6: " + inositol, thiamin, biotin, β-alanine (without pyridoxine)
7: " + inositol, biotin, β-alanine, pyridoxine (without thiamin)
8: " + thiamin, biotin, β-alanine, pyridoxine (without inositol)
9: " + thiamin, β-alanine, inositol, pyridoxine (without biotin)
10: " + inositol, thiamin, biotin, pyridoxine, pantothenic acid (without β-alanine)
11: " + inositol, thiamin, biotin, pyridoxine, β-alanine (without pantothenic acid)
12: " + as in 11, + 0.2 mg./cc. of crude extract of liver.

(race of yeast employed: "Gebrüder Mayer")

Biotin is highly beneficial (see No. 9 without biotin). Beta-alanine and pantothenic acid are also clearly active. For this particular race of yeast vitamins B_1 and B_6 are ineffective, probably because they are synthesized by this organism; however, this is not the case with certain other races ("Old Process" and Fleischmann bakers' yeast). There are still other unidentified factors required by this yeast as evidenced by the fact that the extract of liver is superior to any of the combinations of known vitamins employed (compare curve No. 12 with 10 or 11).

Other factors. — *Bios V.* — According to Miss FARELL an unidentified factor with bios activity can be obtained from tomato juice by precipitation with tannin. To this unknown she gave the name bios V. This precipitate has been enriched and concentrated by Miss ELDER. It is particularly beneficial to *Saccharomyces hanseniaspora valbyensis,* but, according to Miss ELDER, is not required by *Saccharomyces cerevisiae galactosus.* Its action cannot be replaced by any of the following sugars: *d*-ribose, xylose, *l*-arabinose, *d*-arabinose or rhamnose. Recently MARCHANT (1942) has identified bios V as thiamin.

Bios VII and VIII. — Miss ELDER showed that *S. hanseniaspora valbyensis* requires two factors not essential to *S. cerevisiae,* one of these, bios VII, is contained in the crude bios II*b* solution and the other, bios VIII, in the crude bios II*a* solution. MARCHANT found that bios VII can almost be replaced by pyridoxine. Since the latter is slightly less effective than bios VII he concluded that this bios is multiple and that one of the constituents is pyridoxine. Experiments by the same author indicated that the growth of *S. hanseniaspora valbyensis,* Yeast 2335, and *S. galactosus* is more abundant when supplied with crude bios II*a* than when supplied with β-alanine plus *l*-leucine, hence this fraction of bios contains an unknown constituent essential to at least three yeasts. Its identity remains to be determined.

Sterol. — According to DEVLOO a sterol important for the growth of yeast is present in the filtrate of the phosphotungstic precipitate remaining after preparation of biotin. This biosterol can be replaced by other sterols: sitosterol, ergosterol or calciferol. Ergosterol is pro-vitamin D, and is changed to vitamin D_2 (calciferol) upon irradiation by ultra violet light.

To summarize the constituents of the bios complex the following list (which is still incomplete) may be presented:

bios I: *i*-inositol; precipitated by lead acetate.

bios II*a*: β-alanine; not precipitated by lead acetate (*l*-leucine), not adsorbed by animal charcoal.

bios II*b*: *biotin;* adsorbed by animal charcoal; obtained from the product liberated from the adsorbate, precipitated by phosphotungstic acid.

bios III: replaceable by *thiamin,* but not adsorbed, or only partially adsorbed by animal charcoal.

bios IV: terminology not used.

pantothenic acid

pyridoxine (*vitamin* B_6)

possibly sterol.

If we assume that the principal bios factors can all be extracted from the same original material we have the following scheme providing we consider only the chief characteristics of each substance:

Original material
|
+ tannin — precipitation → *Bios V*
|
filtrate
|
+ lead acetate — precipitation → Bios I
|
+ animal charcoal
/ \
filtrate — adsorbate
/ \
Bios IIa — elution
/ \
Bios III — + phosphotungstic acid → Bios IIb
(Other factors?) — \ filtrate
\ *biosterol*

The amino acids. — Beta-alanine and *l*-leucine must undoubtedly be considered as growth factors. However, these substances are on the borderline between growth factors and ordinary nutrients. NIELSEN *et al.* have made a very careful study of the assimilation of amino acids by yeast. Thirty-four amino acids were added to a medium composed of beer wort from which the growth factors were removed by biological adsorption (the medium, before the amino acids were added, had previously been agitated in the presence of yeast which adsorbed the active factors). Only six amino acids were found to be active: β-alanine, asparagine, aspartic acid, glutamic acid, lysine and arginine. When used singly these amino acids have no beneficial effect, in fact, they may even be toxic. When used in the proper combination β-alanine is the most active of these acids in promoting the growth of yeast, when used alone it is toxic. Its maximal action is exhibited in the presence of asparagine, glutamic acid and thiamin. Its action is not due to an impurity, because NIELSEN and HARTELIUS (1938*b*) were able to extract an active β-alanine (or an asparagine) from alanyl-glycine and from glycyl-aspartic acid, both of which are inactive. JANKE has attempted to explain the toxic effect of β-alanine, when used alone, in the following manner: in the absence of asparagine, other amino acids must be diverted from their original destination (the production of proteinaceous materials) in order to be combined with β-alanine. This results in an inhibition of development.

The most interesting observation of NIELSEN and HARTELIUS is that the activity of β-alanine is possible only in the presence of biotin, thiamin, asparagine and glutamic acid. The last two substances can, however, be replaced by organic acids. Citric acid in particular is highly active and behaves in such a way that the optimal dosage of β-alanine is reduced to about one-tenth of the usual size. This represents a kind of "sparing" action of β-alanine under the influence of citric acid. Not all the amino acids behave in the same manner nor do they belong to the same category. Some of them may be assimilated as sources of nitrogen, *viz.*, asparagine, arginine and glutamic acid. The latter two can function in this capacity only when ammonium sulphate is not in excess. The amino acids, with the exception of β-alanine and *l*-leucine, are required in concentrations which are too high (1:10,000 and 1:20,000) to regard them as growth factors. According to the opinion of the author they are micro-nutrients. On the other hand, β-alanine and *l*-leucine, which are not assimilated as a source of nitrogen, can be considered as growth factors, especially since it is known that β-alanine acts as a precursor of pantothenic acid. To distinguish categorically between the two rôles played by amino acids (as material for the construction of proteins, and as true growth factors) is not easy. Further investigations concerning this aspect of nutrition are necessary (see p. 145, amino acids for other microorganisms).

Organisms dependent upon various constituents of bios. Variation in the extent of loss of capacity for synthesis. — The common yeast, *Saccharomyces cerevisiae,* is not the only one which is dependent on bios when grown on a synthetic medium. Other yeasts, bacteria, and fungi also require various bios factors.

Among the bacteria, *Staphylococcus pyogenes aureus,* which requires thiamin or its components and nicotinic acid (KNIGHT), increases its growth 700 per cent upon the addition of 0.005γ of biotin (KÖGL and V. WAGTENDONK, 1938). Recently it was shown by WEST and WILSON (1939) that biotin (coenzyme R) is the principal growth factor of *Rhizobium trifolii* (see p. 160).

Pantothenic acid has been studied extensively in connection with the bacteria. ORLA-JENSEN *et al.* have claimed that this substance acts as a growth factor for certain lactic bacilli but they presented no evidence to support their claim. In 1938, SNELL, STRONG, and PETERSON found that pantothenic acid acts as a growth factor for various bacteria: *Bacillus lactis acidi, Lactobacillus arabinosus, Lactobacillus pentosus, Lactobacillus Delbrückii, Bacillus brassicae, Streptococcus lactis, Leuconostoc mesenteroïdes,* and *Propionibacterium pentosaceum.* Similar results were obtained by KRAUSKOPF, SNELL and MCCOY with various propionic bacteria: *Propionibacterium pentosaceum, Pr. Thonii, Pr. Jensenii, Pr. Zeae, Pr. arabinosum, Pr. Shermanii,* and *Pr. Freudenreichii.* Many pathogenic bacteria of the colon-dysentery and typhus groups, various aerobes and non-spore forming anaerobes do not react to pantothenic acid. The streptococci and certain lactic bacilli require pantothenic acid and, in addition, riboflavin. It should be noted, however, that pure pantothenic acid was not used in any of the above mentioned experiments. The substance used was a synthetic product obtained by combining β-alanine with an unidentified natural portion. MÖLLER has shown for various lactic bacilli that the action of such a preparation was due not to the pantothenic acid itself but instead to a group of factors occurring with the pantothenic acid as impurities. Certain reservations must be made in considering results obtained with pantothenic acid. It seems very probable, however, that this substance acts as a growth factor for *Streptococcus lactis* —125 (WILLIAMS and collaborators).

In the case of *Corynebacterium diphtheriae,* pantothenic acid acts through its β-alanine component which is an essential growth factor for this microorganism.

Much more definite results have been obtained by SUBBAROW and RANE (1939) with a hemolytic streptococcus (Dochez NY 5 strain). In this case an analogue of pantothenic acid was found to be active. This substance was made by the conjugation of β-alanine ethyl ester with the acetylchloride of acetylated 2,5-dihydroxyvaleric acid. However, the material so prepared was needed in larger amounts than was pantothenic acid. The bios substances are just now being studied in connection with bacteria and much remains to be done in this field.

Three of the bios fractions were studied rather extensively in connection with a certain number of fungi by FRIES (1938). The growth of *Nematospora Gossypii* is increased only slightly by inositol, thiamin and biotin, when used singly. The combination inositol + biotin has a marked effect which is further amplified by the addition of thiamin. The combination inositol + biotin plays the rôle of the limiting factor. The growth of *Melanconium betulinum* is markedly increased by the combination biotin + thiamin.

When inositol is added to the combination a further increase in growth is obtained. The three factors used together are as active as yeast extract[1].

Valsa pini and *Hypoxylon pruinatum* grow almost as well when supplied with the combination thiamin + biotin as when supplied with yeast extract. The addition of inositol as the third factor of the combination has only a slight effect.

The most effective combination of factors for *Trichophyton interdigitalis* comprises thiamin + inositol + pantothenic acid + riboflavin. The strange thing is that each of the factors, when used singly is clearly beneficial (phenomenon of replacement?) (MOSHER and collaborators, 1936)[2].

A few substances related to the bios factors mentioned above are also able to function as bios. Mannitol when used in conjunction with a preparation of thiamin (torulin of PETERS) acts as a growth factor for *Streptothrix corallinus* (*Actinomyces corallinus*).

These observations show that the need for growth factors, in other words the loss of capacity for synthesis, varies considerably even in closely related organisms. This is further exemplified by the yeasts. LASH MILLER and his students have shown that the requirements of yeasts differ widely. *Saccharomyces cerevisiae* (race utilized by MILLER and coworkers) requires bios I (inositol) and bios II*b* (impure) and reacts strongly to β-alanine and leucine (bios II*a*). *Saccharomyces mandshuricus* and *Zygosaccharomyces mandshuricus* do not react to inositol. Differences in growth factor requirements exist between races of yeast within the same species. This is clearly demonstrated by WILLIAMS *et al.* (p. 130) in their work with *Saccharomyces cerevisiae* and the constellation pantothenic acid + β-alanine + biotin + inositol + pyridoxine. The "Gebrüder Mayer" race does not respond to pyridoxine; the same is true of the "Old Process" race. The race known as "Fleischmann bakers' yeast" responds very weakly to thiamin added to the constellation inositol + biotin + β-alanine + pyridoxine. When thiamin is substituted in place of inositol in this constellation of four factors the activity remains unchanged. Thiamin apparently is able to replace inositol. The phenomena involved are extremely complex. Usually when a factor is inactive the assumption is that the factor is synthesized; however, if the factor occurs as a member of a complex constellation there is always the possibility that replacement phenomena are operative (p. 208).

Changes in the capacity for synthesis may sometimes occur in a most unexpected manner. For example, *Saccharomyces cerevisiae* ("Gebrüder Mayer" race), which requires pantothenic acid can by a long period of "training", become self sufficient in this respect. The only possible explanation (providing the phenomenon of replacement is omitted) is that the capacity for synthesis, which perhaps has not been completely lost, can be restored to normal.

The fact that pantothenic acid is made up of two components, one of which, β-alanine, is effective by itself in some cases as a

[1] [2] See Additions, p. 271.

growth factor, suggests that the loss of capacity for synthesis involves only one component (the one which acts as a growth factor). If pantothenic acid is actually a universal factor, one can assume that an organism which requires only β-alanine is able to synthesize the lactone component and to utilize the two components in the biosynthesis of pantothenic acid. This calls to attention the case of *Corynebacterium diphtheriae* which reacts very strongly to β-alanine. A more difficult situation to explain is one in which β-alanine and pantothenic acid are effective only when both are present. In any event β-alanine can be considered as the precursor of pantothenic acid.

Specificity of action of bios substances. — Investigations in this connection are, for the most part, very limited.

Bios I (*i*-inositol). — LASH MILLER (1930) and his students at the time they discovered bios I, observed that *i*-inositol could not be replaced by quebrachitol (methyl ether of inositol), by scyllitol (stereoisomer of inositol) or by quercitol. KÖGL and VAN HASSELT noted the inactivity of a series of several polyatomic aliphatic alcohols: *l*-arabitol, adonitol, dulcitol, *d*-sorbitol, *d*-mannitol, as well as cyclic alcohols: *l*-inositol, scyllitol.

The specificity of this factor is therefore very pronounced, if only the optically inactive form of inositol is to be considered as active and the stereoisomers inactive.

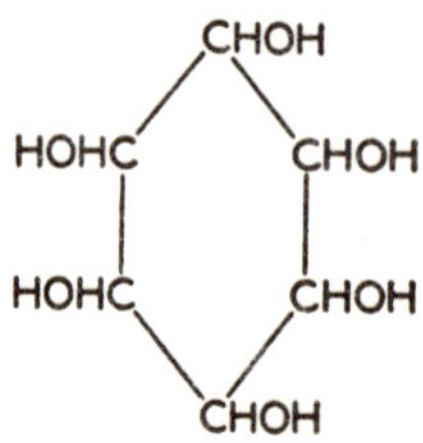

INOSITOL

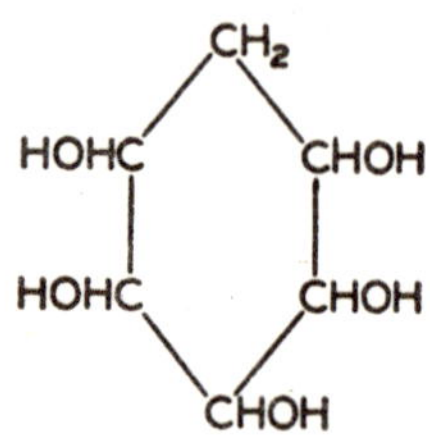

QUERCITOL

Mannitol, which in conjunction with thiamin (torulin of PETERS) acts as a growth factor for *Streptothrix corallinus,* cannot be replaced by inositol (READER).

Biotin. — No investigation has been made.

Thiamin. — The numerous substitution products of thiamin and its components have not yet been studied concerning their effects on yeasts. However, it was observed by SCHULTZ, ATKIN and FREY (1938) that thiamin, as a bios supplement for *Saccharomyces cerevisiae,* could be replaced by pyrimidine and thiazole. In the fermentation test of the above named authors pyrimidine alone (2-methyl-4-amino-5-amino-methylpyrimidine) could be replaced by 2-methyl-5-ethoxy-methyl-6-aminopyrimidine. It would be very interesting to study the numerous substitution products of this vitamin regarding their effects on yeasts.

Sterol. — Its position among the bios substances is still uncertain. It can be replaced by ergosterol, sitosterol and calciferol (irradiated ergosterol, better known as vitamin D_2).

Pantothenic acid. — Some very interesting results have been obtained regarding the specificity of this factor. We have already seen that pantothenic acid (+) possesses strong activity, that the racemic mixture is only half as active, and the (—) form is inactive. Various analogues of pantothenic acid have been synthesized by combining β-alanine with various lactones. The specificity of action has been studied with *Streptococcus lactis* (see lactic bacteria, p. 154).

Functions of bios components. — The only member of this group which has been studied in detail is thiamin (see p. 189). Although biotin, which is essential for yeast and other organisms has received considerable attention, it is not known how this factor acts in the metabolism. Its universal distribution and its presence in quantity in seeds permits the supposition that this bios substance must take part in a very general function. Its presence in the ovarian follicles of the chicken, in which the content increases with the bird's development, is one proof of its importance. The biotin content increases from 439 US (weight of follicle, 0.7 g.) to 1562 US (weight of follicle, 18.1 g.) (KÖGL and VAN HASSELT). It is necessary to mention in passing the presence of large quantities of biotin in the concentrates of the antigray factor of A. MORGAN. The coexistence of these factors gives rise to various suppositions, but does not permit the assumption that the two substances are actually identical. Furthermore, it should be recalled that biotin plays an important rôle in the growth of embryos deprived of their cotyledons and that it occurs in abundance in the buds and young leaves of *Quercus Robur* and *Salix fragilis* (DAGYS).

Pantothenic acid has been investigated to some extent regarding its function. WILLIAMS and PRATT found that pure pantothenic acid in small dosage favors respiration of yeast ("Gebrüder Mayer's" race and Fleischmann's races). Likewise, fermentation is stimulated by pantothenic acid in a medium containing dialyzed macerated yeast. This evidence, although fragmentary, suggests that pantothenic acid takes part in one or several general functions. HELLINGA succeeded in demonstrating that substances produced by certain fungi (*Fusarium*, for example) growing as parasites in the potato plant accelerate the respiration of the host tissues. Inasmuch as biotin, thiamin, riboflavin, inositol, ascorbic acid and heteroauxin are inactive, the possibility that pantothenic acid is the active agent is not excluded.

REFERENCES:-

1871: L. PASTEUR. — Note sur mémoire de M. LIEBIG relatif aux fermentations. C. r. Acad. Sc. Paris 73:1419. Oeuvres 2:363.

1871: J. v. LIEBIG. — Sur la fermentation et la source de la force musculaire. Ann. de Chim. et Physique 23:5.

1901: E. WILDIERS. — Une nouvelle substance indispensable au développement de la levure. La Cellule 18:313.

1921: M. IDE. — The bios of WILDIERS and the cultivation of yeast. J. Biol. Chem. 46:521.

1923: E. J. FULMER, W. W. DUECKER and V. E. NELSON. — The multiple nature of bios. J. Amer. Chem. Soc. 46:723.

1924: G. H. W. LUCAS. — The fractionation of bios, and comparison of bios with vitamins B and C. J. Physic. Chem. 28:1180.

1928: E. V. EASTCOTT. — The isolation and identification of bios I. J. Physic. Chem. 32:1094.

1929: A. M. COPPING. — The effect of bios on the growth and metabolism of certain yeasts. Biochem. J. 23:1050.

1930: W. LASH MILLER. — Bios. J. Chem. Education 7:257.

1931: M. IDE. — Signification du bios. Rev. belge Sci. méd. 3, no. 4.

1932: W. LASH MILLER, E. V. EASTCOTT and E. M. SPARLING. — The fractionation of bios II. Trans. Roy. Soc. Canada III, 26:165.

1933: W. LASH MILLER, E. V. EASTCOTT and J. E. MACHONACHIE. — Wildiers Bios: the fractionation of bios from yeast. J. Amer. Chem. Soc. 55:1502.

1933: R. J. WILLIAMS, C. M. LYMAN, G. H. GOODYEAR, J. H. TRUSDAIL and D. HOLIDAY. — "Pantothenic acid" a growth determinant of universal biological occurrence. J. Amer. Chem. Soc. 55:2912.

1934: W. H. SCHOPFER. — Sur la synthèse d'un facteur de croissance par un microorganisme. C. r. Acad. Sc. Paris 199:1656.

1935: W. VAN HASSELT. — Onderzoekingen over het bios-vraagstuk. Dissert. Utrecht, 114 p.

1935: W. LASH MILLER. — Wildiers bios. Trans. Roy. Soc. Canada III, 29:163.

1936: W. LASH MILLER. — Wildiers bios. Preparation of crude bios fraction from tomato juice. Trans. Roy. Soc. Canada III, 30:99.

1936: F. KÖGL und B. TÖNNIS. — Über das Biosproblem. Darstellung vom kristallisierten Biotin aus Eigelb. Zeitschr. f. Physiol. Chem. 242:43.

1936: R. J. WILLIAMS and E. ROHRMAN. — Beta-alanine and bios. J. Amer. Chem. Soc. 58:695.

1936: W. A. MOSHER, D. H. SAUNDERS, L. B. KINGERY and R. J. WILLIAMS. — Nutritional requirements of the pathogenic mold *Trichophyton interdigitale*. Plant Physiol. 11:795.

1938*a*: N. NIELSEN und V. HARTELIUS. — Wuchsstoffwirkung der Aminosäuren gegenüber Hefe. C. r. Trav. Lab. Carlsberg, série Physiol. 22:250.

1938*b*: N. NIELSEN und V. HARTELIUS. — Untersuchungen über die Wuchsstoffwirkung von β-Alanin, β-Alaninglycin, Asparaginsäure, Glycyl-Asparaginsäure und verwandte Stoffe auf Hefe. C. r. Trav. Lab. Carlsberg, série Physiol. 22:271; Biochem. Zeitschr. 296:359.

1938: F. KÖGL und W. J. VAN WAGTENDONK. — Über die Bedeutung von Biotin für das Wachstum von *Staphylococcus pyogenes aureus*. Rec. Trav. chim. Pays-Bas 57:747.

1938: N. FRIES. — Über die Bedeutung von Wuchsstoffen für das Wachstum verschiedener Pilze. Symbolae botanicae Upsalienses III, Bd. 2.

1938: A. S. SCHULTZ, L. ATKIN and C. N. FREY. — Thiamin, pyrimidine, and thiazole as bios factors. J. Amer. Chem. Soc. 60:490.

1939: W. H. SCHOPFER. — Vitamine und Wachstumsfaktoren bei den Mikroorganismen. Erg. Biol. 16:1-172.

1939: R. E. EAKIN and R. J. WILLIAMS. — Vitamin B_6 as a yeast nutrilite. J. Amer. Chem. Soc. 61:1932.

1939: Y. SUBBAROW and L. RANE. — Pantothenic acid as a growth factor for the Dochez NY 5 strain of hemolytic *Streptococcus*. J. Amer. Chem. Soc. 61:161.

1939: R. J. WILLIAMS, H. H. WEINSTOCK, E. ROHRMAN, J. H. TRUSDAIL, H. K. MITCHELL and C. E. MEYER. — Pantothenic acid III. Analysis and determination of constituent group. J. Amer. Chem. Soc. 61:454.

1940*a*: R. J. WILLIAMS, H. K. MITCHELL, H. H. WEINSTOCK and E. SNELL. — Pantothenic acid VII. Partial and total synthesis studies. J. Amer. Chem. Soc. 62:1784.

1940*b*: R. J. Williams, R. E. Eakin and E. E. Snell. — The relationship of inositol, thiamin, biotin, pantothenic acid and vitamin B_1 to the growth of yeast. J. Amer. Chem. Soc. 62:1204.

1940*c*: R. J. Williams and R. T. Major. — The structure of pantothenic acid. Science 91:246.

1940: P. György, Catharine S. Rose, K. Hoffman, D. B. Melville, V. du Vigneaud. — A further note on the identity of vitamin H with biotin. Science 92:609.

1941*a*: F. Kögl und Th. de Man. — Zur Chemie des Biotins. Zeitschr. physiol. Chem. 269:81.

1941*b*: F. Kögl und L. Pons. — Zur Chemie des Biotins. Nachweis eines Harnstoffrings. Zeitschr. physiol. Chem. 269:61.

1941: K. Hoffman, D. B. Melville and V. du Vigneaud. — Characterization of the functional groups of biotin. J. Biol. Chem. 141:207.

1941: G. B. Brown and V. du Vigneaud. — The effect of certain reagents on the activity of biotin. J. Biol. Chem., 141-85.

1941: M. Landy. — Vitamin H, biotin and coenzyme R. A brief review of the literature. Research Laboratories S. M. A. Corporation, Chagrin Falls, Ohio, p. 1-13.

1942: Cosmo Marchant. — The identification of bios V as vitamin B_1 and of a constituent of bios VII solution as vitamin B_6; their effect upon the reproduction of *Saccharomyces hanseniaspora valbyensis*, yeast 2335, and *Saccharomyces galactosus*. Can. Jour. Res. 20:21-31.

1942: V. du Vigneaud.— The structure of biotin. Science 96:455.

1942*a*: W. J. Robbins and R. Ma. — Vitamin deficiencies of *Ceratostomella*. Bull. Torrey Bot. Club 69:184.

1942*b*: W. J. Robbins and R. Ma. — Specificity of pyridoxine for *Ceratostomella ulmi*. Bull. Torrey Bot. Club 69:342.

1942*c*: W. J. Robbins and R. Ma. — Vitamin deficiencies of *Ceratostomella* and related fungi. Amer. J. Bot. 29:835.

1942: W. H. Schopfer et S. Blumer. — Recherches sur le besoin en facteurs de croissance vitaminiques et le pouvoir de synthèse d'un *Trichophyton*. Le problème du conditionnement des pouvoirs de synthèse. C. r. Soc. Phys. Hist. Nat. Genève 59:106.

1942*a*: W. H. Schopfer. — Recherches sur le besoin en facteurs de croissance et le pouvoir de synthèse de *Rhizopus suinus*. C. r. Soc. Phys. Hist. Nat. Genève 59:101.

1942*b*: W. H. Schopfer. — Les facteurs de croissance pour *Rhizopus suinus*. Déterminisme et relativité des pouvoirs de synthèse. Actes Soc. Helv. Sc. Nat., Sion, 1942.

Chapter XIII.

NICOTINIC ACID AND ITS AMIDE.

Introduction and general considerations concerning the action of this factor on *Staphylococcus.* — Nicotinic acid (and its amide) is now well established as a vitamin in the field of animal physiology, representing one of the important constituents of the heat-stable group of B vitamins sometimes referred to as the vitamin B_2 complex. It is widely distributed in plant and animal tissues where it functions as a constituent of an enzyme and as a proenzyme. This vitamin has the property of preventing and curing pellagra in man and its analogue, blacktongue, in dogs. The identification of the antipellagra vitamin (antiblacktongue vitamin) as a nicotinamide was made by ELVEHJEM and his coworkers in 1937 after they had isolated it in crystalline form from calves liver.

The first investigations on this factor, however, began in the field of microbiology with *Staphylococcus pyogenes aureus.* HUGHES (1932) in studying the growth of this bacterium on a synthetic medium observed that its culture is possible only in the presence of unknown natural substances. According to MUELLER, an acid hydrolyzate of casein does not assure growth of the organism; it is necessary to add 0.001 per cent extract of meat. This extract alone has no action when supplied in dosages below 0.5 per cent. By fractionation of the meat extract HUGHES obtained a preparation which when added to the hydrolyzed casein is active in a dosage of 1 part per 50 millions of medium.

Independently KNIGHT (1935) showed that the extract of marmite is an excellent source of the growth factor required by *Staphylococcus aureus.* The medium employed was composed of a hydrolyzate of casein, tryptophane, tyrosine, cystine and glucose. The marmite extract is also a source of the "sporogenes vitamin" required by *B. botulinus* but this factor is not identical with that required by *Staphylococcus aureus.* Beginning with the marmite extract, KNIGHT carried out experiments designed to concentrate and purify the active agent. From 2 kilograms of marmite he obtained 54 grams (2.7% of the weight of the original) of a highly active substance, soluble in absolute ethyl alcohol and not precipitable by mercuric chloride. This substance was further concentrated by distillation in vacuum at 105°C. The product obtained was colorless, viscous, and non-crystalline. The threshold of activity was about 7.7 mg. for 10 cc. of medium or 10^{-6}. It gave a positive reaction to the test for pyridines with 2,4-dinitro, 1-chlorobenzine, thereby suggesting the possibility of a pyridine nucleus. It was suspected of being a primary or secondary aliphatic amine. KNIGHT then observed that a preparation of cozymase (corresponding to factor V of the hemophilic organisms) containing nicotinamide was active in a dosage of 0.2γ per cc. He found that nicotinic acid was similarly active and able to replace completely the factor which he had concentrated from marmite. The action of nicotinic acid was obviously not due to an impurity (KNIGHT, 1937). This substance must therefore be regarded as one of the members of a complex group of factors required by *Staphylococcus aureus*. Various strains of *S. aureus* and *S. albus* react to the factor concentrated by KNIGHT.

The growth factor requirements of *Staphylococcus aureus* must be rather complex because pure nicotinic acid alone is inactive. It is active only in the presence of thiamin. Nicotinamide when added to a medium containing the hydrolyzate of gelatin, is active in a dosage as low as 0.16γ per 10 cc. of medium (the supplementary factor being present in the hydrolyzate).

From these observations the conclusion can be drawn that at least two growth factors, nicotinamide and thiamin, are essential to the growth of this bacterium. Nicotinic acid exhibits activity in a dosage as low as 0.032γ/10 cc. of medium (2.6×10^{-8} M.) in the presence of thiamin (2×10^{-7} M. for 10 cc.) after 46 hours. Thiamin displays activity in a dosage as low as 5×10^{-10} M. or 0.00015γ per cc. of medium when used in the presence of nicotinamide (1×10^{-5} M. per 10 cc. of medium). Thus thiamin has a much greater activity than nicotinic acid. It can be replaced, as we have seen (p. 108), by its components pyrimidine and thiazole. Nicotinic acid, although not particularly active in itself, completes the action of other factors which are more active.

The action of nicotinamide on *Staphylococcus aureus* was confirmed by LANDY (1938) who made an extensive study of its specificity of action, as well as by KOSER, DORFMAN and SAUNDERS (1938), and KÖGL and v. WAGTENDONK (1938).

KÖGL and v. WAGTENDONK (1938) demonstrated a third factor for *S. aureus*, namely, biotin[1] (bios II*b*, see p. 127). This substance when used in conjunction with suboptimal dosages of thiamin and nicotinic acid has a marked effect on the growth of this organism. When used alone, biotin exhibits some activity but not as great as when used in combination with the other two factors.

TABLE XI. — Action of biotin, thiamin and nicotinic acid on *Staphylococcus aureus* (from KÖGL and v. WAGTENDONK, 1938).

biotin γ/cc. of medium*	0.05γ thiamin 0.05γ nicotinic acid per cc. of medium	5γ thiamin 5γ nicotinic acid per cc. of medium
0	150%	675%
0.005	665%	770%
0.05	690%	800%
0.5	715%	820%
5	745%	870%

*Knight's medium.

[1] Confirmed by PORTER and PELCZAR (1941).

The results are expressed in terms of percentage increase in growth.

It is surprising that the same action amounting to about 670 per cent increase in growth can be exercised by the combination of two factors, thiamin + nicotinic acid (5γ of each) as by three factors, thiamin (0.05γ) + nicotinic acid (0.05γ) + biotin (0.005γ) (see figures italicized in table 11). Apparently biotin permits a more economical use of the other two factors. The phenomenon illustrated here is that of replacement of factors when the dosages are supraoptimal. We merely mention this phenomenon without attempting to explain it, since we do not know what actually takes place. We do not know in what manner the capacity for synthesis of each of these substances is influenced reciprocally by the presence of various dosages of each of the factors. We shall return to this subject later.

Nicotinic acid has been found to act as a growth factor for other organisms. It is the only factor required by *B. proteus*, hence this organism can serve as a simple test object for the assay of this anti-pellagra vitamin. The dysentery bacillus also requires this factor (KOSER, DORFMAN and SAUNDERS). The minimum requirements are: Flexner strain 0.004γ per cc., Hiss Y strain 0.01γ per cc., Strong strain 0.01γ per cc. (growth period, four days). These results were obtained on a basal medium containing glucose, inorganic salts and 15 amino acids.

MUELLER (1937) considers nicotinic acid as an accessory growth factor of *Corynebacterium diphtheriae* (Fig. 11). The liver extract

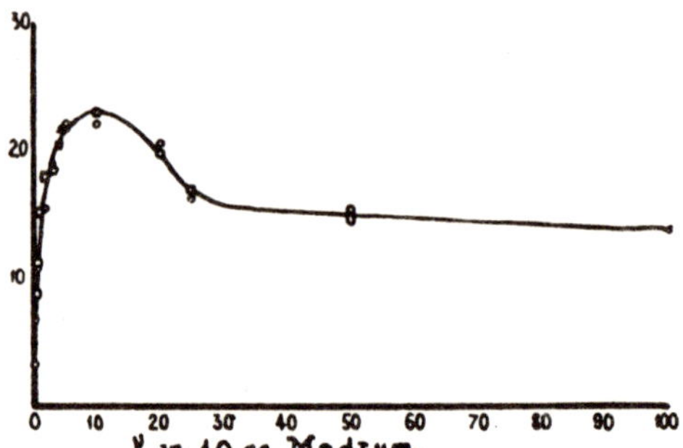

FIG. 11. — Action of nicotinic acid on the development of the diphtheria bacillus (from J. H. MUELLER, 1937).
○ = crystalline preparation from liver extract.
□ = nicotinic acid.

(crystalline preparation) (10 mg. from 300 kg. of liver) has the same activity as pure nicotinic acid. The optimal dosage is 10γ per 10 cc. of medium. Nicotinic acid is only one of a group of factors affecting the growth of this organism, the others being pimelic acid and beta-alanine with the latter indispensable for growth.

Nicotinic acid is also utilized by the lactic bacteria (SNELL, STRONG, and PETERSON).

Cozymase (diphosphopyridine nucleotide or coenzyme I) which acts as a growth factor for certain hemophilic organisms, contains nicotinic acid. It is active on the organisms which require nicotinic acid as a growth factor. It has not been proven that the two substances play the same rôle and that the one is only the precursor of the other, although with animal pellagra the deficiency of the blood in codehydrogenase can be compensated by oral ingestion of nicotinic acid.

In the animal body, nicotinic acid is transformed into the amide and fixed in the cell as a constituent of more complex substances. The two coenzymes (coenzyme I and coenzyme II or Warburg's coenzyme) although they contain nicotinamide as the active constituent for hydrogenation, have only a *weak* vitamin action on animals. It might therefore be assumed that the antipellagra action is due to other compounds containing nicotinamide. This amide may occur in still other compounds. It may be recalled that the pyridine nucleus can occur as a constituent of the heterovitamins of B_1 which undoubtedly are found in nature. This problem of relationship between coenzymes I and II (codehydrogenases) and nicotinic acid in plants and microorganisms has not yet been solved.

Specificity of action of nicotinic acid. — In animals the antipellagra vitamin (nicotinic acid) takes part in two fundamental phenomena, (1) the formation of hemin, and (2) the utilization of proteins. The specificity of action in animals has not yet been studied thoroughly. It is known, however, that plants have an antipellagra action by virtue of another substance, trigonelline (internal salt of N-methyl-pyridine-β-carbonic acid), which gives rise to nicotinic acid and has the same activity as the latter. A reciprocal transformation of the two substances into one another is possible in animals.

In plants, more particularly the microorganisms, the specificity of action has been studied extensively by KNIGHT (1937*a*), KNIGHT and MCILWAIN (1938) (*Staphylococcus aureus*), by LANDY (1938) (*Staphylococcus aureus*), and also by KOSER, DORFMAN and SAUNDERS (1938) (dysentery bacilli). According to LANDY, nicotin-

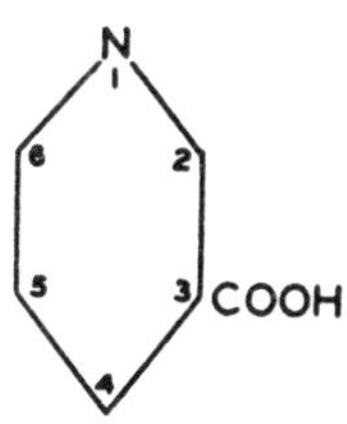

NICOTINIC ACID
(PYRIDINE-3-CARBOXYLIC ACID)

amide has ten times the activity of nicotinic acid; the ethyl amide and the sodium and ammonium salts of nicotinic acid are 1/5 times as active, and the methyl ester 1/25 as active as nicotinic acid.

Trigonelline (chloride and methyl sulphate), despite its activity on animals, is inactive for *Staphylococcus* (KNIGHT and McILWAIN, 1938).

The following table summarizes the activity of the principal products employed.

The following pyridines, methyl and ethyl esters and the amide of nicotinic acid, are active on *Staphylococcus* and the dysentery bacillus *Shigella*. Other pyridines have been tested by KNIGHT and found inactive.

TABLE XII. — Specificity of action of nicotinic acid (from LANDY, 1938, modified).

	Staphylococcus aureus	dysentery bacillus
Pyridine 3-Carboxylic Amide	+	+
Pyridine 3-Carboxylic Ethyl Amide	+	
Pyridine 3-Carboxylic Diethyl Amide (Coramine)	—	
Pyridine 2,3-Dicarboxylic Acid	—	
Sodium Pyridine 3-Carboxylate	+	
Ammonium-Pyridine-3-Carboxylate	+	
Ethyl Pyridine 3-Carboxylate	+	+
2-Methyl Pyridine	—	
Pyridine 2-Carboxylic Acid (Picolinic Acid)	—	
4-Methyl Pyridine	—	
Pyridine 4-Carboxylic Acid (Isonicotinic Acid)	—	

DORFMAN, KOSER and SAUNDERS (1938), working with the dysentery bacillus, studied a series of other derivatives of nicotinic acid. In addition to nicotinic acid and nicotinamide, they found that methyl nicotinate was active (comparable to veal infusion broth) at a concentration of $M \times 10^{-7}$; trigonelline amide, ethyl nicotinate, nicotinuric acid, and ethyl nicotino-acetate were active at $M \times 10^{-6}$, nicotinic acid N-methyl amide active at $M \times 10^{-5}$, and nicotinonitrile active at $M\ 10^{-4}$. Picolinic and quinolic acids were somewhat active at a dosage of $M \times 10^{-4}$, but the samples probably were contaminated by traces of nicotinic acid.

The specificity of this vitamin is not absolute, *i.e.*, a few substitutions are possible. Moreover, the degree of specificity decreases as the dosage is increased.

Growth of Staphylococcus aureus under anaerobic conditions; uracil as a growth factor. — When cultured under aerobic conditions on the basal medium described above, *S. aureus* requires three growth factors: thiamin, nicotinic acid, and biotin. Under anaerobic conditions, this organism can not grow unless biotin is present and, in addition, it is favored by pyruvic acid and an unknown substance functioning as a growth factor. This substance occurs in marmite and in nucleic acid from yeast. Experiments on the degradation of the molecule showed the structure to be that of a pyrimidine. This factor was identified by RICHARDSON (1936) as uracil (2, 6-dihydroxypyrimidine). The synthetic product exhibits activity (at the end of 68 hours) in a concentration as low as M/2,500,000. This dosage is small enough to be characteristic of a true growth factor. RICHARDSON tested 31 other substances, but found no activity in any of them. The compounds included: thymine (5-methyluracil), 4-methyluracil, 1,3-dimethyluracil, 1,3,4-trimethyluracil, xanthine, theobromine, cytostine, isocytosine, guanine, adenine. The fact that these compounds so closely related to uracil are inactive shows that the specificity of the latter is very pronounced. The only compounds having any activity were the following when used in combination: cytosine, isocytosine, guanine, and adenine. These produce only a small amount of growth which is augmented considerably by the addition of uracil.

It can be assumed that *Staphylococcus aureus* is able to synthesize a substance under *aerobic* conditions which it can not synthesize under *anaerobic* conditions. This assumption was substantiated in the following manner: aerobically grown bacteria were hydrolyzed with a molar solution of HCl and the products of hydrolysis were neutralized and added to an anaerobic culture; in another series the filtered and sterilized culture medium was added. Both of these preparations, but particularly the bacterial hydrolyzate, proved effective in supporting anaerobic growth of *S. aureus.* The indispensable factor can therefore be synthesized aerobically. This synthesis can also be accomplished by *Bact. thyphosum.*

S. aureus is the only organism requiring this pyrimidine base as a growth factor.

Uracil is an important constituent of living matter. LEVENE considers it as a primary product and not as a product of deamination of cytosine. This substance must be widely distributed in bacteria as an eventual constituent of nucleotides. Whenever a bacterium is unable to synthesize this portion of the nucleotide molecule, it must be supplied as a growth factor.

Amino acids and the development of Staphylococcus aureus. — The addition of a hydrolyzate of casein to the basal medium of *S. aureus* always introduces an unknown. The hydrolyzate of casein is added to the culture medium of this bacterium for the purpose of supplying the indispensable amino acids which the organism cannot synthesize from the mineral medium. FILDES and his students have investigated the possibility of making up a truly synthetic medium by replacing the hydrolyzate of casein by an appropriate mixture of amino acids. They (FILDES, RICHARDSON, KNIGHT and GLADSTONE, 1936) found that the following amino acids* may be utilized for the synthetic medium: alanine (s)**, valine (s), leucine (s), glycine (s), *l*-proline, *l*-hydroxyproline, aspartic acid (s), *d*-glutamic acid, methionine (s), phenylalanine (s), *l*-tyrosine, *d*-arginine, *l*-histidine, lysine (s). This medium

*The methionine was employed in a concentration of M/50,000; the cystine M/5,000-M/10,000; the tryptophane M/20,000; and the other amino acids M/1,500 and M/1,000.

**s = synthetic preparation.

is suitable for either aerobic or anaerobic cultures. Cystine can be replaced by another sulphur containing organic compound, for example sodium mercaptoacetate.

The above mixture is required by the "untrained" strains. Exhaustive studies by GLADSTONE (1937) revealed, however, that the bacterium can be "trained" progressively to do without these amino acids and to satisfy its nitrogen requirements with ammonium as the sole source of nitrogen. Actually it is not a matter of the organism getting along without these amino acids; instead it has acquired the ability to accomplish their synthesis. The number of amino acids required decreases as follows: 12, 9, 6, 3, 2 and finally only cystine is represented; even it can be eliminated through replacement by mercaptoacetic acid. It is very probable that the capacity for synthesis of each of these amino acids had not been lost completely.

The list of amino acids required by *Cl. sporogenes* is very similar to that required by *S. aureus* (FILDES and RICHARDSON, 1935), the significant difference being that the former requires, among others, tryptophane. This factor apparently plays an essential rôle, since it is absolutely indispensable. It is required in a final concentration of M/10,000.

The action of amino acids on bacteria presents a theoretical problem of a general nature. It is identical with that presented in connection with the amino acids and yeast. The question is—should these compounds be considered as true growth factors? The author's definition stipulates that one of the characteristics of a true growth factor of vitamin nature is that it should be a coenzyme or a fragment of a coenzyme. If, in *Staphylococcus,* the amino acids are required strictly for the production of specific proteins they cannot be considered as growth factors. Since they are effective in relatively small dosages they can be considered as "micronutrients" according to JANKE's definition of the term. The dosages required are smaller than those of ordinary nutrients but larger than the usual dosage of growth factors. This problem of placing the amino acids in the proper category cannot be solved until the function of each of these acids is known.

The case of tryptophane is particularly strange. It is absolutely indispensable for *Cl. sporogenes* and *B. botulinus* (in conjunction with the "sporogenes vitamin") and acts in very small dosage (the smallest of all the amino acids). Its necessity is due to a loss in capacity for synthesis. Despite these characteristics, it is not advisable to consider it as a true growth factor until more information is available.

This problem of interpreting the activity of amino acids is encountered whenever these substances are employed in the preparation of a strictly synthetic medium. The organisms most frequently concerned with this problem are the bacteria, the lactic bacteria in particular (for the amino acids and the bacteria see M. STEPHENSON, 1939).

In summarizing what has been said regarding nicotinic acid we may say that it is a true growth factor, and is required by several bacteria. It is not known to be required by any fungus, but it is required by the higher green plants (roots, see p. 68). It can be considered as a precursor of cozymase, but this has not yet been definitely established.

REFERENCES:-

1932: T. P. HUGHES. — Growth requirements of Staphylococci. J. Bacter. 23:437.

1933: B. C. J. G. KNIGHT. — A vitamin necessary for the growth of *B. sporogenes*: its relation to auxin and other growth factors. British Journ. Exp. Pathol. 14:122.

1935: A. M. PAPPENHEIMER. — The nature of the "sporogenes vitamin", an essential growth factor for *Cl. sporogenes* and related organisms. Biochem. J. 29:2057.

1935: B. C. J. G. KNIGHT. — An essential growth factor for *Staphylococcus aureus*. British Journ. Exp. Pathol. 16:315.

1935: P. FILDES. — The tryptophan and "sporogenes vitamin" requirements of *B. botulinus*. British Journ. Exp. Pathol. 16:309.

1935: P. FILDES and G. M. RICHARDSON. — The amino-acids necessary for the growth of *Cl. sporogenes*. British Journ. Exp. Pathol. 16:326.

1936: G. M. RICHARDSON. — The nutrition of *Staphylococcus aureus*. Necessity for uracil in anaerobic growth. Biochem. J. 30:2184.

1936: P. FILDES, G. M. RICHARDSON, B. C. J. G. KNIGHT and G. P. GLADSTONE. — A nutrient mixture suitable for the growth of *Staphylococcus aureus*. British Journ. Exp. Pathol. 17:481.

1937: G. P. GLADSTONE. — The nutrition of *Staphylococcus aureus*: nitrogen requirements. British Journ. Exp. Pathol. 18:322.

1937: P. FILDES and G. M. RICHARDSON. — The nutrition of *Staphylococcus aureus*: sulphur requirements. British Journ. Exp. Pathol. 18:292.

1937: J. H. MUELLER. — Nicotinic acid as a growth accessory substance for the diphtheria bacillus. J. Bacter. 34:429.

1937*a*: B. C. J. G. KNIGHT. — The nutrition of *Staphylococcus aureus*: nicotinic acid and vitamin B_1. Biochem. J. 31:731.

1937*b*: B. C. J. G. KNIGHT. — The nutrition of *Staphylococcus aureus*. The activities of nicotinamide, aneurine (vitamin B_1) and related compounds. Biochem. J. 31:966.

1938: B. C. J. G. KNIGHT and H. McILWAIN. — The specificity of aneurin and nicotinamide in the growth of *Staph. aureus*. Biochem. J. 32:1241.

1938: P. FILDES. — The growth of *Proteus* on ammonium lactate plus nicotinic acid. British Journ. Exp. Pathol. 19:239.

1938: M. LANDY. — Effect of nicotinic acid, its isomers and related compounds upon nutrition of *Staphylococcus aureus*. Proc. Soc. Exp. Biol. and Med. 38:504.

1938: S. A. KOSER, A. DORFMAN and F. SAUNDERS. — Nicotinic acid as an essential growth-substance for dysentery bacilli. Proc. Soc. Exp. Biol. and Med. 38:311.

1938: A. DORFMAN, S. A. KOSER and F. SAUNDERS. — The activity of certain nicotinic acid derivatives as growth-substances essential for the dysentery bacillus. J. Amer. Chem. Soc. 60:2004.

1938: F. KÖGL und W. J. VAN WAGTENDONK. — Über die Bedeutung von Biotin für das Wachstum von *Staphylococcus pyogenes aureus*. Rec. Trav. chim. Pays-Bas 57:747.

1939: M. STEPHENSON. — Bacterial Metabolism. Longmans, Green & Co., London, New York, Toronto, 391 p.

1941: J. R. PORTER and M. J. PELCZAR Jr. — The nutrition of *Staphylococcus aureus*. J. Bact. 41:173.

Chapter XIV.

GROWTH FACTORS OF THE LACTIC BACTERIA: RIBOFLAVIN, PYRIDOXINE, ETC.

Riboflavin. — The introduction of riboflavin (also known as lactoflavin, vitamin B_2 and vitamin G) and pyridoxine (adermin, vitamin B_6, rat antipellagra vitamin) into the field of plant physiology occurred as a result of studies with the lactic bacteria.

Riboflavin is widely distributed in all groups of plants including the microorganisms. The lactic bacterium, *Bac. Delbrückii,* contains 115 mg. of this factor per kilogram of dry substance; the butyric bacterium, *Clostridium butyricum,* contains 136 mg. The fungus *Eremothecium Ashbyii* synthesizes large quantities of a flavin which crystallizes in the vacuole (GUILLIERMOND, 1936). It has been identified by various methods—spectroscopic, chemical, histological and biological. Its action is identical with that of riboflavin (GUILLIERMOND, FONTAINE, RAFFY, MIRIMANOFF) (Fig. 12). SCHOPFER has shown that riboflavin, as well as lumiflavin and

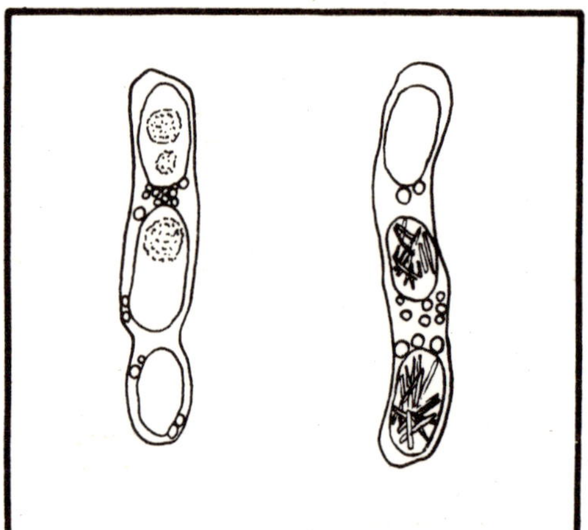

FIG. 12. — Production of riboflavin by *Eremothecium Ashbyii* Guill. (from GUILLIERMOND, 1936). The crystals of flavin appear in the vacuoles.

lumichrome, may accumulate in the vacuole of various plant cells, particularly in the cells of the upper epidermis of the bulb scales of *Allium.* These substances exhibit a characteristic fluorescence and their accumulation in this transparent tissue can be observed by a fluorescent microscope. The riboflavin can be crystallized in the vacuole of the epidermal cells. (C. r. Soc. phys. Hist. nat. Genève, 1941, T. 58). This flavin is produced in abundance by

Aspergillus niger when there is a deficiency of magnesium (LAVOLLAY and LABORAY).

The lactic bacteria are difficult to culture in a synthetic medium. They require, in addition to the usual mineral salts and glucose, cystein and a series of other amino acids, and also growth factors (vitamins). These organisms have become heterotrophic for several of these factors, and their habitat is in part determined by their growth factor requirements.

J. G. DAVIS *et al.* (1929) were the first to demonstrate that lactic organisms isolated from red spots of Cheddar cheese require accessory factors in addition to carbohydrates and proteins. At the time of this report it was impossible to identify growth factors because the culture methods and media were not sufficiently refined. However, the factor required by this lactic organism was shown to be present in concentrates of vitamin B_1. It was then established that vitamin B_1 itself is inactive, hence the essential factor occurred as an impurity in the concentrate. Further investigations of DAVIS, KOSER and SAUNDERS, RAHN and collaborators (1938), and EAGLES and collaborators (1938) definitely established the fact that the lactic bacteria require vitamin factors. But it was by virtue of the investigations of ORLA-JENSEN and his collaborators (1936) that the problem entered into its final experimental stage. ORLA-JENSEN made the fundamental observation that the true lactic bacteria of milk are unable to develop if the milk has previously been treated with activated animal charcoal (the criterion for growth was the proportionate increase in acidity accompanying growth).

The following bacteria require the growth factors contained in milk: *Streptobacterium plantarum, Bacterium bifidum, Propionibacterium technicum, Streptococcus mastiditis, S. faecium, S. glycerinaceus, S. liquefaciens, Thermobacterium lactis,* and *T. helveticum.*

Not only the true lactic bacteria of milk, but also the true lactic bacteria in general require the growth factors of milk. On the other hand, the lactic bacteria of the coli-aerogenes group thrive very well in milk treated with animal charcoal. The bacilli of hay (*Bacillus subtilis*) and of potato (*B. mesentericus*) grow even better in milk which has been treated with animal charcoal. Evidently some of these organisms are able to synthesize their own vitamin factors while others are unable to do so. This fact was demonstrated by the usual method: extracts of the organisms autotrophic for vitamins—*Bacterium fluorescens liquefaciens* and of *Bact. pyocyaneum,* are able to serve as sources of growth factors for *Thermobacterium lactis 10* and *Streptobacterium casei 11,* which are heterotrophic for the growth factors present in milk.

The substance adsorbed by the activated charcoal can easily be liberated by means of pyridine and methyl alcohol. It has precisely the same characteristics as the *riboflavin* present in milk. The true lactic bacteria require various dosages of this vitamin, the rod shaped forms requiring 0.5 mg. per liter, the streptococcus forms much less. At least one factor has been identified (ORLA-JENSEN 1936). It acts, however, in relatively high dosage.

The product eluted from the activated charcoal by pyridine and methyl alcohol contains still another factor which ORLA-JENSEN regarded as pantothenic acid. He also assumed that pyridoxine (vitamin B_6) is an important factor in the eluted product. These assumptions await verification.

The action of riboflavin has been confirmed by WOOD, ANDERSON and WERKMAN who found that the growth of *Lactobacillus mannitopoeus, L. lycopersici,* and *Streptococcus paracitrovorus* in a medium containing casein is strongly accelerated by the addition of riboflavin. *L. pentoaceticus,* on the other hand, is not affected by this vitamin. SNELL and STRONG studied 11 species of lactic bacteria growing under carefully controlled conditions and found riboflavin to be required by the following species: *B. Delbrückii, Leuconostoc Gayoni, Streptobacterium casei,* and *B. lactis acidi.* The following species grow very well without riboflavin: *Leuconostoc arabinosus, L. pentosus, B. brassicae, L. pentoaceticus, L. mannitopoeus, L. mesenteroides, Streptococcus lactis.* These species were shown actually to synthesize a flavin (determination of lumiflavin). KRAUSKOPF, SNELL and McCOY (1939) confirmed the need of riboflavin by the lactic bacteria and by a certain number of streptococci. On the other hand, a great number of organisms do not require this factor, *e.g.,* bacteria of the colon-dysentery-typhoid group and the aerobic and anaerobic spore forming bacteria. The data of ORLA-JENSEN and those of STRONG and SNELL are not in complete accord. This is due in part to the fact that the media employed by these workers are different. SNELL and STRONG made the remarkable observation that lactic bacteria which ordinarily require riboflavin are able to dispense with it in the presence of certain amino acids. This apparent replacement of this factor can be explained by assuming that certain amino acids aid in the synthesis of essential vitamins and thus render the organism autotrophic for these substances. This observation requires further confirmation.

Pyridoxine (adermin, vitamin B_6). — The physiology of the lactic organisms is much more complicated than it would appear from the foregoing discussion. *Bacterium acetylcholini* (*Streptococcus* of ORLA-JENSEN) requires two unidentified factors in addition to riboflavin when it is cultured on a medium containing mineral salts (among others, Mn, NH_4 or Na-acetate), glucose, and at least 10 amino acids. One of these factors is thermolabile, soluble in ether, sensitive to acids and alkalis, and contains nitrogen (SNELL, STRONG, and PETERSON, 1937), while the other is thermostabile and resistant to acids and alkalis. The latter is present in peptones, yeast, and urine. This thermostabile factor was identified by MÖLLER (1938) as pyridoxine. He found that the pure crystalline vitamin B_6 is fully as active as a concentrate of the thermostabile factor. The following dosages are required: *Bacterium acetylcholini* strain No. 10, 10^{-6}; *B. acetylcholini* strain No. 3, 10^{-6}; *Streptobacterium plantarum*, 10^{-6}; *Bacterium utile* (HENNEBERG), 10^{-6} g./cc. MÖLLER also found that certain yeasts require pyridoxine, a fact which has since been confirmed.

The thermolabile nitrogenous factor was found by SNELL to be replaceable by a highly purified preparation of pantothenic acid (calcium pantothenate preparation of R. J. WILLIAMS). MÖLLER points out, however, that the casein used by SNELL as a source of amino acids contained an appreciable quantity of other growth factors for the lactic bacteria. By utilizing a synthetic medium based on amino acids, MÖLLER (1939) showed that *Streptobacterium plantarum* requires three fractions which he called *F*, *G* and *H*. The purification and concentration of these fractions may lead to the identification of other chemically pure factors.

The *F* fraction corresponds closely to the thermolabile nitrogenous factor of SNELL. SNELL, STRONG and PETERSON (1938) were able to replace this fraction by pantothenic acid. A preparation containing 80 per cent pantothenic acid is active at a concentration of 3×10^{-8} g./cc. on *Streptobacterium casei* and also on lactic and propionic bacteria. This vitamin is also required as a growth factor for yeast (R. J. WILLIAMS *et al.*, 1940) and for numerous bacteria. In addition to the lactic and propionic bacteria, its action on pathogenic streptococci has been pointed out by HUTCHINGS and WOOLLEY, and by SUBBAROW and RANE and on *Corynebacterium diphtheriae* by MUELLER. The latter organism requires only the beta-alanine constituent of pantothenic acid. It is able to synthesize the other component and condense the components into a molecule of pantothenic acid. Pantothenic acid is known to contain (see p. 129) in addition to beta-alanine, a lactone of butyric acid (dihydroxy-dimethylbutyro-lactone). Recent investigations suggest the presence of an impurity in the calcium pantothenate prepared from natural products. WILLIAMS *et al.* (1940), in determining the constitution of pantothenic acid have synthesized this vitamin and have also produced various lactones. The synthetic product (lactone condensed with beta-alanine) is definitely active on *Streptococcus lactis* - 125, hence the pure pantothenic acid is responsible for the activity.

The *G* fraction occurs in various natural products, *e.g.*, casein, peptone, globin, zein, vitellin of egg yolk, liver and yeast. It is not absolutely indispensable to *Streptobacterium plantarum* but it does have a clearly beneficial effect on growth. Its nature has not yet been definitely established.

The *H* fraction is highly active and has furnished results of unusual interest. MÖLLER (1939-1940) has described fraction H as a group of substances contained in a preparation of natural pantothenic acid subjected to rigorous purification. This group has been found to consist of five factors, nicotinic acid, pyridoxine, biotin, and two unidentified factors H′ and J.

Biotin. — A highly purified concentrate of this factor acts in a dosage as low as 26 Saccharomyces Units per cc. (1 US = 1/25,000γ). This vitamin is responsible for about half of the activity of fraction H. A saponified concentrate of biotin is more active than an unsaponified concentrate. The saponified methyl ester of biotin is much more active than the unsaponified ester. According to MÖLLER, the form of biotin found in the lactic bacterium *Staphylococcus pyogenes aureus* is not the saponified methyl ester (Fig. 13).

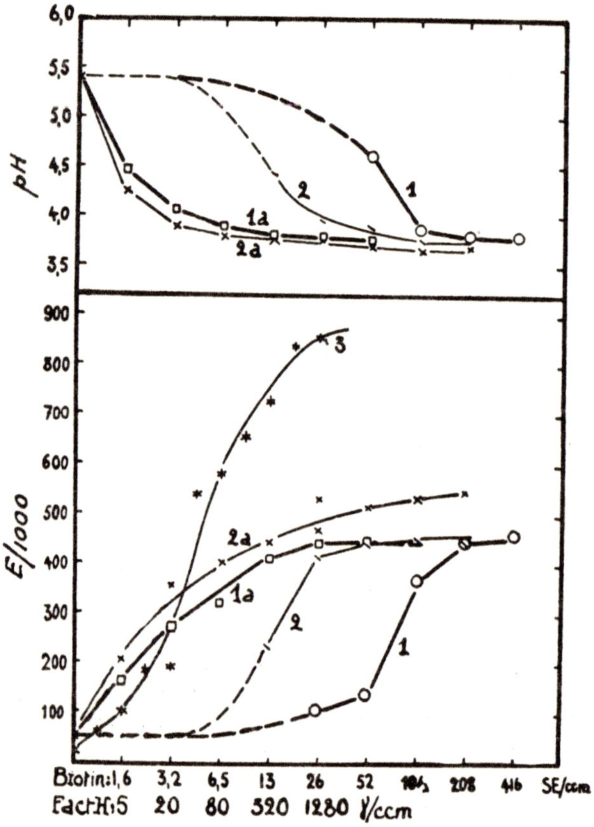

FIG. 13. — Growth of *Streptobacterium plantarum* (strain 10S) in a medium containing biotin and fraction H, but lacking fractions G and H' (from MÖLLER, 1939). — 1. biotin-methylester; 1*a*, the same saponified; 2, concentrate of biotin; 2*a*, the same saponified; 3, fraction H. — Above - increase in the acidity of the medium. — Below - growth of the culture, E/1,000 measured by LANGE's photoelectric colorimeter.

Streptobacterium plantarum, unlike *Staphylococcus aureus,* does not exhibit the "replacement" phenomenon, *i.e.,* the presence of biotin neither increases nor decreases the need for other factors.

The optimal biotin dosage established for *Streptobacterium plantarum* is approximately the same as that found by NILSSON *et al.* for *Bacterium radicicola.* However, MÖLLER suggests that the biotin concentrates employed probably contained another factor, *viz.*, the unknown and unidentified H′.

Nicotinic acid and adenine. — The first demonstration of the need for nicotinic acid by the lactic organisms was made by SNELL, STRONG and PETERSON (1938) with *Streptobacterium casei* and *Leuconostoc arabinosus.* Shortly thereafter MÖLLER found this vitamin had a definite action on certain strains although it was not consistently active and for some strains was completely inactive.

The same vitamin action is exhibited by adenine. This is of particular interest because of the relationship between this substance and nicotinic acid. The data of SNELL *et al.* show that these two factors when used in combination, under the conditions defined, have an optimal dosage of about 0.5×10^{-6} grams per cc. When used separately they have a much lower activity. Both of these substances are constituents of cozymase (see formula p. 53) which is known to be required by the bacillus of Pfeiffer (LWOFF). In the case of *Streptobacterium plantarum* the preparations of cozymase exert an effect almost equal to the combination nicotinic acid + adenine. Apparently the need for these two substances is due to the inability of the organism to synthesize the components of cozymase. Adenine can be replaced only by guanine. No activity is present in xanthine, hypoxanthine and various pyrimidines. Nicotinic acid, on the other hand, is easily replaceable. The action obtained with an optimal dosage of this factor (10^{-7} grams per cc.) can be duplicated by *i*-inositol in a dosage 1000 times stronger. It is again a matter of replacement of which we know nothing as already mentioned concerning other bacteria.

Thiamin. — This vitamin is also a necessary factor for certain species of lactic bacteria. *Streptobacterium plantarum* was found to require thiamin by WOOD, ANDERSON and WERKMAN. This was confirmed by MÖLLER, who showed that this vitamin has an optimal action at a dosage of about 10^{-8} grams per cc.

In addition to thiamin and other factors mentioned above, ascorbic acid exerts a stimulating effect on certain species, but the required dosage is so great that its action is probably not that of a true growth factor (see p. 39). It is entirely possible that this action is exercised on the medium, external to the organism.

Other factors contained in definite fractions have not yet been identified. MÖLLER (1940) tested the action of numerous known substances to determine whether they could replace factors H′ and J, but obtained negative results. These factors do not correspond to the unidentified ones of SNELL and PETERSON, since their characteristics are different.

Different investigators work with different media thereby making it difficult to compare their results. Also the criteria used to measure growth give rise to controversies. It is uncertain that a parallelism exists between acid formation (phenomenon of fermentation) and growth. It is known in the case of yeast that factor Z accelerates fermentation without acting specifically on growth. Since acid production by lactic bacteria represents the same phenomenon as fermentation in yeasts it is very probable that acid production and growth proceed independently.

In general it may be said that organisms are very individualistic in their growth factor requirements. Various strains of the same species react differently. The species whose growth factor requirements are fairly well known are: *Streptobacterium plantarum* P24 (original strain of ORLA-JENSEN), *Bacterium acetylcholini* (KEIL) = *Streptobacterium plantarum* 10S of ORLA-JENSEN (various strains), *Bacterium cucumeris fermentati* B_1, *Bacterium utile* B_3, *Bacillus lactis acidi, Betacoccus arabinosus* (ORLA-JENSEN), *Streptococcus lactis* M13, and certain other related strains.

Amino acids. — The problem of supplying the essential amino acids becomes rather delicate when one is attempting to culture yeast, staphylococci and certain other bacteria on a strictly synthetic medium. A complete list of amino acids required by the lactic bacteria has been compiled by MOLLER (1940). ORLA-JENSEN pointed out the following as indispensable for *Streptobacterium plantarum* (strain 10S): glutamic acid, leucine, valine, aspartic acid, isoleucine, and methionine, in dosages of 2 to 4 $\times$ 10^{-5} grams per cc.; the following are utilized with variable effects: cysteine, tryptophane, phenylalanine, and alanine in dosages of 1 to 5 $\times$ 10^{-4} grams per cc.

Specificity of action of growth factors on the lactic organisms. — Despite extensive investigations on this subject the specificity of some of the essential growth factors of these organisms is poorly understood. This is true particularly of nicotinic acid. Although this vitamin can be replaced by *i*-inositol in high dosage the mechanism of this replacement remains unexplained.

The specificity of synthetic pantothenic acid on *Streptococcus lactis* has been studied by R. J. WILLIAMS *et al.* (1940). Various lactones were synthesized and condensed with β-alanine. The lactone of pantothenic acid is α-hydroxy-β, β-dimethyl-γ-butyrolactone. Three other lactones were found to have a weak but definite activity on this bacterium, *viz.*, α-hydroxy-γ-*n*-valerolactone, α-hydroxy-β-methyl-γ-butyrolactone, α-hydroxy-α-methyl-γ-butyrolactone. WOOLLEY working with *Streptococcus zymogenes* found that pantothenic acid can be replaced by α, delta-dioxyvaleric acid; with certain strains, α, epsilon dioxycapronic acid is active. SUBBAROW and RANE (1939) showed that N-(α, delta-dihydroxyvaleryl)-β-alanine is active in the growth of the Dochez NY5 strain of hemolytic streptococcus. None of the above compounds, however, possessed more than a fraction of 1% of the activity of pure pantothenic acid. "Hydroxypantothenic acid" (N-(α-hydroxy-β,-β'-dimethylolbutyryl)-β-alanine), on the other hand, was shown (MITCHELL, SNELL, and WILLIAMS, 1940) to possess striking biological activity. Its effectiveness, however, varies with the microorganisms and the testing conditions. The specificity does not seem to be very marked, but the data are not sufficient to warrant a definite conclusion.

The specificity of riboflavin has been much studied by SNELL (1939).

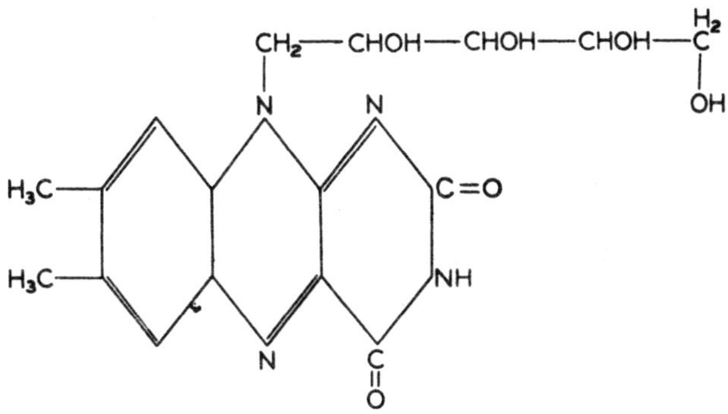

RIBOFLAVIN: 6,7-DIMETHYL-9-D-RIBOFLAVIN

This vitamin, in order to be fully active on animals, must have two methyl groups in the sixth and seventh positions, but can be replaced by tri- and tetramethylene rings (cyclopropane and cyclobutane, respectively). One of the methyl groups can be replaced by an ethyl group. The substitution of a methyl group in the fifth or eighth position destroys the vitamin activity. The NH group in the third position must be free. The sugar group must have the alcoholic structure.

The results obtained with *Streptobacterium casei* and *B. lactis acidi* are presented in the following table.

TABLE XIII. — Specificity of action of flavins on lactic bacteria (from MÖLLER, 1940).

	S. casei		*B. lactis acidi*		growth of rat
	growth	conc. 10^{-7} g/cc.	growth	conc. 10^{-7} g/cc.	
6,7-dimethyl-*d*-riboflavin	+ + +	0.5	+ + +	0.5	+ + +
6-methyl-*d*-riboflavin	+	3	+	0.5	+ + +
7-methyl-*d*-riboflavin	+ +	1	+ +	1	+ + +
6-ethyl-7-methyl-*d*-riboflavin	+ + (+)	0.5	+ +	1	+ + +
6,7-dimethyl-*d*-araboflavin	0	10	0	10	+ + or 0
6,7-dimethyl-*l*-araboflavin	0	10	(+)	10	+
6-ethyl-7-methyl-*l*-araboflavin	0	10	0	10	0
6,7-dimethyl-*d*-glucoflavin	0	10	0	10	0
9-*d*-riboflavin	0	10	0	10	0
5,6-benzo-riboflavin	0	10	0	10	0
6,7,9-trimethyliso-alloxazine (lumiflavin)	0	300	0	300	0
6,7-dimethyl-alloxazine	0	300	0	300	0
tetracetate of riboflavin	0	1	0	1	+ + +

Growth of the bacteria was determined on the basis of acid formation. The data in the last column were furnished by KUHN (see MÖLLER, 1940).

SNELL and PETERSON point out that with a particular test a certain activity is present in 6,7,dimethyl-araboflavin (*d*- or *l*-form), 6-ethyl-7-methyl-*l*-araboflavin, and also 5,6-benzo-riboflavin.

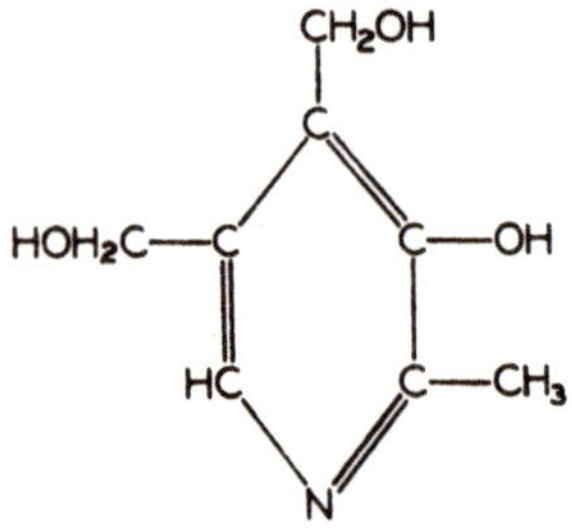

PYRIDOXINE, $C_8H_{11}O_3N$;
IN THE HYDROCHLORIDE FORM,
$C_8H_{12}O_3NCL$

In general the specificity of action of riboflavin on microorganisms is comparable with that on animals, although it is slightly more pronounced. Increasing the dosage does not bring about a loss of specificity.

The specificity of action of pyridoxine has been studied by MÖLLER (1940).

TABLE XIV. — Specificity of action of pyridoxine on the lactic bacteria (from MÖLLER, 1940).

	Streptobacterium plantarum (10S)		Rat	
	Action	Conc. 10^{-6} g/cc.	Action	Conc. day/rat
Pyridoxine (hydrochloride) 2-methyl-3 hydroxy-4,5-dihydroxymethylpyridine hydrochloride	+ + +	0.5-1	+ + +	7.5-10
Pyridoxine (base) 2-methyl-3-methoxy-4,5-dihydroxy-methylpyridine	+ + +	0.8	+ + +	6-8
methyl-ether of pyridoxine I 2-methyl-3-methoxy-4,5-di-hydroxymethylpyridine	+ + +	510	+ + +	5000
methyl-ether of pyridoxine II 2-methyl-3-hydroxy-4-methoxymethyl-5-hydroxy-methylpyridine hydrochloride	+ + (+)	64-128		
Acetate of pyridoxine 2-methyl-3-hydroxy-4,5-diacet-oxymethylpyridine hydrochloride	+ +	16-128	+ + +	10-13
Isopyridoxine 2-hydroxymethyl-3-hydroxy-4-methyl-5-hydroxymethylpyridine hydrochloride	+ + (+)	128-256	0	200
4-desoxy-pyridoxine 2-methyl-3-hydroxy-4-methyl-5-hydroxymethylpyridine	+ +	16-128	0	200
4,5-bi-dehydroxy-pyridoxine 2-methyl-3-hydroxy-4,5-dimethylpyridine	0	256	0	200

The specificity of action of pyridoxine on *Streptobacterium* approaches that on the animal, but it is less pronounced.

Except for riboflavin, the growth factor requirements of the lactic bacteria closely resemble those of yeast. In both cases these needs are multiple. Despite extensive research the list of essential factors is not complete and there may still be some surprises in store.

The propionic bacteria. — The propionic bacteria, since they are closely related to the lactic organisms, will furnish additional data of the same general nature although the growth factor requirements of these two groups of bacteria are slightly different.

These microorganisms are extremely difficult to culture and as a consequence it is not easy to determine whether a vitamin is absolutely essential as a growth factor.

FROMAGEOT and TATUM (1933), and TATUM, PETERSON and FRED have pointed out the fact that certain crude extracts from plants, particularly those from potato and wheat, contain substances which favor the growth of certain propionic bacteria. WOOD, TATUM and PETERSON (1936) obtained from yeast an acid fraction, soluble in ether, which contains factors essential to certain of these organisms.

Potato extract is beneficial to certain strains of the following species: *Propionibacterium pentosaceum, P. Freudenreichii, P. Thönii, P. Jensenii, P. Zeae, P. arabinosum.* These organisms also require a second factor (TATUM, WOOD, and PETERSON, 1936) which is found in the hydrolytic products of protein but is not an amino acid or a fraction of the protein molecule. The same factor can be obtained from caseinogen, egg white, yeast extract, powdered milk, and other substances. It can be extracted most readily with acetone. The extracts are listed below in order of decreasing activity: acetone extract of powdered milk, alcoholic extract of powdered milk, alcoholic extract of caseinogen. The active substance contained in these matérials cannot be replaced by any of the following: inositol, heteroauxin, ascorbic acid, nor nicotinamide. Pantothenic acid is only slightly active. Under the conditions of culture employed by TATUM, WOOD, and PETERSON, crystalline thiamin is able to replace the acetone extract of caseinogen, in fact, it is even more active than the latter. The optimal dosage of thiamin varies with the strain of bacterium from 0.15 to 0.005γ per cc. This vitamin can therefore be included among the factors required by certain species of *Propionibacterium. Propionibacterium pentosaceum* has been investigated extensively in connection with thiamin and cocarboxylase (see p. 189). Strangely enough this organism can adapt itself to live without thiamin, *i.e.*, it regains its ability to synthesize this factor (SILVERMAN and WERKMAN, 1939). Thiamin represents only one member of a group of factors which undoubtedly is very complex. The ether extract of yeast, which must be present in the basal medium, has not yet been analyzed from the viewpoint of growth factors. The work of KRAUSKOPF, SNELL and McCOY (1939) indicates that pantothenic acid (impure sample) is necessary for the propionic bacteria.

The butyl alcohol bacteria. — Certain bacteria producing butyl alcohol (*Clostridium butylicum, C. acetobutylicum*) are likewise dependent upon one or several growth factors. McDaniel, Woolley and Peterson (1939) have found that they require a factor contained in liver, yeast, malt, maize, and the commercial preparations of vitamin concentrates. It can be extracted by means of ether and can be adsorbed on animal charcoal. Elution by alcohol and pyridine is incomplete. This factor can be combined into a copper salt which has a definite stimulating action in a dosage of 0.01 to 1.0γ per cc. It cannot be replaced by any of the following substances: riboflavin, heteroauxin, nicotinamide, thiamin, pyridoxine, pimelic acid, beta-alanine, pantothenic acid, or a combination of amino acids from a natural source. Apparently the growth factor required by *Cl. acetobutylicum* is biotin.

REFERENCES:-

1933: C. Fromageot und E. L. Tatum. — Über einen Aktivator des Stoffwechsels der Propionsäurebakterien. Biochem. Zeitschr. 267:360.

1936: E. E. Snell, E. L. Tatum and W. H. Peterson. — Growth factors for bacteria, III. Some nutritive requirements of *Lactobacillus Delbrückii.* J. Bacter. 33:207.

1936: A. Guilliermond. — L'*Eremothecium Ashbyii,* nouveau champignon parasite des capsules du cotonnier. Rev. de Mycol. 1:115.

1936: H. G. Wood, E. L. Tatum and W. H. Peterson. — Growth factors for bacteria, IV. An acidic ether-soluble factor essential for propionic acid bacteria. J. Bacter. 33:227.

1936: S. Orla-Jensen, N. C. Otte und Agnete Snog-Kjaer. — Der Vitamin-bedarf der Milchsäurebakterien. Zbl. Bakter. II, 94:434 & 447.

1936: E. L. Tatum, H. G. Wood and W. H. Peterson. — Vitamin B_1, a growth stimulant for propionic acid bacteria. Biochem. J. 30:1898.

1937: S. Orla-Jensen. — Les microorganismes et les vitamines. Annales des Fermentations 3:1.

1937: E. E. Snell, F. M. Strong and W. H. Peterson. — Growth factors for bacteria, VI. Fractionation and properties of an accessory factor for lactic acid bacteria. Biochem. J. 31:1789.

1938: B. A. Eagles, O. Okulitch and A. S. Kadzielawa. — Wildier's bios and the lactic acid bacteria. The relation of bios to the water soluble B-vitamins. Canad. J. Res. B. 16:46.

1938: E. F. Möller. — Vitamin B_6 (Adermin) als Wuchsstoff für Milchsäurebakterien. Zeitschr. f. physiol. Chem. 254:285.

1938: E. E. Snell, F. M. Strong and W. H. Peterson. — Pantothenic acid and nicotinic acid as growth factors for lactic acid bacteria. J. Amer. Chem. Soc. 60:2825.

1938: O. Rahn and C. P. Hegarty. — Accessory factors in lactic acid fermentation. Proc. Soc. Exper. Biol. and Med. 38:218.

1939: E. F. Möller. — Das Wuchsstoffsystem der Milchsäurebakterien. Zeitschr. f. physiol. Chem. 260:246.

1939: E. E. Snell and F. M. Strong. — The effect of riboflavin and of certain synthetic flavins on the growth of lactic acid bacteria. Enzymologia 6:186.

1939: J. G. Davis. — The nutritional requirements of the lactic acid bacteria. J. Dairy Res. 10:186.

1939: L. E. MacDaniel, D. W. Woolley and W. H. Peterson. — Nutrient requirements of certain butyl alcohol producing bacteria. J. Bacter. 37:259.

1939: M. Silverman and C. H. Werkman. — Thiamin effects in bacterial metabolism. Iowa State College Journ. of Science 13:365.

1939: E. J. KRAUSKOPF, E. E. SNELL and E. McCOY. — Growth factors for bacteria. A survey of the pantothenic acid and riboflavin requirements of various groups of bacteria. Enzymologia 7:327.

1939: Y. SUBBAROW and L. RANE. — Pantothenic acid as a growth factor for the Dochez NY5 strain of hemolytic streptococcus. J. Amer. Chem. Soc. 61:1616.

1940: H. K. MITCHELL, H. H. WEINSTOCK jr., E. E. SNELL, S. R. STANBERRY and R. J. WILLIAMS. — Pantothenic acid, V. Evidence for structure of non-beta-alanine portion. J. Amer. Chem. Soc. 62:1776.

1940: E. F. MÖLLER. — Nährstoffe und Wuchsstoffe der Milchsäurebakterien. Angewandte Chemie 53:204.

1940: H. K. MITCHELL, E. E. SNELL and R. J. WILLIAMS. — Pantothenic acid, IX. The biological activity of hydroxypantothenic acid. J. Amer. Chem. Soc. 62:1792.

Chapter XV.

THE NITROGEN FIXING BACTERIA (RHIZOBIUM AND AZOTOBACTER) AND THE GROWTH FACTORS REQUIRED BY THEM. COENZYME R (BIOTIN).

The nitrogen fixing bacteria play a highly important rôle in the cycle of living matter. For this reason their culture and their capacity for synthesis should be given special consideration.

Rhizobium. — These bacteria live symbiotically in the roots of legumes, furnishing them with atmospheric nitrogen which they are able to fix. Numerous experiments have been carried out with cultures on synthetic or natural media for the purpose of determining whether these bacteria are able to fix nitrogen when separated from the plant. Although the early results concerning the ability of these organisms to grow in artificial culture were conflicting, it was soon recognized that they could be cultured on a medium of more or less definite composition.

ALLISON (1927) was the first to report that *Bacterium radicicola* (*Rhizobium trifolii*) isolated from *Trifolium pratense* could be cultivated on a synthetic medium containing saccharose, ammonium sulphate or potassium nitrate, with a small quantity of extract from straw. The same observation was made simultaneously by THORNTON. Even at an earlier date BEIJERINCK employed aqueous extracts from legumes, enriched with asparagine and saccharose for the isolation and culture of these bacteria. ALLISON successfully employed more than 26 different extracts from members of the *Leguminosae* and other families (*e.g.*, lettuce, cabbage, beet, spinach, tomato and potato) as sources of unknown accessory factors. According to BJÄLFVE and NILSSON (1938) the extract from straw is excellent for this purpose.

ALLISON, HOVER and BURK (1933) carried out extensive investigations concerning the unknown substances required for the culture of the root nodule organisms. They established the fact that these substances are widely distributed in animal and plant cells, and also in bacteria. The substances are found in commercial preparations of saccharose and in the cells of *Azotobacter* (see p. 167 for the vitamin content of *Azotobacter*).

Since there was no parallelism between the action of extracts on *Rhizobium* and on *Saccharomyces* the premature conclusion was drawn that the factor required by *Rhizobium* was different from bios (bios, as known at that time). This unknown factor, since it was supposed to take part in the respiratory processes, was called "coenzyme R", a name which still persists.

The basal medium required for the culture of *Rhizobium* is relatively simple, consisting of mineral salts and a carbon source (mannitol). To this medium are added accessory substances, *e.g.*, yeast extract.

THORNE, NEAL and WALKER (1936) determined the respiratory quotient (volume of CO_2 produced / volume of O_2 consumed) for various species of *Rhizobium* (*R. meliloti, R. trifolii, R. leguminosarum, R. japonicum*). The quotient varies with the composition of the medium as follows:

Basal medium	respiratory quotient
mannitol + yeast extract	0.5–0.6
glucose + yeast extract	0.7
mannitol + KNO_3	1.0

These figures must be accepted with reservation, since it has been noted by PIETZ (1938) that, in the presence of KNO_3, gaseous nitrogen may be produced and, during the decomposition of sugar, hydrogen may be formed. These gaseous products interfere with the manometric measurement of carbon dioxide and oxygen leading to values for the former which are too high and values for the latter which are too low. However, since the figures obtained by THORNE *et al.* are relatively constant, they can be considered as characteristic of the metabolism of the bacteria under the conditions described.

Cultures of *Rhizobium* are very sensitive to the redox potential (ALLYN and BALDWIN). When the medium has been too strongly oxidized with H_2O_2 or with potassium permanganate, no growth can take place. The unfavorable action of these oxidizing agents can be counteracted by adding cysteine or thioglycolic acid both of which are capable of dehydrogenation in their disulphite form.

The bacteria containing cells of the root nodules have an *r*H of 9-11 at *p*H 7, and an *r*H of 4-6 at *p*H 6 in an atmosphere of nitrogen. Hence this tissue possesses a strong capacity for reduction. Cultures of these bacteria likewise produce very low redox potentials. In the presence of air the *r*H reaches 22-24.

In the nodules there occur "red bodies" which are derivatives of dihydroxyphenylalanine.

Cultures of *Rhizobium* will grow only under aerobic conditions or in the presence of oxidized substances, *e.g.*, yeast extract or nitrates. The action of free oxygen can also be replaced by weakly oxidized dihydroxyphenylalanine (corresponding to the "red bodies"). Thus the "red bodies" play an essential rôle. They are related to dihydroxyphenylalanine and to the melanins in the following manner:

$$\text{tyrosine - dihydroxyphenylalanine - red bodies - melanin}$$
$$\text{(reduced)} \quad \rightleftarrows \quad \rightleftarrows \quad \text{(oxidized)}$$

In a culture growing under aerobic conditions the dihydroxyphenylalanine is rapidly oxidized into "red bodies", then into melanin. The same culture is capable of reducing the "red bodies". Thus the reaction is reversible, and by alternating the cultural conditions between the aerobic and anaerobic type, the reaction, "red bodies" ⇄ leucobase, can be controlled at will, providing the conditions are such that the oxidation does not exceed the "red body" stage. According to FRIEDHEIM, the redox system, "red body" ⇄ leucobase, has an *r*H of about 15. When the *r*H is below 15 the "red bodies" occur in their reduced form. In the presence of air the nodules and also the cultures of *Rhizobium* have always an *r*H of 24. Even at this redox potential, the "red bodies" are not oxidized to melanin. Apparently these microorganisms must live within the limits of *r*H 15 and 24 (PIETZ, 1938). It is thus obvious that the "red bodies" play an important rôle in the life of these bacteria by enabling them to grow under anaerobic conditions.

These observations are the first to furnish a possible clue concerning the physiology of "coenzyme R". THORNE, NEAL and WALKER (1936) find from their work that the growth of the root nodule bacteria is favored by strong reducing agents. For example, yeast extract seems to act as an effective donator of hydrogen. The question might be asked whether the unknown factor, contained in the natural extracts, does not act simply by adjusting the *r*H to the proper level for the culture. Before attempting to answer this question we must first consider the chemical nature of the growth factors required by *Rhizobium.*

NILSSON, BJÄLFVE, and BURSTRÖM (1938 *a, b, c*) made extensive investigations concerning all aspects of the nutrition of the root nodule bacteria. They found that *R. trifolii* was unable to develop on a strictly synthetic medium. No improvement in growth was obtained upon the addition of reducing substances such as cysteine, glutathione, and glyceric aldehyde to a natural medium. On this point the ideas of these authors do not agree with those of ALLISON. However, they were able to confirm the early observations of ALLISON concerning the action of the extracts of various natural products (yeast, potato, and straw). It was shown that the action of the natural extracts is not due to the nitrogen which they furnish. Although the extract acts as a very favorable nitrogen source, the action is due to other accompanying substances.

NILSSON *et al.* state that yeast loses much of its activity upon treatment with ether. The ether extract by itself has only a weak action, but in combination with the water soluble residue, its action is maximal and comparable with that of the untreated yeast extract.

Vitamin B_1 appears to be able to replace partially the yeast extract. Already in 1927 WERKMAN had studied the action of concentrates of thiamin and arrived at the conclusion that it was not a question of a vitamin action, but of an effect produced by accompanying substances which were easily assimilable. This hypothesis could then be supported by the fact that the vitamin employed was not pure. The investigations of CLARK and those of ALLISON *et al.* do not permit the conclusion that a vitamin is necessary. Nevertheless, it is now known that vitamin B_1 when used alone does exert a rather weak action and when it is used in conjunction with hydrolyzed casein or with an ether extract of yeast this vitamin is very active. Accordingly, the conclusion can be drawn that thiamin is capable of replacing the factor which is soluble in water but insoluble in ether. The ether extract can be replaced by an amyl alcohol extract but the latter exerts an action even when used alone. The effect of vitamin B_1 and the ether extract from yeast used singly and in combination is readily demonstrated by the following data:

	Number of cells per cc. at the end of 11 days.
Basal medium containing KNO_3 (control)	7 millions
Basal medium + 0.08γ vitamin B_1 per cc.	35 millions
Basal medium + 0.01 cc. of ether extract per cc.	45 millions
Basal medium + 0.08γ vitamin B_1 + 0.01 cc. of ether extract	600 millions

Thus the combination, vitamin B_1 + ether extract, acts in much the same manner as the combination, crude aqueous extract (following extraction with ether) + ether extract. This is true irrespective of the presence or absence of nitrogen in the medium. It should be pointed out, however, that even the combination vitamin B_1 + ether extract is less effective than the complete extract from yeast or from straw. The figures in the latter case amount to 1000—2000 millions of bacteria per cc. of medium.

Riboflavin, inositol, beta-alanine, nicotinic acid and pimelic acid (the last three used in combination) have no noteworthy effect.

An unusual observation can be made if the number of bacteria produced is compared with the turbidity coefficient (absorption of light) of a culture containing only amyl alcohol extract and one containing amyl alcohol extract + thiamin. The extract alone has a marked effect and the number of bacteria produced is large. Additions of this extract beyond the optimal dosage are ineffective, except for a slight possible reduction in the number of cells. An addition of thiamin does not produce a very great increase in the number of bacteria (the dosage of amyl alcohol extract being optimal), although it greatly increases the opalescence and the turbidity coefficient. This vitamin is much more effective when the dosage of the amyl alcohol extract is small.

A microscopic examination of the cells reveals that thiamin in conjunction with the amyl alcohol extract brings about a marked increase in the size of the cells which acquire the form of bacteroids. The strange thing is that when thiamin is used alone (without any amyl alcohol extract) an increase in the volume of the cells occurs but the characteristic branching of the bacteroids is lacking. This type of effect produced by vitamin B_1 has not been found in any other organism.

These data seem to contradict those of WEST and WILSON (1938*b*), who reported that *Rhizobium* can be cultured in the absence of vitamin B_1 and that the bacterium is able to synthesize this vitamin. By using *Staphylococcus aureus* as a test object, WEST and WILSON (1938*a*) found the thiamin content of *Rhizobium* to be 19.6γ per gram of dry matter. The validity of this test for thiamin is open to question since *Staphylococcus* requires biotin as an additional factor. This fact was not yet known or, at any rate, was not taken into consideration by WEST and WILSON. But now it is known that after several passages on a vitamin-free medium the *Rhizobium* culture becomes definitely dependent on an exogenous source of thiamin.

The ether extract and the amyl alcohol extract still remain to be considered. Because of the fact that the latter is active when used alone it can be assumed that it contains a little thiamin but, from all indications, an additional factor is also required by this bacterium. This factor appears to be extremely active, much more active than thiamin. It is soluble in ether, thermostabile and relatively stabile in alkalis. Experiments designed to concentrate and isolate the active substance (NILSSON, BJÄLFVE and BURSTRÖM, 1939) led to a crystalline product. The crystals, however, were less active than the mother liquor thereby suggesting the activity of an impurity on the crystals. By chemical analysis the crystals were identified as succinic acid. A sample of purified commercial succinic acid, however, was found to be inactive, hence it was certain that the action was due to an impurity of unusual activity and present in weak dosage. A concentrated preparation of the ether soluble factor (5γ) used in conjunction with thiamin (0.09γ per cc.) was found to produce 1400 millions of bacteria per cc. of medium.

Because of the unusual activity and stability of the unknown factor the action of biotin was suspected. Attempts to replace the ether soluble fraction with pure biotin were successful for three strains of *R. trifolii*. The action of biotin, although it is very strong, may fail to be observed because of the fact that a period of induction is required for initiating growth of the culture (NILSSON, BJÄLFVE and BURSTRÖM, 1939).

Biotin alone is only slightly active, but when used in conjunction with thiamin it is extremely active. These facts are obvious from the following data:

	Number of bacteria at the end of 21 days millions/cc.
Control (vitamin-free basal medium)	5
Basal medium + 0.07γ thiamin per cc.	60
Basal medium + 0.0005γ biotin per cc.	7
Basal medium + 0.0005γ biotin +0.07γ thiamin per cc.	1500

The growth promoting action of biotin on *Rhizobium* can be observed when this vitamin is used in a concentration as low as 1:100,000,000,000. The maximal effect is reached with a dosage of 1:2,000,000,000. These figures are similar to those obtained with other organisms.

The combination, biotin + thiamin, is just as effective as the crude yeast extract. However, it has not yet been proven that the ether soluble factor and biotin are identical. It is very probable that these two substances are the same since their general characteristics are alike. It can be assumed that biotin is slightly soluble in ether or amyl alcohol.

WEST and WILSON (1939) arrived at the same conclusion as NILSSON *et al.* did regarding the action of biotin although they attacked the problem from a different point of view. Their work represents a renewal of investigations to determine whether the growth factor requirements of *Rhizobium* are different from those of *Saccharomyces.* It will be recalled that early observations on the growth factor requirements of *Rhizobium* led to the hasty conclusion that *the factor* required by this bacterium is different from *that* (bios) required by *Saccharomyces.*

WEST and WILSON attempted to determine the exact nature of coenzyme R. They tested the action of concentrates of biotin (prepared according to the method of KÖGL and TÖNNIS) on the growth of *Rhizobium trifolii* and observed a highly beneficial effect. The biological activity of the concentrate was determined at each stage of its preparation and was found to increase parallel with the increase in biotin content. This factor replaced completely coenzyme R. None of the following were able to replace this coenzyme: thiamin, cocarboxylase, riboflavin, nicotinic acid, nicotinamide, adenine, cozymase, pyridoxine (vitamin B_6), beta-alanine, uracil, inositol, vitamin C, ergosterol, heteroauxin, pantothenic acid (concentrate), and the sporogenes vitamin.

Contrary to previous assumptions, it was observed that a close parallelism exists between the action of concentrates of coenzyme R (or simply of extracts of yeast) on *Rhizobium* and its action on *Saccharomyces.* That the two organisms have one character in common is demonstrated by experiments concerned with inactivation, extraction, oxidation and adsorption of the active principle. The only difference in their nutrition is that *Saccharomyces* requires pantothenic acid and beta-alanine whereas *Rhizobium* does not. It was finally shown by WEST and WILSON, working with *Nematospora Gossypii* (requiring biotin as a growth factor) that the preparations of coenzyme R contained biotin.

When it was learned by NILSSON and his collaborators that pure biotin can act as a growth factor the investigations which have just been analyzed assumed their full significance.

In summarizing the available information on the culture of *Rhizobium trifolii,* two categories of factors must be considered: (1) Physico-chemical factors, particularly the redox potential. — An unfavorable *r*H prevents the growth of the culture. (2) Growth factors of vitamin nature. — This organism requires at least two growth factors. Biotin is essential for growth. Vitamin

B_1 is beneficial, but is less active than biotin; its effect on this bacterium is somewhat different from that which it exercises on other microorganisms. Biotin replaces completely the ether soluble fraction of NILSSON and collaborators. It is identical with coenzyme R. There may exist still another thermostabile factor for which WEST and WILSON propose to reserve the term coenzyme R (*sensu stricto*). This hypothesis, however, is not supported by the observations of NILSSON and collaborators, *viz.*, the combination biotin + thiamin is fully as active as crude yeast extract.

From all indications it seems likely that the action of the redox potential is distinct from that of growth factors.

Some investigators were inclined to attribute the activity of the growth substances of *Rhizobium* to the specific action of certain minerals. THORNE and WALKER have shown that iron in a concentration of 1.0 per cent strongly accelerates the growth of the culture in the complete absence of organic growth factors. The importance of mineral elements for the culture of *Rhizobium* and *Azotobacter* has been pointed out by STEINBERG (1938). Despite their importance they do not remove the necessity of growth factors of vitamin nature.

The possible rôle of growth factors in the fixation of nitrogen is still unknown.

Azotobacter. — *Azotobacter chroococcum* has been employed as a suitable source of the growth factor "coenzyme R" for *Rhizobium trifolii.* Apparently *Azotobacter* is able to synthesize this factor and is therefore self-sufficient in this respect.

The physico-chemical conditions required by this organism are very precise. RIPPEL and B. LEHMANN (1936) showed that the addition of very small quantities of agar to the culture medium exerted a very favorable effect on growth and on the fixation of nitrogen. This effect is not due to an organic growth factor, as might be suspected, but to the action of the agar on the viscosity and the surface tension of the culture medium.

Mineral elements play an important part in the nutrition of this bacterium. Iron when used in conjunction with agar acts very favorably (RIPPEL, 1936). Molybdenum and vanadium are essential to the growth of *Azotobacter* and to the fixation of atmospheric nitrogen by this bacterium (BORTELS, 1937).

It has not been demonstrated with certainty that this organism requires organic growth factors. WERNER, however, states that organic growth factors are utilized since extracts from yeast and certain algae (*Hantzschia, Prasiola, Chlorococcum*) accelerate growth. These investigations require confirmation. They fail to eliminate the possible action of mineral elements contained in the extracts.

It must be recalled that the problem of growth factors of *Azotobacter* was raised by MOCKERIDGE as early as 1917 when he showed that the extracts of bacterial infusions favor considerably the fixation of atmospheric nitrogen by *Azotobacter chroococcum* and *Bacillus radicicola* (*Rhizobium*). The action of a true organic growth factor was by no means demonstrated and these investigations must be repeated from the modern viewpoint.

Little is known concerning the ability of *Azotobacter* to synthesize growth factors. BONNER and GREENE, employing the *Phycomyces* test, found the thiamin content of cells of this bacterium to be very high, *viz.*, 140 mg. per kg. of air dry material. The capacity for synthesis cannot be discussed until more information is available.

REFERENCES:-

1927: F. E. ALLISON. — The growth of *Bacillus radicicola* on artificial media containing various plant extracts. J. Agr. Res. 35:915.

1933: F. E. ALLISON, S. R. HOVER and D. BURK. — A respiration coenzyme. Science 78:217.

1934: F. E. ALLISON and S. R. HOVER. — An accessory factor for legume nodule bacteria. J. Bacter. 27:561.

1936: D. W. THORNE, O. R. NEAL and R. H. WALKER. — Physiological studies on *Rhizobium*, VII. The respiratory quotient. Arch. Mikrobiol. 7:477.

1936: A. RIPPEL und B. LEHMANN. — Über die Wirkung von geringen Mengen Agar auf Wachstum und Stickstoffbindung von *Azotobacter* und auf andere mikrobiologische Vorgänge. Arch. Mikrobiol. 7:210.

1936: A. RIPPEL. — Eisen-, Agar- und Humuswirkung bei *Azotobacter*. Arch. Mikrobiol. 7:590.

1938: J. PIETZ. — Beitrag zur Physiologie des Wurzelknöllchenbakteriums. Zentralbl. Bakter. II, 99:1.

1938: G. BJÄLFVE und R. NILSSON. — Stroh als Nährboden für *Bacterium radicicola*. Ann. Landwirtschaftl. Hochschule Schwedens 5:71.

1938*a*: R. NILSSON, G. BJÄLFVE und D. BURSTRÖM. — Über Zuwachsfaktoren bei *Bacterium radicicola*, I. Ann. Landwirtschaftl. Hochschule Schwedens 5:291.

1938*b*: R. NILSSON, G. BJÄLFVE und D. BURSTRÖM. — Über Zuwachsfaktoren bei *Bacterium radicicola*, II. Ann. Landwirtschaftl. Hochschule Schwedens 6:299.

1938*c*: R. NILSSON, G. BJÄLFVE und D. BURSTRÖM. — Über Zuwachsfaktoren bei *Bacterium radicicola*, III. Ann. Landwirtschaftl. Hochschule Schwedens 7:51.

1938: R. A. STEINBERG. — Applicability of nutrient solution purification to the study of trace-element requirements of *Rhizobium* and *Azotobacter*. J. Agric. Res. 57:461.

1938*a*: P. M. WEST and P. W. WILSON. — Biological determination of vitamin B_1 (thiamin) in *Rhizobium trifolii*. Science 88:334.

1938*b*: P. M. WEST and P. W. WILSON. — Synthesis of growth factors by *Rhizobium trifolii*. Nature 142:397.

1939: P. M. WEST and P. W. WILSON. — Biotin as a growth stimulant for the root nodule bacteria. Enzymologia 8:152.

1939: R. NILSSON, G. BJÄLFVE und D. BURSTRÖM. — Über Zuwachsfaktoren bei *Bacterium radicicola*, IV. Ann. Landwirtschaftl. Hochschule Schwedens 7:302.

1939: R. NILSSON, G. BJÄLFVE und D. BURSTRÖM. — Biotin als Zuwachsfaktor für *Bacterium radicicola*. Naturwiss. 27:389.

Chapter XVI.

THE GROWTH FACTOR REQUIREMENTS OF THE HEMOPHILIC ORGANISMS. FACTORS X AND V.

General introduction. — Practically all the organisms considered in previous chapters are very similar with respect to their growth factor requirements, the two factors most frequently encountered being thiamin and biotin. The hemophilic bacteria lead us into another domain and new factors appear among their vitamin requirements.

We shall begin our discussion of this group of organisms with the influenza bacillus (*Bacillus influenzae*). PFEIFFER showed in 1892 that *B. influenzae* grows luxuriantly on natural media based on blood. Hence it was assumed that the latter contains substances which are highly beneficial, perhaps indispensable to the life of this microorganism. The blood serum deprived of corpuscles does not support satisfactory growth whereas the washed red corpuscles are perfectly suitable for the culture of the bacillus. This observation led PFEIFFER to attribute a determining rôle to hemoglobin. He stated, without being able to demonstrate it experimentally, that the iron of hemoglobin was the essential growth promoting principle. The problem took on an entirely different complexion when it was learned that the necessary growth promoting substances were also present in plant and other animal tissues. GRASSBERGER was able to show that other bacteria were capable of furnishing the Pfeiffer bacillus the substances which it required. If the culture is contaminated with *Staphylococcus*, growth is strongly stimulated. The "satellite phenomenon" described later by DAVIS (corresponding to the "giant colonies" of GRASSBERGER and to the "nurse colonies" of NEISSER) is a new expression of this stimulation. The presence of microorganisms of the most diverse types exerts a stimulating action, *e.g.*, *Staphylococcus aureus* and *S. albus*, various strains of *Pneumonococcus*, the meningococcus, the diphtheria and pseudo-diphtheria bacilli, the colon bacillus, the typhoid and paratyphoid bacilli, the gonococcus, Friedländer's bacillus, *Sarcina lutea* and others. These organisms, in accelerating the growth of the Pfeiffer bacillus, exhibit the phenomenon of unilateral stimulation which, in the case of other organisms, has been responsible for the isolation of a growth factor, *e.g.*, the stimulation of the growth of *Polyporus* by bacteria (FRIES).

It is not necessary to have the bacteria in the living condition in order to stimulate the development of colonies of *B. influenzae;* dead bacteria, incorporated in the medium, produce the same effect.

This action, which may be favorable or inhibitory within the same group of hemophilic bacilli, has been used to differentiate certain species.

An explanation of certain discrepancies in the results of various investigators was published by DAVIS (1917). He found a favorable action on the hemophilic organisms resulting from the addition of sterile fragments of plants to the culture medium; the polishings of rice and of wheat are active, especially if they come from seeds which have commenced to germinate. DAVIS, along with others, has shown that blood contains two growth factors which differ in their properties. One of them, the so-called "X factor", is thermostabile; the other, the "V factor", is much less thermostabile (THJÖTTA and AVERY, 1921). The name "V" was chosen in order to designate the vitamin characteristics of this factor (vitamin in the older sense). The term vitamin should not be reserved exclusively for the V factor, since the X factor likewise has characteristics typical for vitamins as we shall see. Each factor alone is inactive; the two factors must be used in combination. Their presence prolongs the life of the culture and shortens the incubation period. Several investigators immediately confirmed the presence of two factors in blood (RIVERS and POLE, 1921, FILDES, 1923), which have since been conclusively demonstrated.

The general characteristics found during the early investigations were as follows: the X factor is relatively thermostabile, withstanding a temperature of 120°C., but not longer than 20 minutes; it is active at very high dilution. The V factor is thermolabile, not resistant to boiling for longer than 10 minutes, and is active only in large dosage. These characteristics were ascertained through the use of crude substances and impure extracts then available.

Numerous investigators have studied the distribution of these two factors. The V factor is abundant in potatoes, bananas, and many plant extracts. This may be demonstrated by a very simple experiment - blood, when heated, is rendered ineffective for the culture of the Pfeiffer bacillus, the V factor having been destroyed; it again becomes effective if various plant extracts are added to it.

Attempts to separate the two factors have been made, particularly by KOLLATH. These were successful only in part since they did not succeed in isolating the factors in a highly purified state. The investigations concerning the two factors must be discussed separately.

Factor X. — At first the hypothesis was advanced that the activity of the X factor, contained only in the corpuscles and not in the plasma, must be due to iron. This view was advanced by the Pfeiffer school, particularly KOLLATH. According to them the X factor was simply an organic compound containing iron, hence the Pfeiffer bacillus was not hemophilic, but instead *siderophilic.* KOLLATH (1925) performed the following experiment and obtained results which seemed to support the hypothesis of active iron. Bacteria were cultivated on a medium containing the oxide of ammoniacal iron and their ash was then tested for its action on *B. influenzae.* The ash was active only when it contained iron; the ash from some of the bacteria was devoid of iron and hence was inactive. These results indicated that the assimilation of iron and its incorporation in an organic compound controls the production of an active factor by the bacterium.

It was chiefly the work of OLSEN (1921) which directed the attention definitely toward hemin, suspected as the active growth factor. Compounds of known chemical composition were employed. The following substances were found to be active: oxyhemoglobin, methemoglobin, carboxyhemoglobin, and hemin, whereas the following were inactive: hematoporphyrin, hemocyanin, bilirubin, pyrrole from chlorophyll, peroxidase of WOLFF. The fact that hematoporphyrin is devoid of any activity seemed to prove that iron is the active element since it is absent from this molecule. It was thought possible to obtain growth of *B. influenzae* in the presence of certain salts of iron alone, for example, the active iron of BAUDISCH. The belief also prevailed that a correlation exists between the peroxidase activity of the factor and its activity as a growth factor.

These contradictory results were partially cleared up by the work of LWOFF (1934). He studied some of the flagellate parasites (trypanosomides): *Strigomonas oncopelti, S. fasciculata,* and *Leptomonas ctenocephali.* The first is parasitic in the latex of *Asclepias syriaca,* from where it passes into an hemipterous insect, *Oncopeltus;* the second lives as a parasite in the intestine of *Culex pipiens;* and the third is a parasite of the intestine of the dog flea, *Ctenocephalus canis.*

The growth of these flagellates, like that of the hemophilic bacteria, is dependent on blood added to the culture medium. The action of the blood is strictly quantitative; it can be measured by the intensity of respiration. In the presence of blood (and peptone) the flagellate synthesizes a respiratory enzyme system which otherwise can not be built up. The activity of the blood is pronounced with respect to growth (cell division) and the intensity of respiration in these flagellates, *e.g.,* 1×10^{-12} grams of blood permits 6.19 cell divisions. For respiration, 1 gram-atom of iron transports 4.83 gram molecules of oxygen per second at 28°C.

Hemophilus influenzae (*Bacillus influenzae*) is just as sensitive to the effect of blood as are the flagellates. This bacterium responds to blood in dilution up to 1:4-5 millions (concentration of hemin, 1×10^{-9}).

Since hemin is undoubtedly the active factor it was possible to study the specificity of its action and to compare its growth promoting action with its influence on respiration and peroxidase activity. The results obtained with *Strigomonas fasciculata* (from A. LWOFF, 1934, 1938) are presented in the following table.

TABLE XV. — The action of hemin and its derivatives on *Strigomonas fasciculata* (from A. LWOFF, 1938).

	reaction of peroxidases	influence on respiration	influence on cell division
Hemin	+	+	+
Deuterohemin	+	0	0
Protohemin	+	+	+
Hematohemin	+	0	
Mesohemin	+	0	0
Pyrrohemin	+	0	
Pheophorbide-a-hemin	+	0	
Pheophorbide-b-hemin	+	0	0
Protoporphyrin	0	+	+
Cytochrome C	+	0	0
Deuteroporphyrin	0	0	
Hematoporphyrin	0	0	0
Mesoporphyrin	0	0	0
Aetioporphyrin	0	0	
Pyrroporphyrin	0	0	
Pheophorbide a	0	0	
Rhodin (of chlorophyll b)	0	0	
Peroxidase of WOLFF	+	0	0
Active iron of BAUDISCH	+		0

The results obtained with protoporphyrin are particularly clear. Since it contains no iron, it has no peroxidase activity but it is active as a growth factor.

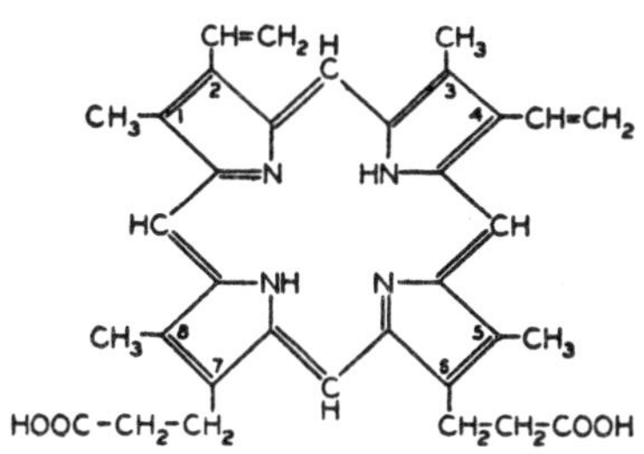

PROTOPORPHYRIN

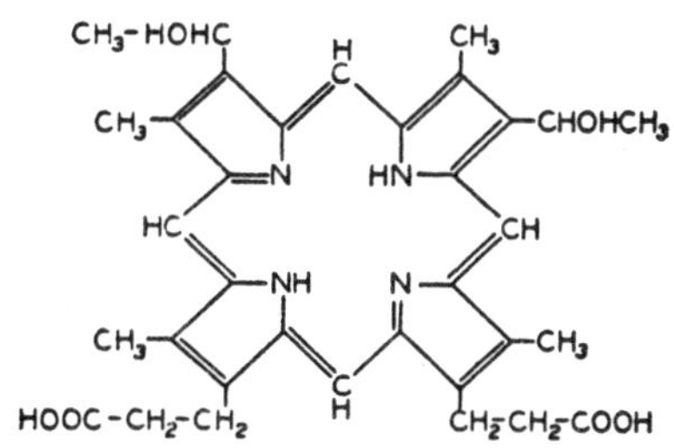

HEMATOPORPHYRIN

The fact that hematoporphyrin is inactive can not be explained by the absence of iron, but by its structure. Iron need not necessarily be present in the active molecule since protoporphyrin does not contain it and still functions as a growth factor. There is no longer any support for the hypothesis regarding a possible relationship between growth promotion and peroxidase activity. The introduction of iron into the molecule of protoporphyrin transforming it into protohemin does not alter its growth promoting activity.

The group in the second position is what determines the activity. Protoporphyrin has a vinyl ($CH_2 = CH$) group in that position. The substitution of the vinyl group by CH_3CHOH (hematoporphyrin), by H (deuteroporphyrin) or by CH_3CH_2 (mesoporphyrin) nullifies the growth promoting activity of the substance.

In the final analysis, the X factor activity of the blood is due to hemin. The fact that an organism requires blood does not necessarily indicate, however, that there is a need for hemin; it may be due to a need for cholesterol, present in the serum, or a need for ascorbic acid as well as the V factor present in the corpuscles.

The problem of the X factor is thus solved for the present. It does not imply that iron is entirely useless; on the contrary, it must be present in the culture medium. The rôle of hemin is to participate in the constitution of respiratory enzymes, cytochrome and others. Iron is the active element of these enzymatic systems. All the early investigations indicate that iron is indispensable. M. LWOFF (1933*b*, complete literature up till 1933) concluded from the works of KOLLATH and others that iron is the active element in hemin. He was able, by stabilizing the cultural conditions by a protective colloid, to replace blood in the culture of Trypanosomides. Moreover, he observed that protoporphyrin (devoid of iron) is able to replace blood and hemin in the culture of these organisms. He concluded that "it appeared uncertain that iron, particularly the active iron of BAUDISCH, was able to replace hemin".

The fact that protoporphyrin is active is contrary to expectations. If the organism can be satisfied with protoporphyrin (devoid of iron) is must be able to make protohemin by means of the iron which is present. Even this does not solve the problem unless the protohemin enters into the constitution of cytochrome. According to KEILIN protohemin exists in the free state in the cells of aerobic organisms and takes part in the construction of the prosthetic group of various respiratory enzymes (cytochrome, oxidase, catalase, and peroxidase). Hemin is necessary only under aerobic conditions. The strains of hemophilic organisms which are able to live anaerobically are content with the V factor but under aerobic conditions these same strains require hemin. This is explained by the rôle played by hemin in oxidation phenomena. However, it is necessary to mention that, according to SNYDER and BROH-KAHN (1939), hemin can be replaced by cystine in aerobic culture of *H. influenzae.* These investigators concluded that the function of hemin is to destroy H_2O_2 formed during the course of aerobic respiration.

In addition to *Hemophilus influenzae,* in which the need for hemin has been demonstrated by OLSEN and studied by A. LWOFF (1937), numerous other organisms have been found incapable of synthesizing completely their respiratory enzymes. The following are dependent on hemin as a growth factor: *Hemophilus conjunctivitidis* (P. FILDES), *H. canis* (RIVERS, 1922), *H. ducreyi* (A. LWOFF and PIROSKY), and in addition the following flagellates according to M. LWOFF: *Strigomonas muscidarum, S. culicidarum* var. *anophelis, S. fasciculata, Leptomonas ctenocephali, Leishmannia tropica, L. donovani, L. agamae, L. ceramodactyli, Schizotrypanum cruzi* (*cf.* A. LWOFF, 1938). For the present this completes the list of organisms known to require the X factor. The capacity to synthesize cytochrome, whose occurrence is universal, appears to have been lost only by these hemophilic organisms. However, this fact does not preclude the possibility that hemin is involved in regard to other organisms which are difficult to culture.

Factor V. — The existence of the V factor, with properties distinct from those of the X factor, was recognized by DAVIS (1917). Its name was supplied by THJÖTTA and AVERY. Its properties have been determined by numerous investigators; its thermolabile nature and its destruction by alkalis was recognized by FILDES. A. and M. LWOFF (1936) recognized the nature of the V factor and were able to show that it functions as a coenzyme.

The test object employed for this factor is *Hemophilus parainfluenzae* (strain 4101 of the Lister Institute). This organism does not develop in a water solution of peptone unless it is supplied with factor V in the form of a decoction of bakers' yeast. This yeast product, without being concentrated, is active in a dosage of 1:10,000.

The properties of this factor are as follows: not precipitated by barium acetate at *p*H 8.5, nor by lead acetate at *p*H 6.8; precipitated by lead acetate at *p*H 9.5 and by an excess of mercuric nitrate at *p*H 7, and also by alcohol; resistant to desiccation in vacuum (A. and M. LWOFF, 1936). These properties correspond to those of the coenzyme of WARBURG and CHRISTIAN (triphosphopyridine nucleotide). A. and M. LWOFF observed that a concentrated preparation of this coenzyme extracted from erythrocytes of the horse, has a V factor activity in a dilution of 1 to 30 millions. This represents a 1:100 million dilution of the coenzyme since the degree of purity of the preparation was 0.3. The coenzyme and the V factor activity of yeast extract are identical. There are still other characteristics which indicate a marked similarity between the coenzyme of WARBURG and the V factor: destruction in alkaline medium, resistance to boiling in acid medium, stability of the reduced form in alkaline medium.

It can be demonstrated experimentally that the dehydrogenase activity of bacteria deficient in the V factor is more than 20 times weaker than that of bacteria cultured with an excess of this factor. The time required to reduce methylene blue by bacteria deficient in factor V is reduced considerably upon the addition of factor V (codehydrogenase) when these organisms are cultured in the presence of various substances: glucose, pyruvate, fumarate, malate, ethyl alcohol, asparagine, *d*-valine, *l*-valine. The addition of the codehydrogenase to the medium of these bacteria brings about a ten fold increase in respiration and a twenty fold increase in aerobic and anaerobic glycolysis. It is therefore certain that factor V is required by *Hemophilus parainfluenzae* because of the inability of this organism to synthesize its complete dehydrogenase system.

The study of the chemical constitution of the codehydrogenases has already given us a good understanding of the rôle which they play.

The active substance in the cozymase of HARDEN and YOUNG is codehydrogenase I of v. EULER. The latter is composed of one molecule of the purine base adenine, two molecules of pentose, two molecules of phosphoric acid, and nicotinamide. The codehydrogenase II of WARBURG has the same structure except for the presence of an additional molecule of phosphoric acid. The structural formula of codehydrogenase I is as follows:

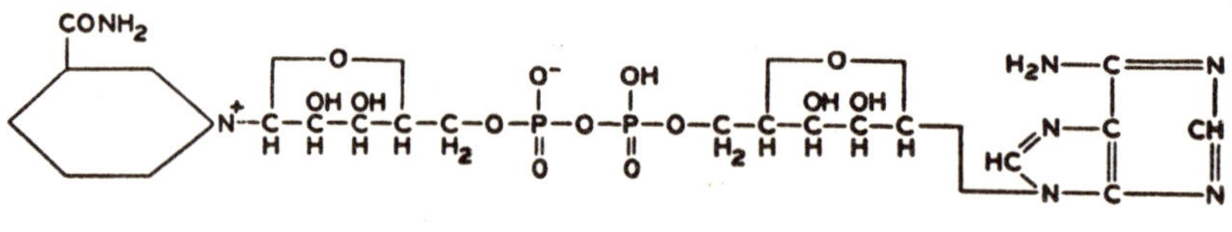

The acid hydrolysis of cozymase (codehydrogenase I) liberates the nicotinamide, the adenine, and two molecules of pentose-phosphoric acid. The alkaline hydrolysis furnishes nicotinamide and the triphosphoric acid of adenosine (adenylic acid). The enzymatic hydrolysis frees the nucleoside of nicotinamide (nicotinamide + one molecule of pentose) and the adenosine (adenine + one molecule of pentose).

Experiments show that nicotinic acid and its diethylamide and also adenylic acid when added alone have no growth promoting activity. Thus the bacterium is unable to synthesize the complete molecule of cozymase from the components of this factor. On the other hand, the red corpuscles of man are able to synthesize this factor from nicotinic acid (KOHN and KLEIN, 1939).

Nicotinamide represents the active group in the molecule and takes part in the phenomena of hydrogenation and dehydrogenation (see p. 53).

The cozymase of v. EULER (codehydrogenase I) is a diphosphopyridine nucleotide, whereas codehydrogenase II of WARBURG is a triphosphopyridine nucleotide. By a dephosphorylation of codehydrogenase II it is possible to obtain codehydrogenase I (cozymase). This transformation can be effected either by purely chemical means or by enzymatic action. Both of the codehydrogenases are involved in the metabolism of *Hemophilus parainfluenzae.* Under certain conditions when the codehydrogenases are not united with their specific apozymases a reciprocal transformation of the two codehydrogenases can be brought about by this bacterium. The nicotinamide is attached in the same manner in both molecules.

It is highly probable that when nicotinic acid is required as a growth factor for an organism, this compound participates in the construction of the molecule of codehydrogenase. This has not been proven, however. We have already seen that codehydrogenase makes up part of the vitamin B_2 complex. The physiological relationships existing between codehydrogenase and nicotinic acid will be discussed later (see p. 192).

Hemophilus parainfluenzae is the only organism known to require factor V. Furthermore, this factor is the only one required by this bacterium. It is not impossible that other microorganisms which are difficult to culture also require this factor.

The study of the hemophilic microorganisms has introduced an entirely new set of growth factors, composed of two very different factors. However, when the part played by these substances in the enzymatic systems is discussed, it will be found that the two factors have features which are alike physiologically. In certain cases, however, these factors do not act alone; they act in conjunction with some of the better known vitamins.

The growth factor requirements of some of the hemophilic organisms (A. and M. LWOFF) may be summarized as follows:

Bacteria. — *Hemophilus canis* requires only hemin; *Hemophilus parainfluenzae* requires only cozymase; *Hemophilus influenzae* requires both hemin and cozymase.

Flagellates. — *Strigomonas oncopelti* requires only thiamin; *Strigomonas fasciculata* and *S. culicidarum* require thiamin and hemin; *Schizotrypanum cruzi* requires hemin and ascorbic acid.

REFERENCES:-

1917: D. J. DAVIS. — Food accessory factors (vitamins) in bacterial culture with special reference to hemophilic bacilli. J. Inf. Dis. 21:392.

1921: TH. THJÖTTA and O. T. AVERY. — Growth accessory substances in the cultivation of hemophilic bacilli. J. Exp. Med. 34:97.

1921: T. M. RIVERS and A. K. POLE. — Growth requirements of influenza bacilli. Johns Hopkins Hosp. Bull. 32:202.

1921: O. OLSEN. — Über die Bedeutung des Blutes für das Wachstum des Pfeiffer'schen Influenzabazillus. Zentralbl. Bakter. I, 85:12.

1922: T. M. RIVERS. — *Bacillus haemoglobinophilus canis.* J. Bacter. 7: 579.

1923: P. FILDES. — The classification of hemoglobinophilic bacteria based upon their relation to blood pigment and to the vitamin factor. Brit. J. Exp. Path. 4:265.

1925: W. KOLLATH. — Vitaminähnliche Substanzen in ihrer Wirkung auf das Wachstum der Influenzabazillen, III. Die Rolle des sogenannten X-Faktors für die Biologie der Influenzabazillen. Zentralbl. Bakter. I, 95:279.

1933*a*: M. LWOFF. — Recherches sur la nutrition des Trypanosomides. Ann. Inst. Pasteur 51:55.

1933*b*: M. LWOFF. — Remarques sur la nutrition des Trypanosomides et des bacteries parahémotropes. Ann. Inst. Pasteur 51:107.

1934: A. LWOFF. — Die Bedeutung des Blutfarbstoffes für die parasitischen Flagellaten. Zentralbl. Bakter. I, 130:498.

1936: A. and M. LWOFF. — Studies on codehydrogenase, I. Nature of growth factor "V"; II. Physiological function of growth factor "V". Proc. Roy. Soc. Lond. B, 122:352.

1937: A. LWOFF. — Rôle physiologique de l'hémine pour *Hemophilus influenzae* Pfeiffer. Ann. Inst. Pasteur 59:129.

1938: A. LWOFF. — Les facteurs de croissance pour les microorganismes. Rapport présenté au ler Congrès intern. des Microbiologistes de Langue française, Paris, p. 58.

1938: TH. L. SNYDER and R. H. BROH-KAHN. — Substitution of cysteine for protohemin as the "X" factor for growth of *H. influenzae.*

1939: H. I. KOHN and J. R. KLEIN. — The synthesis of cozymase and factor V from nicotinic acid by the human erythrocytes. J. Biol. Chem. 130:1.

Chapter XVII.

CERTAIN INDIVIDUAL FACTORS: ASCORBIC ACID, VITAMIN D AND CHOLESTEROL, PIMELIC ACID, THE SH GROUP.

Ascorbic acid. — We have seen that ascorbic acid can be present in the green flowering plants in rather large amounts. It is not certain that all the ascorbic acid in such plants can be considered as a growth factor of vitamin nature. The data concerning the ascorbic acid content and its rôle in microorganisms are even more fragmentary.

At one time non-green plants, particularly the lower cryptogams, were regarded as unsuited for the synthesis of vitamin C and ascorbic acid. This assumption has since been shown to be untenable. BERNHAUER found a reducing substance in *Aspergillus niger* closely related to ascorbic acid. FUKOMOTO and SHIMOMURA made the same observation concerning *Aspergillus cellulosae, A. fugimatus, A. nidulans,* and *A. melleus.* A similar reducing substance was found in *Penicillium glaucum* and *P. luteum.* We may recall, however, that vitamin C and ascorbic acid are not necessarily equivalents. For example, the presence of reducing substances in fungi suggests the existence of a precursor of ascorbic acid which does not exercise any antiscorbutic action on animals (guinea pigs).

In certain lactic streptococci, *Str. lactis, Str. cremoris, Str. citrovorus,* and also in various lactic bacilli, the ability to synthesize ascorbic acid or a related substance has been assumed. The results obtained with *B. prodigiosus* are particularly interesting. When cultured on *d*-xylose (the substance from which ascorbic acid is synthesized chemically) this microorganism synthesizes a substance giving a positive reaction for ascorbic acid (Tillmann's reaction), which may actually be ascorbic acid or a related substance. The same synthesis has been assumed for *B. bifidus* and also in Kombucha symbionts (a yeast + an acetic bacterium of Javanese tea).

The results concerning the ability of higher fungi to synthesize ascorbic acid are contradictory. SCHEUNERT and RESCHKE obtained negative results with various edible fungi. Although many microorganisms produce reducing substances giving a positive Tillmann's reaction (2,6-dichlorophenol-indophenol) which are possibly related to ascorbic acid, it has not been proven that these microorganisms synthesize vitamin C which is active on animals.

We must now examine another aspect of the problem, *viz.,* the question of whether the loss of the ability to synthesize ascorbic acid makes it necessary to add this substance as a growth factor.

Bacteria. — Ascorbic acid is recognized generally as having a favorable action on the growth of obligate anaerobes. These organisms are enabled to grow in the presence of oxygen upon the addition of ascorbic acid (*Clostridium Welchii,* for example, according to KLIGER and GUGGENHEIM). It may be recalled that cysteine is able to bring about the same effect. KLIGER and GUGGENHEIM mention one case in which it was more than probable that the action was external to the organism and that ascorbic acid was effective because of its adjusting the redox potential of the medium.

MORELL, working with *B. coli,* was able to observe a relationship between the oxidizing processes and ascorbic acid.

There are indications in a certain number of cases that ascorbic acid is required as a growth factor. This was first shown by RAHN and HEGARTY (1938) working with *Streptococcus lactis* when they observed that exhausted cultures are definitely stimulated by the addition of ascorbic acid. SARTORY and collaborators also report that certain cocci react favorably to the presence of ascorbic acid. The tubercular bacillus has yielded contradictory results. *B. coli, Aerobacter aerogenes,* and *B. subtilis* (under aerobic conditions) are inhibited by ascorbic acid (GUHA and GUPTA).

Sometimes it is possible to follow the utilization and the transformation of ascorbic acid in microorganisms. During the course of lactic fermentation, ascorbic acid is transformed into its oxidized form (NOWOTELNOW and WADOWA). *Bacterium Leichmannii* (*Thermobacterium acidophilum*) is able to hydrogenate the dehydrogenated ascorbic acid into ascorbic acid (TKATSCHENKO). The hemolytic strains of *B. coli* and the paratyphoid bacilli of the B type (*Salmonella Schottmülleri*) break down ascorbic acid by oxidation (STEPP and SCHROEDER); the same is true of *Streptococcus mucosus* and *Enterococcus.* In many cases the spore-forming aerobic bacteria bring about a reversible oxidation of ascorbic acid. We will be able to understand better these cases when we learn the rôle of ascorbic acid in oxidation-reduction.

Fungi. — Only a few fungi have been investigated concerning their response to the presence of ascorbic acid. *Aspergillus* and *Saccharomyces* (various species) are known to respond favorably to this substance, although it is not an indispensable growth factor.

Green algae and flagellates. — In these chlorophyll containing organisms ascorbic acid seems to play an important rôle as a growth factor (ONDRATSCHEK, 1940 *a, b, c*).

Hematococcus pluvialis has been studied in great detail and has been shown to require vitamin B_1 and ascorbic acid as growth factors. With potassium nitrate as a source of nitrogen the effect is as follows (expressed as the number of cells per mm^3 of solution after maximal growth has occurred):

control (without acetate in the light, without vitamin supplements)	350
without acetate in the light + ascorbic acid	2800
with acetate in the light + ascorbic acid	3400
with acetate in the dark + ascorbic acid	1700

With asparagine as the nitrogen source the figures are as follows:

control (without acetate in the light, without vitamin supplements) 300
without acetate in the light + ascorbic acid 4000
with acetate in the light + ascorbic acid 4200
with acetate in the dark + ascorbic acid 2100

The effect of ascorbic acid is more pronounced when used with asparagine (mixotroph culture) than when used with potassium nitrate. Ascorbic acid apparently compensates for a deficit by acting either externally or internally (Fig. 14). In the presence

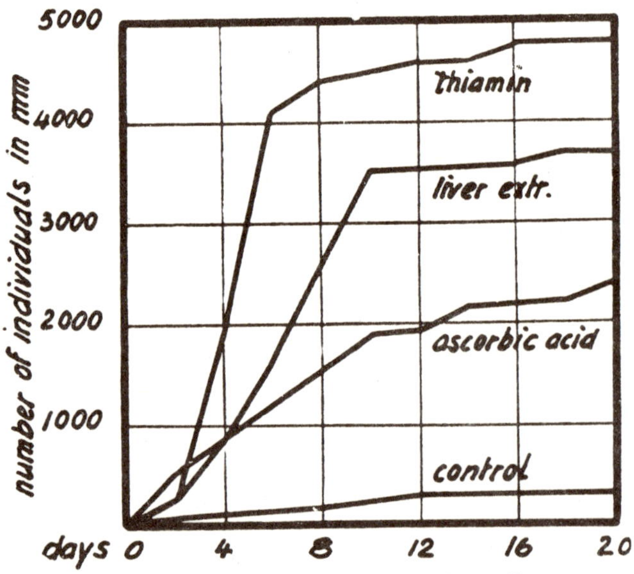

FIG. 14. — Action of ascorbic acid and thiamin on *Hematococcus pluvialis* cultured in the dark with asparagine and acetate (from ONDRATSCHEK, 1940*a*).

of asparagine and acetate the effect of ascorbic acid begins to appear with a dosage of 10γ per 100 cc. of solution; the maximal action is obtained with a dosage of 1 mg. (ONDRATSCHEK, 1940*a*). This dosage is relatively high and is above the limit assigned to true growth factors. If 1 mg. is accepted as the optimal dosage for ascorbic acid, it is found that one cell of this organism requires 0.0025γ of this vitamin providing all the ascorbic acid is absorbed and utilized. Even this dosage represents a very large number of molecules.

The striking fact is that sterilization in the autoclave diminishes only slightly the action of this vitamin on these microorganisms. Its effect on animals, on the other hand, is destroyed by oxidation and by heat. This apparently confirms the assumption that it does not have the same vitamin action on microorganisms which it has on animals.

But another fact supports the possibility that ascorbic acid acts as a true growth factor (vitamin) on these microorganisms, *viz.*, the alga *Hormidium flaccidum* does not require vitamins for its growth. According to our view this signifies that this alga is able to synthesize the necessary vitamins when cultured in a purely mineral solution, *i.e.*, it is completely autotrophic with respect to vitamins. Furthermore, an extract of a natural product such as plant tissue rich in both known and unknown factors has no effect. Extracts from *Hormidium* grown on a strictly mineral medium contain 8.4 to 68.5γ of ascorbic acid per gram of dry matter (titration by methylene blue according to MARTINI and BONSIGNORE). Thus extracts made from vitamin synthesizing organisms can serve as sources of growth factors for those species which are unable to synthesize all the necessary factors (ONDRATSCHEK, 1940*b*). Ascorbic acid favors the growth of cultures of *Euglena gracilis* and *E. viridis* (Eugleninae), *Chlamydomonas dorsiventralis, C. humicola, C. orbicularis, Chlorogonium euchlorum, Hematococcus pluvialis* (Chlamydomonadidae), *Hormidium Barlowi, Uronema Barlowi, U. gigas* (Ulotrichales). The control cultures on the vitamin-free medium in each case undergo an appreciable development, indicating that ascorbic acid is not absolutely essential; it merely favors growth. *Euglena gracilis* shows a definite need of vitamins for its metabolism. When the cultural conditions are *autotrophic* additional thiamin beyond the optimal dosage is ineffective. On the other hand, additional ascorbic acid beyond the optimal dosage exerts an additional auxogenic effect. When the cultural conditions are *mixotrophic,* additions of thiamin beyond the optimal dosage have a weak action; additional ascorbic acid has a very strong action. When the cultural conditions are *heterotrophic* ascorbic acid in excess of the optimal dosage exerts almost no effect whereas excessive thiamin acts strongly. It is clearly demonstrated herewith that the growth factor requirement, which may also be spoken of as the capacity for synthesis, is dependent upon the general metabolism. Ascorbic acid is a growth factor when the medium is strictly mineral. Thiamin is an essential factor when the medium contains nitrogen and carbon in the organic form and the culture is grown in the dark. Under cultural conditions which permit the simultaneous utilization of the CO_2 of the air (autotrophic) and of organic substances, thiamin and ascorbic acid act with equal intensity.

It is surprising to note that widely distributed species such as *Euglena gracilis,* which appear to be completely autotrophic in nature, can be rendered partially or completely heterotrophic for vitamins. This species was found by DUSI (1939) to respond favorably to thiamin. All the species of algae and flagellates mentioned in the preceding pages of this chapter respond to thiamin which acts in conjunction with ascorbic acid (ONDRATSCHEK). In certain cases the effects of thiamin and ascorbic acid are additive and of the same magnitude (*Euglena*), whereas in others the simultaneous action of the two vitamins is not strictly additive (*Chlamydomonas* and *Uronema*).

Despite certain reservations, ascorbic acid must be considered as a growth factor in the case of the green algae listed above as organisms requiring this vitamin. We shall place ascorbic acid on the borderline between true growth factors and substances functioning as nutrients.

The use of ascorbic acid as a nutrient and as a source of carbon has been considered. It apparently is unable to replace sodium acetate as a source of carbon for *Hematococcus pluvialis* since attempts to make such a substitution bring about a cessation of growth after the second passage.

Heterotrophic flagellates. — The first microorganism shown to require ascorbic acid was *Schyzotrypanum cruzi* (M. LWOFF, 1938). This pathogenic trypanosome is easily cultured on media based on blood. It would be reasonable to suspect that this organism, like *Strigomonas,* might require hemin as a growth factor. However, hemin alone does not permit the culture to undergo a series of passages. Obviously a second factor is required. Since the results obtained with red blood corpuscles were very inconsistent, M. LWOFF assumed that the second factor was labile and believed it to be ascorbic acid. This factor is known to be destroyed rather rapidly in serum. Experiments show that ascorbic acid when used together with hemin and serum produces cultures comparable to those produced with whole blood. Hence ascorbic acid is indispensable to this species of trypanosome. An unknown factor which is present in the serum and may possibly be cholesterol is designated provisionally as the TS factor by M. LWOFF. Unfortunately nothing is known concerning the dosage required and we do not know whether this substance should be considered as a true growth factor in the sense that we now use the term.

Heterotrophic flagellates of the group Tetramitides also require ascorbic acid (CAILLEAU, 1937, 1940). Three species, *Trichomonas columbae, T. foetus* and *Eutrichomastix colubrorum* require ascorbic acid and also cholesterol as growth factors. A culture of one of the above species can be grown on a broth base supplemented with veal liver extracted by alcohol and acetone. The liver extract, after 14 months in storage, is no longer active but it can be reactivated by the addition of cholesterol and ascorbic acid. Since these investigations of CAILLEAU do not indicate the dosage of ascorbic acid added, it is necessary to make the same reservation as in the case of *Schizotrypanum cruzi* regarding the action of vitamin C as a true growth factor of vitamin nature.

A study concerning the specificity of action has been started with *Eutrichomastix colubrorum* (R. CAILLEAU, 1939). The results obtained are as follows: *l*-ascorbic acid, *d*-arabo-ascorbic acid (isoascorbic acid), and *d*-gluco-hepto-ascorbic acid are active on flagellates and on vertebrates. The reductone and *d*-gluco-ascorbic acid are inactive on vertebrates but are active on flagellates. The activity on vertebrates, is strictly dependent upon the stereochemical configuration of the vitamin molecule, but this is not the case with the flagellates. Hence the specificity of vitamin C activity is not very marked in *Eutrichomastix.*

At present no other microorganisms are known to require ascorbic acid.

Cholesterol and vitamin D. — Although vitamin D is very important in the animal kingdom, it appears at present to play only a secondary rôle in the plant kingdom. The reason for this situation is that it has hardly been studied from the botanical point of view. Plants contain the precursor of vitamin D_2 called provitamin D_2. This precursor is a sterol, known as mycosterol or ergosterol, and is abundant in fungi, particularly the yeasts. Provitamin D_2 is transformed into the active vitamin D_2 by irradiation with ultraviolet light.

Bacteria. — PRICKETT and MASSENGALE (1931) have studied the action of ergosterol on the growth of various species of *Mycobacterium, M. leprae, M. phlei, M. tuberculosis, M. avium, M. berolinensis,* and *M. smegatis* (9 strains representing 6 species). When these species of bacteria are cultured on an agar medium with 5 per cent glycerol, they do not synthesize ergosterol; its addition to the medium definitely favors growth and the production of pigments (studied by INGRAHAM and STEENBOCK). Strange, indeed, is the fact that the irradiated (activated) ergosterol does not accelerate growth; it actually retards growth when used in high concentrations.

It has not been proved that ergosterol acts as a true growth factor. This type of action might be suspected because the first impression seems to indicate that this substance is required because of a loss in capacity for synthesis. Its action, however, is only stimulatory in nature.

Bacteria, for the most part, are devoid of sterols. This is assumed to be true for *Mycobacterium tuberculosis, Corynebacterium diphtheriae* and *Escherichia coli* (BEHRING, CHARGRAFF). CAILLEAU (1937), using a biological assay (test organism—*Trichomonas columbae*), found that the acetone extracts of certain bacteria, which will be listed later, contain no sterols. The lack of sterols in bacteria is in direct contrast to the situation in fungi, in which sterols are widely distributed. Only a very few bacteria contain these compounds. For example, *Azotobacter chroococcum,* according to GREAVES (1935) contains one or several sterols synthesized from a purely mineral medium + glucose.

The absence of sterols in a bacterium need not necessarily indicate a loss of capacity for synthesis accompanied by a need for one of these compounds as a growth factor. Even the fact that a sterol, *e.g.*, ergosterol, favors growth does not necessarily mean that it acts as a true growth factor. The action may be purely of stimulatory nature.

Flagellates. — In contrast to the bacteria, certain flagellates (Trypanosomatidae, Trichomonadidae) of the genus *Trichomonas* require sterols as indispensable factors. These organisms have been studied thoroughly by CAILLEAU (1937).

A culture of *Trichomonas columbae* requires that the basal medium be supplemented with an acetone extract from liver. This extract may be replaced by natural products such as egg yolk, hydrolyzed lecithin, and bile. On the other hand, the hydrochloride of choline does not permit a culture to survive a series of passages. These facts point to cholesterol as the active factor in question. Experiments show that an alcoholic solution of cholesterol permits cultures of *Trichomonas* to undergo a number of passages. The concentration of cholesterol required is relatively high, *viz.*, 1:10,000. This is definitely higher than the usual dosage of growth factors. The action of an impurity might be suspected because of the size of the dosage and because of the fact that the vitamin D effect discovered in cholesterol was found to be due to the contamination of the latter with ergosterol. This hypothesis of an impurity is not supported experimentally since absolutely pure samples of cholesterol have the same activity as impure samples.

It has been demonstrated by CAILLEAU that cholesterol is required by the following: *Trichomonas columbae* (1939*b*), *T. foetus* (1938), *T. batrachorum,* and *Eutrichomastix colubrorum* (1939*a*). All of these organisms, except *Trichomonas batrachorum,* require an additional factor, ascorbic acid. Both of these factors are absolutely indispensable for growth.

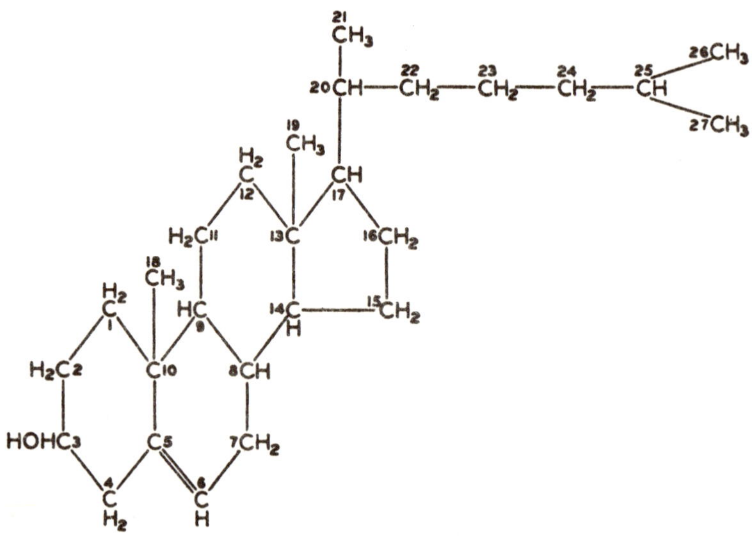

CHOLESTEROL

The specificity of action of cholesterol has been studied very thoroughly with *Trichomonas columbae* by means of 71 products more or less closely related to cholesterol. This specificity is rather pronounced. The following substances are active: cholesterol, cholestanol, epidehydrocholesterol, allo-cholesterol, acetate of cholesterol, palmitate of cholesterol, 22-dihydro-ergosterol, α-ergosterol, γ-sitosterol, γ-sitostanol, cinchol, cis-cholestan-3-4-diol, cis-Δ-5:6-cholestine-3:4-diol, and ergostanol. The following substances are somewhat less active: ergosterol, α-dihydro-ergosterol, the acetate of γ-sitosterol, and pyridino-allocholesterol.

Any satisfactory discussion of the specificity of action of cholesterol presupposes an acquaintance with the chemistry of this complex group of sterols (see FIESER, 1937). Since a discussion of the chemistry of these compounds is beyond the scope of this volume, we simply give the structural formula of cholesterol and point out the substitutions which inactivate the molecule in the case of *Trichomonas columbae.*

According to CAILLEAU the epimerization of the sterol in the third, fourth or fifth position completely inactivates the molecule. The elimination of the double bond between the fifth and sixth carbon atoms does not alter the activity. The same is true for the esterification of the alcohol group in the third position (the acetate and the palmitate are active). On the other hand, the transformation of the alcohol group in the third position into a ketone, and also the elimination of the hydroxyl groups in the third and fourth positions of cis-Δ-5:6-cholestine 3:4-diol, or their replacement by chlorine destroys the activity. Finally the side chain may be altered without reducing the activity (cinchol, ergosterol, and sitosterol are physiologically active), but the complete suppression of the side chain destroys the growth factor action. (For details refer to the original publications of CAILLEAU 1937, 1939*a* and *b*).

The sterols which are active as growth factors seem to constitute a separate group. The carcinogenic substances, the estrogenic hormones (female sex hormones), the male sex hormones, and the anti-rachitic substances (vitamin D_2) have no growth factor action on *Trichomonas.* The inactivity of vitamin D_2 (calciferol) and lumisterol is explained by assuming that the OH radical in the third position has been epimerized.

Practically nothing is known concerning the physiological rôle of cholesterol in the flagellates. In higher animals which require the antirachitic vitamin it is believed that cholesterol increases the permeability of the intestine for calcium. Moreover, complex relationships exist in animals between the action of vitamin D and certain hormones (thyroid and parathyroid). It is impossible to make further comparisons between the reactions of the flagellates and those of vertebrates. However, it is conceivable that the cholesterol might, as a constituent of the ectoplasm, take part in the phenomena of permeability.

In the higher plants the female sex hormones act as growth hormones under certain conditions.

It is difficult on the basis of the available evidence to consider cholesterol as a true growth factor inasmuch as the dosage required for action is large and its physiological rôle is still obscure. Nevertheless its action is very interesting.

The flagellates listed above are the only organisms known to require cholesterol and are the only ones studied extensively in this connection. It has been observed by R. DEVLOO that cholesterol is utilized by yeasts in which it functions as bios but this report lacks confirmation and interpretation.

Pimelic Acid. — This substance has been studied very little. It is known to be active only on *Corynebacterium diphtheriae.* This bacterium when cultured in a synthetic medium containing a carbon source and the usual mineral salts (MUELLER, 1937) requires, in addition, cystine, *d*-lactic acid, as well as traces of heavy metals, iron, manganese, copper and zinc. This medium, moreover, does not support normal growth unless it receives supplements of liver extract. The latter was used by MUELLER and his collaborators for the isolation of growth factors. *Corynebacterium diphtheriae* will be encountered elsewhere in this book since it requires nicotinic acid and beta-alanine as growth factors. The latter is the really essential factor and is active in very small dosage. This bios II*a* component must participate in the synthesis of pantothenic acid. In addition to these two factors, a third substance exercises a favorable action. It is found for instance in the urine of cows, from which material MUELLER extracted and identified the factor as pimelic acid. This acid alone, or nicotinic acid alone, is inactive, whereas beta-alanine alone is active. The maximal effect, however, is exercised by a combination of the three. Pimelic acid must in this case be considered as a co-factor. When the cultural conditions are optimal it acts quantitatively, the maximal effect being produced by a dosage of 0.5γ per 10 cc. of medium. The natural pimelic acid has the same activity as the synthetic product. (The response of the culture to this substance was measured by means of a nitrogen analysis of the bacterial colony).

The specificity of action appears to be very pronounced. No activity was observed in any of the following dibasic acids: oxalic acid (COOH-COOH), malonic acid ($COOH\text{-}CH_2\text{-}COOH$), succinic acid ($COOH\text{-}(CH_2)_2\text{-}COOH$), glutaric acid ($COOH\text{-}(CH_2)_3\text{-}COOH$), adipic acid ($COOH\text{-}(CH_2)_4\text{-}COOH$), suberic acid ($COOH\text{-}(CH_2)_6\text{-}COOH$), and azelaic acid ($COOH\text{-}(CH_2)_7\text{-}COOH$).

Recent experiments of DU VIGNEAUD and coworkers (1942) with the Allen strain of the diphtheria bacillus indicate that pimelic acid is utilized for the synthesis of biotin. These workers found, however, that pimelic acid is unable to replace biotin in its growth-stimulating effect on yeast. Apparently the latter has lost its ability to synthesize the other component or to condense the components into a molecule of biotin.

R. E. and ESTER EAKIN (1942) demonstrated independently the physiological relationship between pimelic acid and biotin. They suspected a possible relation between pimelic acid and biotin when DU VIGNEAUD, HOFFMAN, and MELVILLE (1942) reported the tentative structural formulae for the latter with a side chain—$CH_2CH_2CH_2CH_2COOH$, and accordingly designed a unique procedure for a study of the biosynthesis of biotin. They selected an organism

(*Aspergillus niger*) whose rate of growth is not affected by either biotin or pimelic acid (autotrophic for biotin), cultured it on a biotin-free medium, and demonstrated the activity of pimelic acid in promoting the synthesis of biotin. Cultures supplied with pimelic acid (1 mg./12 cc. of medium) produced 16 to 36 times as much biotin as the controls. These authors extended their studies to the specificity of action and found the lower homologues of pimelic acid (succinic, glutaric, and adipic acids) and an isomer, β-methyl adipic acid, inactive. The higher homologues, suberic and azelaic acids, however, possess activity comparable to that of pimelic acid. Apparently the specificity of action is less pronounced in *A. niger* than in *C. diphtheriae.*

The SH Group. — Various organisms, particularly the lower ones, are unable to reduce oxidized sulphur; they require this element in the reduced form (*Saprolegnia,* for example, according to VOLKONSKY). On purely theoretical grounds, LWOFF considers reduced sulphur as a growth factor but the author disagrees with him on this point since the major part of sulphur must be utilized as food with plastic action. This, however, does not preclude the possibility that a *small part* of this reduced sulphur functions as a growth factor.

Repeatedly cysteine and glutathione (SH form) have been found to have a beneficial action on the aerobic fermentation of *Saccharomyces cerevisiae* and an inhibitory action on respiration and growth. In the case of *Propionibacterium pentosaceum* these compounds accelerate both aerobic and anaerobic fermentation as well as respiration (CHAIX). The influenza bacillus exhibits a growth response to these compounds (KOLLATH).

The most interesting effect produced by compounds containing reduced sulphur is that exerted on cell division (HAMMET). This process is retarded by glutathione. On the other hand a favorable action on the growth of *Aspergillus niger* has been pointed out by GUHA and GUPTA.

The relationship between the SH group and ascorbic acid is well known. The latter protects the former against oxidation to SS. The SH group is able to reduce dehydroascorbic acid. Other relationships exist according to JOYET-LAVERGNE. Glutathione is synthesized in the chondriome where it is found associated with vitamin A. The latter protects the glutathione against oxidation.

Although the position which the SH group should occupy in the system of growth factors is difficult to decide, it is necessary to include it in this discussion. The reduced sulphur probably acts as a precursor of compounds which function as true growth factors.

REFERENCES:-

1931: P. S. PRICKETT and O. N. MASSENGALE. — *Mycobacterium:* Ergosterol content of certain species and effect of ergosterol on their growth. J. Infect. Dis. 49:297.

1935: J. E. GRAEVES. — Can *Azotobacter chroococcum* synthesize vitamin D? J. Bacter. 30:143.

1937: J. H. MUELLER. — Studies on cultural requirements of bacteria, X. Pimelic acid as a growth stimulant for *C. diphtheriae*. J. Bacter. 34:163.

1937: R. CAILLEAU. — La nutrition des Flagellés tétramitidés. Les stérols, facteurs de croissance pour les Trichomonades. Ann. Inst. Pasteur 59:137.

1937: L. F. FIESER. — The chemistry of natural products related to phenantrene. 2nd ed., Monographs of the Amer. Chem. Soc. No. 70, Reinhold Pub. Corp.

1938: M. LWOFF. — L'hématine et l'acide ascorbique, facteurs de croissance pour le Flagellé *Schizotrypanum cruzi*. C. R. Acad. Sci. Paris 206:540.

1938: R. CAILLEAU. — Le cholestérol et l'acide ascorbique, facteurs de croissance pour le Flagellé tétramitidé *Trichomonas foetus*. C. R. Soc. biol. Paris 127:861.

1938: O. RAHN and C. P. HEGARTY. — Accessory factors in lactic acid fermentation. Proc. Soc. Exper. Biol. a. Med. 38:218.

1939*a*: R. CAILLEAU. — L'activité de quelques substances voisines de la vitamine C, envisagées comme facteurs de croissance pour le Flagellé *Eutrichomastix colubrorum*. C. R. Soc. biol. Paris 131:964.

1939*b*: R. CAILLEAU. — L'acide ascorbique, facteur de croissance pour le Flagellé *Trichomonas columbae*. C. R. Soc. biol. Paris 130:319.

1939: H. DUSI. — La pyrimidine et le thiazol, facteurs de croissance pour le Flagellé à chlorophylle *Euglena pisciformis*. C. R. Soc. biol. Paris 130:419.

1940: R. CAILLEAU. — Cholesterol and ascorbic acid, two growth factors for Trichomonads. 3rd Inter. Congress f. Microbiology, Report of Proceedings (1939), p. 493.

1940*a*: K. ONDRATSCHEK. — Experimentelle Untersuchungen über den Einfluss von Wirkstoffen auf die Vermehrung einiger mixotropher Algen. Arch. f. Mikrobiol. 11:89.

1940*b*: K. ONDRATSCHEK. — Experimentelle Untersuchungen über den Zusammenhang zwischen Vitaminbedürfniss und Verlust des Synthesevermögens einiger mixotropher Algen. Arch. f. Mikrobiol. 11:228.

1940*c*: K. ONDRATSCHEK. — Vitamine als vermehrungsbegrenzende Faktoren bei *Hematococcus pluvialis*. Arch. f. Mikrobiol. 11:219.

1942: V. DU VIGNEAUD, K. HOFFMAN and D. B. MELVILLE. — On the structure of biotin. J. Amer. Chem. Soc. 64:188.

1942: V. DU VIGNEAUD, K. DITTMER, E. HAGUE and B. LONG. — The growth-stimulating effect of biotin for the diphtheria bacillus in the absence of pimelic acid. Science 96:186.

1942: R. E. EAKIN and ESTER EAKIN. — A biosynthesis of biotin. Science 96:187.

Chapter XVIII.

FUNCTIONS OF VITAMINS (GROWTH FACTORS); THEIR ACTION AS COENZYMES.

The fact that vitamins are required in very small dosages presents an interesting problem regarding the mechanism of their action. What are the functions of thiamin, riboflavin, ascorbic acid, hemin and cozymase which are required for the growth of certain organisms?

Vitamin A and carotene. — Little is known concerning the rôle of carotene in plants. This is true at least for the chemical mechanism of its action or actions. We have seen that carotene is formed in the chondriome and that it must play a role in cellular oxidations. Carotene possesses properties of oxygenases. In the presence of peroxidase it activates molecular oxygen. The system, carotene → pyrogallol → peroxidase, is inhibited by oils, the less saturated ones being the more effective (BORODINE). It has been impossible to determine whether carotene acts as a growth factor, and, except for the particular case of carotinoid factors affecting the sexual processes of *Chlamydomonas,* the chemical mechanism of its action remains shrouded in obscurity.

Thiamin. — The B vitamins are much better known with respect to the chemistry of their action on metabolism. Thiamin has been known for a long time to participate in the metabolism of carbohydrates. A deficiency of this vitamin in animals results in an accumulation in the blood of methylglyoxal (pyruvic aldehyde) produced as an intermediate product of carbohydrate metabolism.

It is from the study of alcoholic fermentation that the precise rôle of thiamin was learned. This process consists of a series of steps: phosphorylation of glucose and production of hexose-diphosphoric acid, and the splitting of this molecule into two molecules of triose phosphoric acid; these in turn produce α-glycerophosphoric acid and phosphoglyceric acid; the phosphoglyceric acid is hydrolyzed into pyruvic acid and phosphoric acid; the pyruvic acid is split enzymatically by *carboxylase* into acetaldehyde and CO_2, the latter being released as a product of fermentation. The acetaldehyde reacts with one molecule of triose phosphoric acid in accordance with an oxidation-reduction reaction in which the triose phosphoric acid is oxidized to phosphoglyceric acid and the acetaldehyde is reduced to ethyl alcohol as the second product of fermentation.

The step in carbohydrate metabolism which requires the action of thiamin is the one in which pyruvic acid is split. The rôle of this vitamin was elucidated by a series of discoveries beginning in 1937. LOHMANN and SCHUSTER (1937) extracted from yeast a cocarboxylase which was highly active under certain conditions (*p*H 6.2, in the presence of Mg). They determined the structure of this cocarboxylase and found two atoms of phosphorus attached in the form of an ester to a group whose composition corresponds to that of thiamin. Cocarboxylase can therefore be considered as a dipyrophosphoric ester of thiamin. In the metabolism of yeast (decarboxylation of pyruvic acid) this phosphoric ester of thiamin is able to replace the cocarboxylase extract.

The formation of cocarboxylase (pyrophosphate of thiamin) is tied up with the cycle of adenylic acid (LIPMANN):

adenosinetriphosphate + thiamin ⇄ adenylic acid + cocarboxylase

It has even been assumed that the active form of vitamin B_1 is not cocarboxylase, but *acetylated cocarboxylase.* The biosynthesis of cocarboxylase in the presence of thiamin is easily accomplished by microorganisms in general, and certain propionic bacteria are able to synthesize acetyl-cocarboxylase in the presence of thiamin.

The action of cocarboxylase on pyruvic acid in the presence of washed yeast in alkaline solution containing Mg, is stimulated by free thiamin. The same stimulation can be observed in the presence of orthophosphate of thiamin, or simply pyrimidine (OCHOA). Similar observations were made with acetyl-thiamin by WESTENBRINK. The stimulating action of thiamin depends in part on the manner in which the yeast containing the apocarboxylase has been washed. The explanation of this fact given by WESTENBRINK (1940) is as follows: thiamin does not act directly on the system, pyruvic acid + carboxylase + pyrophosphate of thiamin (cocarboxylase) + Mg. Instead it acts upon the enzyme, phosphatase which brings about a dissociation of pyrophosphate of thiamin in two separate steps. The intermediate product is orthophosphate of thiamin. The two steps are carried out by the same phosphatase (optimal *p*H 3.7). This phosphatase, by destroying the cocarboxylase, stops the action of the latter on pyruvic acid. The action of this phosphatase is inhibited by the addition of either thiamin or the pyrimidine component. Hence the apparent stimulation of thiamin on cocarboxylase is in reality an inhibition of the enzyme destroying the cocarboxylase.

It is beyond the scope of this chapter to discuss in detail the enzymological aspects of the thiamin problem. It is of interest to us, however, to know the chemical mechanism of the action of thiamin demonstrated by means of plants. Only the microorganisms enter into this aspect of the subject.

The first demonstration was made by means of *yeast.* In this organism cocarboxylase (pyrophosphate of thiamin) accelerates the decarboxylation of pyruvic acid, the criterion for decarboxylation being the volume of CO_2 released (LOHMANN and SCHUSTER).

HILLS, working with *Staphylococcus aureus,* studied the mechanism of action of thiamin on a medium based on pyruvic acid. The volume of oxygen absorbed was measured. In the absence of thiamin oxygen consumption was small. No appreciable increase was obtained with a suboptimal vitamin dosage, but with an optimal dosage the consumption was much increased. The combination, pyrimidine + thiazole, has the same effect as thiamin. When these thiamin components were employed, a certain lag period was observed; this is probably due to a preliminary synthesis of the molecule of thiamin. When used singly, the components of thiamin were found to be inactive.

According to KREBS, *Staphylococcus aureus* growing on a medium based on pyruvic acid under anaerobic conditions produces acetic and lactic acid in the following manner:

2 molecules of pyruvic acid + H_2O → acetic acid + lactic acid + CO_2.

The acetic acid is transformed into acetoacetic acid. Pyruvic acid may also undergo a dismutation which results in β-hydroxybutyric acid, lactic acid, and CO_2 as follows:

3 molecules of pyruvic acid + H_2O → β-hydroxybutyric acid + lactic acid + $2CO_2$.

KREBS' explanation of the mechanism of action of *Staphylococcus aureus* and of the consumption of oxygen is based on the preceding observations according to which oxygen is utilized strictly for the oxidation of lactic acid.

SILVERMAN and WERKMAN (1939) working with three species of bacteria, *Propionibacterium pentosaceum, P. Petersonii,* and *Lactobacillus mannitopoeus* observed that the addition to thiamin markedly accelerates CO_2 production by cultures growing on a medium based on sodium pyruvate. An experiment of a somewhat different nature was performed by HAAG who cultured *Phycomyces Blakesleeanus* on a medium based on glucose and supplied with suboptimal doses of thiamin. Under these cultural conditions he observed an accumulation of pyruvic acid which can be detected colorimetrically.

All observations just described agree in one respect, namely, that thiamin, as a constituent of cocarboxylase, participates in the decarboxylation of pyruvic acid.

The reaction mechanism in *Bacillus Delbrückii* is somewhat different. LIPMANN (1937) observed that this bacterium does not bring about the decarboxylation of pyruvic acid into CO_2 and acetaldehyde. Instead an acetone extract of the organism causes a dehydrogenation of pyruvic acid. He believes that the dehydrogenase and the carboxylase utilize different protein carriers but that the active group in both cases is identical with cocarboxylase.

This short examination indicates the lack of complete accord concerning the mechanism of action. Apparently thiamin is a constituent of a cocarboxylase, of a dehydrogenase, and finally of an anaerobic dismutase (see R. R. WILLIAMS and SPIES, 1939).

The mechanism of action of cocarboxylase on pyruvic acid may be represented according to the following simplified scheme:

(a) action of cocarboxylase:

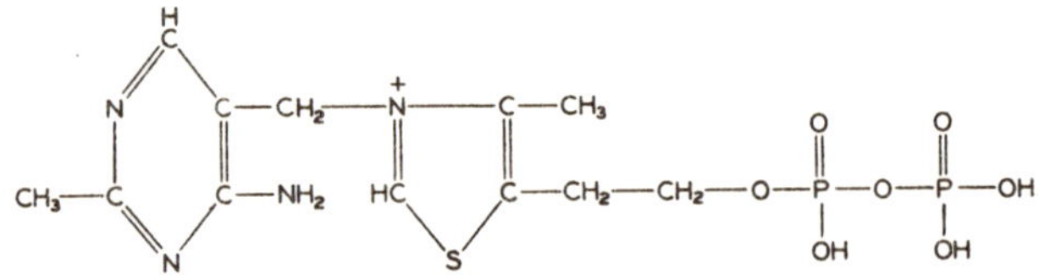

COCARBOXYLASE

The process of decarboxylation requires the presence of the NH_2 group in the fourth position of the pyrimidine portion of the molecule. The formula of the molecule of cocarboxylase may be abbreviated as R——NH_2, in which R represents the remainder of the vitamin.

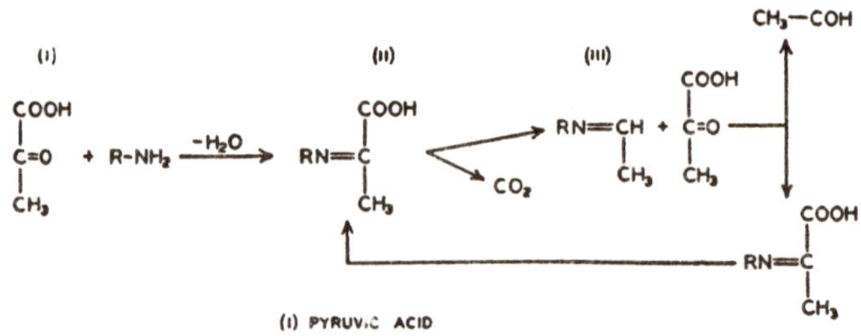

(I) PYRUVIC ACID

(II) KETOIMIDE OF PYRUVIC ACID

(III) ACID DERIVATIVE OF ACETALDEHYDE

The study of the specificity of action of thiamin on various microorganisms has shown that the free NH_2 group is fundamental to the process illustrated above and that all substitutions for this group destroy the vitamin action. A derivative with OH in place of NH_2 produced by GREWE is completely inactive on several of our microorganisms (SCHOPFER and BLUMER).

The decarboxylation process begins with the production of a ketoimide of pyruvic acid which yields CO_2 and an acid derivative of acetaldehyde; the latter reacts with another molecule of pyruvic acid producing a molecule of ketoimide and one of acetaldehyde, and the process is repeated again and again.

(b) action of a dehydrogenase:

When thiamin acts as a dehydrogenase the essential part of the molecule is the H in the second position of the thiazole portion of the molecule (LIPMANN, 1937; LIPMANN and PERLEMANN, 1938). The reduction takes place at the nitrogen atom of thiazole as follows:

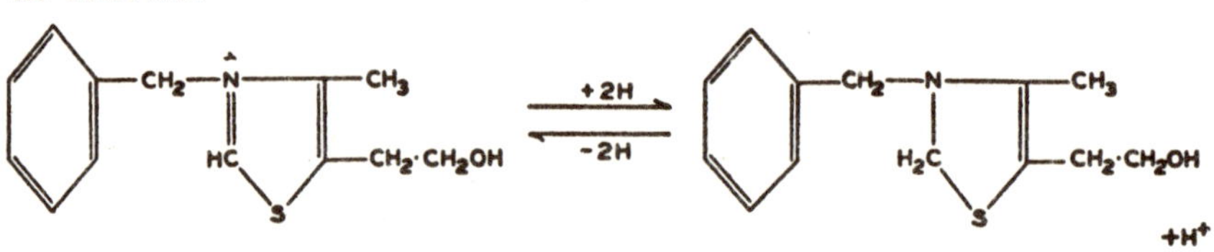

THIAMIN DIHYDROTHIAMIN

In this phenomenon, the hydrogen atom in the second position of thiazole, or more particularly the nitrogen in the third position, is of prime importance. The author, from an extensive study of the specificity of action of thiazole, has demonstrated experimentally that all substitutions for the H destroy the vitamin activity. Only the heterotrophic flagellates of LWOFF tolerate the presence of a CH_3 group in place of H. It must be assumed that these organisms possess the ability to demethylate the thiazole and restore it to its active form.

It is entirely possible to conceive a mechanism with a simultaneous occurrence of two functions, *e.g.*, a reduction and decarboxylation of pyruvic acid and phosphopyruvic acid, respectively. The relationships existing among the components of carboxylase (cocarboxylase, apocarboxylase and magnesium) and pyruvic acid may be demonstrated as follows:

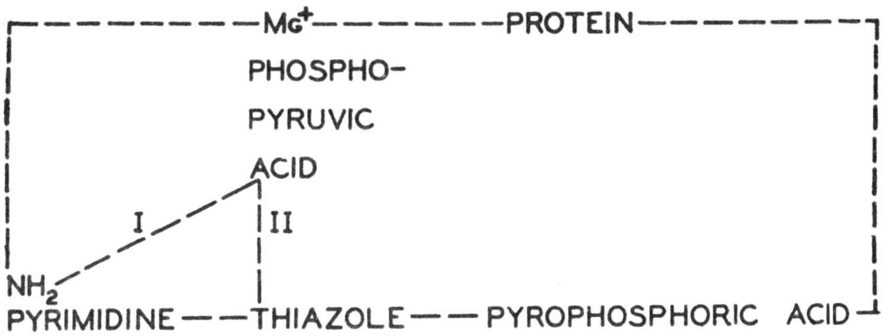

The following reactions may occur: I. a hydrogenation of pyruvic acid, and union with the nitrogen of the pyrimidine, II. *a reduction of the nitrogen of the thiazole* and a simultaneous oxidation and *a decarboxylation of the pyruvic acid,* followed by a reoxidation of the nitrogen of the thiazole by a dehydrogenase system. A. JUNG has proposed a scheme in which the action of cocarboxylase is related to a cycle of oxidation and phosphorylation (fumaric acid - pyruvic acid - diphosphopyridinonucleotide - adenylic acid - Mg). To discuss this interesting concept here would involve too many details (see JUNG, 1940).

In summarizing, we may say that thiamin is an essential constituent of coenzymes; that the loss of the ability to *synthesize thiamin means that vital enzymatic reactions* cannot be carried out unless this vitamin is supplied as an exogenous growth factor.

Riboflavin. — This flavin is found in all animal and plant tissues and occurs in a number of different forms. It occurs in the free state and also in combination with various specific proteins. One such combination is represented by the "yellow enzyme" in which the riboflavin is bound to phosphoric acid and a protein. This enzyme functions as a hydrogen carrier in the biological oxidation of glucose (THEORELL, 1937). The hydrogenation is carried out by the nitrogen atoms in the first and tenth positions of the isoalloxazine portion of the enzyme.

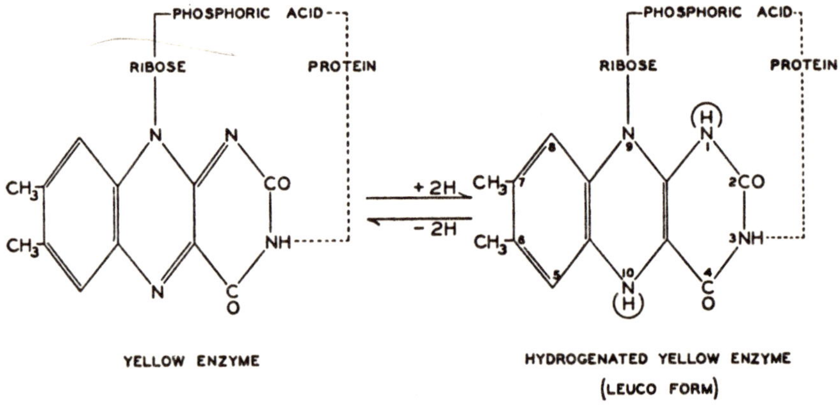

The "yellow enzyme" is transformed by hydrogenation into its leuco form (leuco-riboflavin) which can again be dehydrogenated, the reaction being reversible. The activated hydrogen of the reduced enzyme reacts with oxygen of the air or is transferred to another acceptor. Hence the enzyme functions as an "oxytrope" dehydrogenase.

The fact that riboflavin is required by microorganisms which have lost the capacity to synthesize it (lactic bacteria for example) is easily understandable in view of the importance of the function of the "yellow enzyme" in the transport of hydrogen.

Nicotinic acid and the codehydrogenases. —The amide of nicotinic acid is a component of the two codehydrogenases, coenzyme I and II which are constituted respectively by:

(I) nicotinamide - ribose - phosphoric acid (2 molecules) - ribose - adenine

(II) nicotinamide - ribose - phosphoric acid (3 molecules) - ribose - adenine.

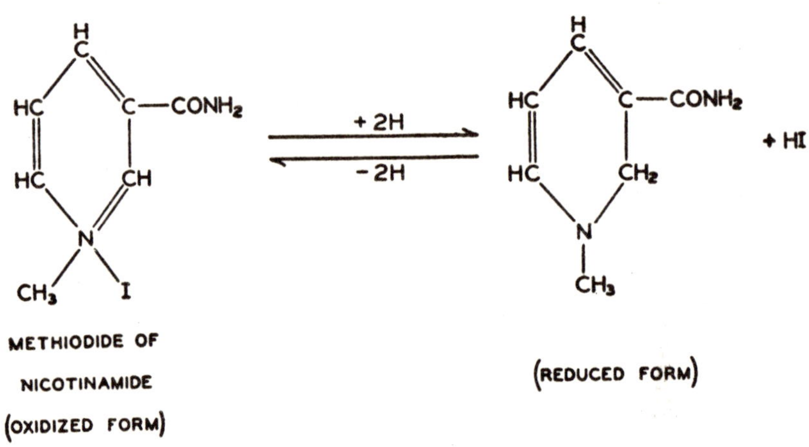

The nicotinamide is the active part of the coenzyme in that it acts as the carrier of hydrogen from one substrate to another. According to KARRER and BENZ all pyridines are able to undergo a reversible hydrogenation by virtue of their pentavalent atom of nitrogen and in this respect they are similar to thiazole.

We have stated, with certain reservations, that nicotinic acid is an essential growth factor in microorganisms and plants in general. This substance occurs in tissues as a constituent of the codehydrogenases (factor V). The importance of these factors is due to their function as coenzymes as in the case of the preceding vitamins.

Ascorbic acid. — We have already stated that it is doubtful whether ascorbic acid functions as a true growth factor. One of its essential functions consists in the establishment of a favorable redox potential of the culture medium. The action is of this nature in all cases in which the organism requires the vitamin in large dosage.

Plants contain an oxidase capable of oxidizing ascorbic acid into dehydroascorbic acid. This oxidase is probably constituted by a copper-protein complex (SZENT-GYÖRGYI). We are not fully informed, however, concerning the rôle played by ascorbic acid in biological oxidations (*cf.* TAUBER 1938, and SZENT-GYÖRGYI, 1939-1940; CARROLL, 1943).

Vitamin K. — The chemical mechanism of this recently discovered vitamin is not yet well known. This problem has been studied by E. L. MCCAWLEY and C. GURCHOT working with vitamin K (natural and synthetic forms), 2-methyl-1, 4-naphthoquinone and phthiocol (2-methyl-3-hydroxy-1,4-naphthoquinone), all of which have vitamin K activity. They state that the hydroquinone form is rapidly oxidized by molecular oxygen and suggest that this vitamin acts as a catalyzer of oxidation-reduction. It is assumed that this vitamin takes part in the synthesis of prothrombin since it influences the prothrombin concentration. It exerts, through its side chain, a favorable action on the activity of *cathepsin* of the liver. The relations between the chemical mechanism of action of vitamin K and the two other above named substances are not known. The suggestion has been advanced by BERNHEIM and BERNHEIM (1940) that the vitamin participates in an oxidation-reduction system involving the oxidation of SH groups to -S-S-linkages. Another hypothesis of similar nature was proposed by BAUMBERGER (1941), according to which the formation of the blood clot involves an oxidation of SH groups of fibrinogen to —S—S— groups in fibrin.

Hemin. — This substance is also known as factor X. According to present knowledge, it is required as a growth factor only by some of the hemophilic organisms (A. and M. LWOFF, 1937*b*).

The hemes (with Fe^{++}) are constituents of reduced cytochrome; the hemins (Fe^{+++}) are constituents of oxidized cytochrome, of catalase, and of peroxidase. Reduced cytochrome is oxidized by a respiratory enzyme, cytochrome-oxidase or indophenol-oxidase (*cf.* KEILIN, 1933).

The absence of hemin, or of protohemin or the inability to synthesize these substances, which are widely distributed in all aerobic organisms, must therefore result in trouble of a fundamental nature in the course of oxidation. The situation may be remedied and growth permitted to occur by the addition of hemin or of protohemin as growth factors.

Vitamins in the enzymatic systems. — We shall now establish the rôle which growth factors play in the transport of hydrogen. In this discussion we shall consider only the substances which we have recognized as necessary growth factors in one case or another.

The codehydrogenases (factor V) (A. and M. LWOFF 1937*a*) receive hydrogen from the substrate and are transformed into dihydrocodehydrogenases. The latter are stabile in air and are not oxidized directly, but they are dehydrogenated to the codehydrogenases by the "yellow enzyme" which functions as the hydrogen acceptor and is transformed into the leuco-form which in turn gives up its hydrogen to the oxygen of the air or to various other acceptors and is thus reoxidized. The leuco-form of the "yellow enzyme" can also be rapidly dehydrogenated by cytochrome C which is again oxidized by the cytochrome oxidase (indophenol oxidase). The dehydrogenase nucleopyridine system and the cytochrome system require another substance, acting as a "carrier" or "go-between" represented by diaphorase. The latter is a dinucleotide based on alloxazine and adenine. Thus the hydrogen taken from the substrate by the codehydrogenases and transferred to various acceptors finally comes in contact with oxygen. These reactions may be represented in the following manner:

$$H_2 \rightarrow \text{substrate} \rightarrow \text{dehydrogenases} — \text{codehydrogenases} \rightarrow \text{yellow enzyme} \rightarrow \text{cytochrome} \rightarrow O_2.$$

Each of these steps, through which the hydrogen passes, involves a growth factor of vitamin nature. If any link in the chain of substances is lacking, the organism is unable to grow unless a substitute can be found. If factor V (codehydrogenase), the first link, is lacking, access to hydrogen will be closed under aerobic and also under anaerobic conditions. If factor V is present, but Factor X (the hemin component of cytochrome) is absent, there still remains an open course because the yellow enzyme can transfer its hydrogen to an acceptor other than cytochrome.

The above example clearly illustrates the real significance of vitamin factors and makes it easy to understand why such substances are absolutely necessary.

The organism can be compared with an extremely complex mechanism composed of an infinite number of gear wheels. If one of these wheels is missing or if it is unable to function, and if it cannot be replaced, the entire mechanism stops. If one pauses to reflect concerning the mechanism of action of this growth factor or of this gearwheel he is amazed at the precision of the mechanism. When the jeweler has to insert a delicate wheel in a watch he must put it in the proper place, the only one which will permit the mechanism to start again.

The vitamin factor which acts often in such small dosages that only a few molecules are required for each cell can not be absorbed in a haphazard fashion by the living material. It must be placed in a definite locus in the extremely complex structure of the cytoplasm, *i.e.*, where it can be integrated normally in this structure. We know that *Phycomyces* adsorbs thiamin with considerable force. The thallus is able to extract thiamin from the substrate on which it is adsorbed. Yeast likewise actively adsorbs biotin (KÖGL). This property was utilized by NIELSEN in separating a growth factor of yeast from a complex medium (biological adsorption). The vitamin molecule, once adsorbed and absorbed, must find by selection the frequently circumscribed region (Haftbezirk of KÖGL) in which its action will be manifested. At first thought it seems striking that living matter is represented as an extraordinarily complex structure in which the biologist can change nothing. The living cell, on the other hand, is able to absorb the few molecules which it lacks and to move them to the proper place.

We are now able to return to our first definition of a vitamin growth factor, *viz.*, that it is *an extremely active substance of organic nature, required by an organism because the latter has lost the capacity to synthesize it.* An additional characteristic can now be added as the result of our study of the mechanism of action, namely that *it functions as a coenzyme or a fragment of a coenzyme.* This new criterion makes it understandable why the substance is extremely active (active in very small dosage). It is due to the manner in which a coenzyme acts (SCHOPFER, 1939).

This definition applies to the principal growth factors which we have studied: thiamin, riboflavin, nicotinic acid, the codehydrogenases, ascorbic acid and perhaps carotene. No precise information is yet available regarding the function of biotin or pantothenic acid.

As we explained the mechanisms involved in the loss of capacity for synthesis of growth factors and the variations of this phenomenon we held that a relationship exists with the adaptive enzymes from which the organism can "learn" to synthesize the necessary enzyme. The loss of the capacity to synthesize an indispensable enzyme, coenzyme or fragment of a coenzyme always causes an organism to become dependent on that substance as a growth factor. Experiments will show whether these views are correct. If an organism lacks a coenzyme, or synthesizes the latter too slowly, an addition of the enzyme or of the coenzyme furnished by another organism should be effective by virtue of the growth factor action.

Our study started with growth factors but led to enzymes and coenzymes. It is possible to proceed in the opposite direction and, starting with enzymes and coenzymes, we may discover new growth factors. There are a number of microorganisms which can not yet be cultured, even by employing all the known growth factors. Apparently unknown factors exist which are members of other enzyme systems. By taking into account the fact that a growth factor is a coenzyme, it may be possible to determine by trial and error whether various unsuspected coenzymes may eventually function as growth factors.

On a theoretical basis we can exclude from the group of true growth factors all substances which are not normal constituents of organisms even if they stimulate growth when used in small dosage. A stimulant has nothing in common with a true growth factor.

In the course of this discussion we have mentioned a few of the enzyme systems and have considered one of the possible paths of hydrogen. This brings up the question of what takes place in an organisms whose growth involves a great number of reactions catalyzed by many enzymes. Why do all organisms regardless of their systematic position in the plant or animal kingdom require approximately the same vitamin factors when the ability to synthesize so many others might equally well have disappeared? The question remains unanswered. The assumption might be made that the needs which became apparent are those which pertain to substances which are absolutely indispensable and irreplaceable. The inability of organisms to synthesize their essential vitamins (auxo-heterotrophism) appears to be a general characteristic of living matter, irrespective of the systematic position of the organisms.

REFERENCES:-

1933: D. KEILIN. — Cytochrome and intracellular respiratory enzymes. Erg. Enzymforschung 2:239.

1937*a*: A. and M. LWOFF. — Studies on codehydrogenases. Proc. Roy. Soc. London B, 122:352.

1937*b*: A. and M. LWOFF. — Rôle physiologique de l'hémine pour *Hemophilus parainfluenzae.* Ann. Inst. Pasteur 59:129.

1937: K. LOHMANN und PH. SCHUSTER. — Untersuchungen über die Cocarboxylase. Biochem. Z. 294:188.

1937: F. LIPMANN. — Die Dehydrierung der Brenztraubensäure. Enzymologia 4:65.

1937: H. THEORELL. — Das gelbe Ferment: seine Chemie und seine Wirkungen. Erg. Enzymforschung 6:111.

1938: F. LIPMANN and G. PERLEMANN. — Hydrogenation of vitamin B_1 and other quaternary thiazoles. J. Amer. Chem. Soc. 60:2574.

1938: G. M. HILLS. — Aneurin (vitamin B_1) and pyruvate metabolism by *Staphylococcus aureus.* Biochem. J. 32:383.

1938: H. TAUBER. — The interrelation of ascorbic acid (vitamin C) with enzymes. Erg. Enzymforschung 7:301.

1939: R. R. WILLIAMS and T. D. SPIES. — Vitamin B_1 (Thiamin) and its use in medicine. New York, The Macmillan Company, p. 331-355.

1939: M. SILVERMAN and C. H. WERKMAN. — Thiamin effects in bacterial metabolism. Iowa State College Journal of Science 13:365.

1939: W. H. SCHOPFER. — Vitamine and Wachstumsfaktoren bei den Mikroorganismen, mit besonderer Berücksichtigung des Vitamins B_1. Erg. Biologie 16:1-172.

1939: A. SZENT-GYÖRGYI. — Les oxydations biologiques au niveau cellulaire. Exposés annuels de Biochimie médicale, 2nd series, p. 1.

1940: A. JUNG. — Die Funktionen der Vitamine des B-Komplexes im Organismus. Zeitsch. f. Vitaminforschung, supplement no. 1, p. 1-114.

1940: H. G. K. WESTENBRINK, D. A. VAN DORP, M. GRUBER and H. VELDMAN. — Aneurinphosphates and yeast phosphatase. Enzymologia 9:73.

1940: A. SZENT-GYÖRGYI. — Lectures on Oxydation, Fermentation, Vitamins, Health and Diseases. 120 pp.

1940: F. BERNHEIM and M. S. C. BERNHEIM. — Action of 4-amino-2-methylnaphthol on the oxidation of certain sulfyhydryl groups. J. Biol. Chem. 134:457-458.

1941: J. P. BAUMBERGER. — Proc. Am. Physiol. Soc. 1941:18.

1943: G. H. CARROLL. — The rôle of ascorbic acid in plant nutrition. Bot. Rev. 9:41.

Chapter XIX.

THE CAPACITY FOR SYNTHESIS AND THE EFFECT OF ITS LOSS ON THE ORGANISM.

The discussion in many of the preceding chapters has been guided by a single general idea, *viz.*, that the need for vitamins is determined by a loss of capacity for synthesis. We have shown how the higher plant synthesizes its necessary vitamins and how it utilizes them. We have shown also how a partial heterotrophism may become established in the higher green plant. And finally we have pointed out the fact that the vitamin requirements with few exceptions, are the same in microorganisms as in the higher animals. It is possible with each group of organisms to demonstrate experimentally that a loss of capacity for synthesis is responsible for the need for an exogenous supply of vitamins. The object of the present chapter is to analyze this loss of capacity for synthesis in the light of the general facts established in the preceding chapters.

Conditioning and relativity of the capacity for synthesis. — The general assumption is that an organism's capacity for synthesis is very stable, and that during the course of a series of experiments this characteristic remains rather constant. This assumption is supported by the work of LASSEN who cultured a strain of *Saccharomyces cerevisiae* for a year and a half on a thiamin-free synthetic medium during which time the organisms capacity for synthesis did not change. Another assumption is that when a loss of capacity for synthesis once appears in a species, this feature remains definitely constant. This is demonstrated by experiments with *Ustilago violacea,* a parasite of *Melandrium album* and *M. rubrum,* requiring thiamin or its two components (pyrimidine + thiazole) as growth factor or factors. S. BLUMER (1940) made a careful study of this smut in which he investigated nine varieties differing morphologically and in their parasitic habitats. He found that all of his 215 single-cell cultures exhibited identical growth factor requirements. The loss in capacity for synthesis is therefore a matter of a genotypically determined character which involves the entire species and not just a strain.

Despite the apparent constancy in this character, variations in the capacity to synthesize vitamins can be produced experimentally.

BEADLE and TATUM (1941, 1942), by X-raying cultures of *Neurospora* prior to meiosis, induced changes in the ability to synthesize vitamins. Among the mutants obtained in this way are: (1) one unable to synthesize pyridoxine, (2) one unable to make the thiazole half of the thiamin molecule and (3) one unable to make *para*-aminobenzoic acid. Each of these differs from the normal by a single gene, and is distinguishable from the normal only by its inability to synthesize a certain vitamin. These facts are consistent with the assumption that each of the genes involved is concerned with the control of one and only one specific chemical reaction. Other mutants not yet investigated by these workers as regards inheritance are characterized by loss of ability to synthesize a growth factor (contained in yeast extract) different from any vitamin known to the authors, loss of ability to utilize fat as a carbon source, and in a number of other ways.

The ideal organism for study is one in which the loss of capacity for synthesis is conditioned genotypically. The deletion or the suppression of the gene bearing this character for synthesis prevents a microorganism from growing on a medium lacking the necessary growth factor. And when growth and cell division are *completely* suppressed the organism dies. Sometimes the organism is able to develop slightly during the first passage on the vitamin-free medium at the expense of growth factor reserves carried by the inoculum from the preceding culture, but by the time it has undergone one or two passages and has used up all its reserves it dies because the capacity for synthesis is completely lacking.

In another case the situation may be different. For example, the red yeast *Rhodotorula rubra* is able to grow to a limited extent on a synthetic vitamin-free medium even when inoculated very lightly, *i.e.*, the controls are not absolutely nil. After several passages of the organism on a vitamin-free medium its reserves become exhausted but the organism still continues to produce a weak growth. Evidently the capacity for synthesis on the medium employed has not been completely lost. A. LWOFF (personal communication) observes that during the course of numerous passages (36 passages in the course of 14 months) of the Lister strain No. 4585, the growth of the culture improves progressively. After a year's "training", cultures of this organism produce three or four times greater turbidity in the medium than that produced by the initial culture. This demonstrates that the capacity for synthesis, which was very weak, has increased. Similar observations on several organisms have been made in the author's laboratory.

A distinction must be made, at least theoretically, between a complete loss of capacity to synthesize a growth factor and a *very marked reduction* of this function. This is a difficult task, even after having performed a great number of experiments. We shall now consider the various possible cases.

The augmentation of a limited capacity for synthesis (*training*). — The case of *Rhodotorula rubra* as communicated by A. LWOFF has already been mentioned in this chapter. This organism, after a long period of "training" on a vitamin-free synthetic medium, develops better than at first. Other cases of this kind are known. *Proprionibacterium pentosaceum* requires among various unidentified factors, thiamin but "learns" to do without it. In other words, the bacterium has "learned" again to synthesize this vitamin (SILVERMAN and WERKMAN, 1939).

A similar case in connection with the amino acids is that of *Staphylococcus aureus*. This organism can be cultured on a synthetic medium providing a complex group of amino acids is supplied. As pointed out previously (see p. 146) the amino acids do not act as true growth factors according to our definition. The point which we want to make at this time is that this bacterium, after a period of adaptation (training), no longer requires these amino acids as constituents of the medium (GLADSTONE, 1937), because it has "learned" again to synthesize them from ammonium. The organism has become autotrophic in regard to this phase of its nutrition.

Reappearance of the (apparently lost) ability to synthesize as a result of a quantitative variation in the medium. — *Pythium Butleri,* under certain cultural conditions, requires pyrimidine (of thiamin) as an indispensable growth factor, but it can be "trained" to do without this factor by reducing the mineral salt content from 16.4 to 1.64 grams per liter. The synthesis of this factor is inhibited by some unknown mechanism when the mineral salt content is too high. The possibility still exists that the fungus may be "trained" by a gradual adaptation to accomplish the synthesis of pyrimidine in the presence of a strong concentration of mineral salts (ROBBINS and KAVANAGH, 1938). Riboflavin likewise is formed by *Aspergillus niger* only when the medium contains little or no magnesium (LAVOLLAY and LABORAY). *Rhodotorula Sanniei,* studied by FROMAGEOT and TCHANG (1938), differs little from *Rhodotorula rubra* and likewise requires pyrimidine as a growth factor when cultured on a medium based on glucose. If this carbon source is replaced by glycerine (purified by distillation) the former red yeast grows normally and becomes pigmented in the absence of growth factors. The capacity for synthesis, which was apparently lost, has been reestablished.

Such cases make it obvious that caution must be exercised in the interpretation of observations concerning the disappearance of the capacity for synthesis. The statement that an organism is heterotrophic is valid only for the cultural conditions given and only when these conditions are under careful physico-chemical control. For a given organism the number of cultural conditions is enormous. Before it can be determined with certainty that heterotrophism is complete and genetically controlled, it is necessary to exhaust all possible cultural conditions.

A good set of rules to follow in dealing with this type of problem is: for each apparently heterotrophic organism there exists probably *one* cultural condition in which the organism is auxo-autotrophic; this condition must be found; exhaustive investigations must be made in which all possible cultural conditions are tested and in which careful consideration is given to all physico-chemical characteristics of the medium, to all quantitative and qualitative variations in the ingredients of the medium, and also to various combinations of all these characteristics.

Although the cases just mentioned involved microorganisms, the same rules hold equally well for the higher plants.

This discussion leads us readily to a consideration of another biological problem, namely, race. A certain strain can be completely heterotrophic for a certain vitamin whereas another strain is only partially heterotrophic or is entirely autotrophic. For example, the embryos of certain races of *Pisum* require more ascorbic acid than others, *i.e.,* they differ in their capacity for synthesis.

The cause of variations in the capacity for synthesis as a function of the medium. — It is difficult to discuss this subject since the chemistry of the biosynthesis of growth factors is poorly known. The requisites for the biosynthesis of a growth factor with a complex molecular structure are: (1) that the precursors be present or capable of being synthesized, and (2) that the enzyme systems required for these syntheses be present and active. An unfavorable medium may prevent the synthesis of the precursors, and also prevent the synthesis of enzymes or bring about their inactivation. Any changes in the medium which remove these inhibitions and obstructions will help render the microorganism autotrophic with respect to growth factors.

If an organism requires the complete molecule of thiamin and cannot utilize the combination pyrimidine + thiazole, it is because the enzyme necessary for the condensation of these components is absent or is unable to function. It may be possible to find a cultural condition which permits the enzyme to act, thus changing the organism from the type requiring thiamin to the type requiring pyrimidine + thiazole.

Analogy with the adaptive enzymes. — An analogy with the above mentioned enzyme category seems to exist; the resemblance, however, may be superficial. An organism possesses certain enzymes which remain constant and which are present irrespective of the cultural condition. The same organism, on the other hand, may possess other enzymes of the adaptive type formed by a more or less prolonged adaptation and originating only on a specific substrate under a definite set of cultural conditions. This phenomenon is well illustrated by the case of *Bacterium aerogenes* studied by KARSTRÖM. This bacterium which lives in cream and possesses the enzymes necessary for the utilization of lactose and its hydrolytic products is normally incapable of fermenting xylose. If, however, ammonium sulphate and yeast water are added to the xylose, the bacteria become adapted to the medium and multiply in it. The enzymes necessary for the utilization of xylose have been formed by virtue of the nitrogen source present (see KARSTRÖM, 1938). A similar observation has been made concerning *Bact. coli* by STEPHENSON and STICKLAND. They found this organism unable to grow in a solution of pure formate unless bouillon is added. The latter permits the formation of a hydrolytic enzyme necessary for cell multiplication. Despite the fact that the origin of enzymes is still shrouded in obscurity, these cases in which enzymes are formed under the influence of the medium are comparable with situations in which the biosynthesis of growth factors is dependent on the medium. The latter probably controls the mechanism by which growth factors are produced. This possibility is strengthened by the fact that the majority of the true growth factors are coenzymes or fragments of coenzymes.

Among the so-called adaptive enzymes, KARSTRÖM distinguishes several groups: (1) those which are adaptive in a general way, *i.e.*, their formation is dependent on the general composition of the nutrient medium, (2) those which are typically adaptive (*sensu stricto*), *i.e.*, their formation is dependent on the presence of a specific substrate, (3) those which are not typically adaptive, which exist in small quantity, and whose formation is strongly augmented by the presence of a specific substrate. The first group of enzymes is undoubtedly analogous with those growth factors whose formation is dependent on the general composition of the nutrient medium. An analogy also exists between the third group and certain growth factors whose normal synthesis is limited but is capable of considerable increase on a specific substrate.

It is not yet possible to compare growth factors and enzymes in every respect. Once the mechanism of the biosynthesis of growth factors is better understood the validity of the analogy proposed here will appear in a different light.

These ideas of the author were developed on the basis of the assumption that, when an auxo-heterotrophic organism is shown to become auxo-autotrophic following a change in the medium, it has regained its ability to synthesize the necessary factors. This raises the question of whether the organism, under the influence of a new set of cultural conditions, has not become adapted to doing without the growth factor which is no longer synthesized. This possibility should be foreseen and taken into consideration but it appears hardly probable. Nevertheless it has been established in the case of the chick that an adaptation to a lack of vitamin B_1 may develop under certain conditions (LANCZOS, 1939). The mechanism of this adaptation is not known.

It is easy, moreover, to find out by biological analysis whether the organism really contains the growth factor or is actually getting along without it. We are assuming that we are dealing with the usual growth factors required by living matter, present in the most diverse kinds of cells, and participating in the same functions irrespective of the kind of organism. This fundamental principle which forms the basis of our discussion in this book permits us to eliminate the hypothesis of a lack of usefulness of the factor which is no longer synthesized.

The author wishes to emphasize the importance of basing experiments on the fact that a growth factor required by the organism is one which can no longer be synthesized by it. This line of reasoning, which is well established experimentally, is valid for several vitamins functioning as growth factors. Theoretically a vitamin which is ineffective when supplied to an organism is one which normally is present by virtue of synthesis. Thiamin is necessary for *Phycomyces* because it cannot be synthesized. Pyridoxine has no effect when supplied to this fungus because the thallus contains an abundance of this factor (JUNG and SCHOPFER, 1939). There are certain exceptions to this principle. For example vitamin E (alpha-tocopherol) has no action on *Phycomyces* but it is not present in the thallus as shown by the potentiometric analysis with gold chloride (SCHOPFER and BLUMER, 1939).

This discussion of the variations in auxo-heterotrophism may be summarized by listing the various causes responsible for the inability of plants to synthesize vitamins:

A. *internal conditions* — synthesis normally lacking or strongly inhibited on all media employed, *i.e.*, the condition is inherent in the organism.

B. *external conditions*
   (1) *conditions of the medium*
      (a) quantitative variations in the constitution of the medium (*Pythium Butleri*).
      (b) qualitative variations in the constitution of the medium (*Rhodotorula Sanniei*).
   (2) *time factor*
   The synthesis may be slowed up to such an extent as to suggest case A. Although this may actually be true, the external factors of the medium are usually responsible. It is sometimes difficult to distinguish between the type induced by the medium and that conditioned by an internal factor.

By applying the facts to a more general scheme we are able to interpret them theoretically according to the outline given below. In this reclassification of conditions influencing the synthesis of growth factors, we assume that the production of a vitamin factor is dependent upon one or several genes which act quantitatively. By a chain of chemical phenomena and reactions a gene leads to the formation of a few molecules of the indispensable vitamin factor. The following possibilities exist.

(1) *The gene is absent* and no synthesis of the growth factor is possible under any cultural conditions. This represents the ideal case of auxo-heterotrophism. When an organism is classified in this category, it is necessarily assumed that it cannot be cultured in the absence of one or more vitamins.

(2) *The gene is inhibited in its action.*

(3) *An inhibition occurs at some point in the chain of reactions* leading to the formation of the vitamin factor. If the inhibition is not complete, a small quantity of growth factor may be formed. This inhibition may be dependent on the medium (external factor) in which case a modification of the medium permits the organism to synthesize its necessary growth factors. If this inhibition is complete, it leads to practically the same result as in

case 1 in which the gene is entirely absent. To distinguish between cases (1) and (3) is very difficult and is possible only after prolonged experimentation.

This discussion remains purely theoretical because no consideration was given to the phenomena of replacement according to which a factor that is suboptimal or absent can be replaced by a supraoptimal quantity of another factor. Practically nothing is known concerning the mechanism of these phenomena.

The classification of organisms with respect to thiamin proposed by FRIES (1938) coincides perfectly with that of the author. FRIES recognizes four groups: (1) organisms not requiring an external supply of thiamin, *Aspergillus niger*. This fungus belongs to the author's auxo-autotrophic group. (2) Organisms dependent on an exogenous source for part of their thiamin requirements, *e.g., Sclerotinia cinerea, Penicillium glaucum,* etc. In these organisms the synthesis is diminished due to an unknown reason. Organisms of this type are able to produce a sparse growth on a vitamin-free medium but they are favored by the addition of thiamin. The time factor enters and retards synthesis. Such retarded syntheses have actually been observed by KNIGHT in *Staphylococcus.* (3) Organisms under some conditions partially, under other conditions wholly dependent on thiamin, the synthesis of this factor being conditioned by external or internal factors. (4) Organisms completely heterotrophic in respect to thiamin, *i.e.,* the capacity for synthesizing this factor has been entirely lost, *e.g., Phycomyces Blakesleeanus, Polyporus adustus.*

Auxo-heterotrophism and heterotrophism in general. Polyphyletic nature of auxo-heterotrophism. — All the evidence indicates that heterotrophism for growth factors is independent of heterotrophism for carbohydrates. The need for a certain growth factor may appear in organisms which are only in part heterotrophic for carbohydrates. Conversely, organisms in which heterotrophism for carbohydrates is more pronounced are often self-sufficient with respect to vitamins. Furthermore there are cases in which one species of saprophytic microorganism is entirely dependent upon thiamin as a growth factor whereas another related species, whose metabolism and nutritive requirements are very similar, is able to synthesize all of its necessary thiamin. For example, in the *Mucoraceae, Phycomyces Blakesleeanus* requires thiamin or its components pyrimidine and thiazole, whereas *Mucor mucedo* is able to dispense entirely with this factor.

On the other hand, heterotrophism for a certain factor, thiamin for example, appears to be entirely independent of the systematic position of the species and of phylogeny. At various phylogenetic levels diverse organisms, both animal and plant, can be found which require this factor. The bacteria, protozoans, phycomycetes, ascomycetes, basidiomycetes and seed plants have representatives which are dependent on exogenous sources of thiamin. In other words, the loss of capacity for synthesis and the needs which arise therefrom are identical in organisms which are widely separated systematically. It is therefore permissible to say that the origin of auxo-heterotrophism is polyphyletic. At all levels in the plant and animal kingdoms, the requirements of living material are identical and the losses in capacity for synthesis are similar.

Complexity of auxo-heterotrophism. Multiplicity of essential factors. — Certain organisms seem to be able to grow on a synthetic medium when only one growth factor is added, the growth obtained being equal to that obtained on the most favorable natural medium. *Phycomyces* seems to be of this type, with thiamin the only factor required. *B. proteus* likewise requires only one factor —nicotinic acid. Experiments with *Phycomyces* (see p. 106) show that other substances are necessary. They probably are not true growth factors (according to our definition), nevertheless they are indispensable.

Most auxo-heterotrophic organisms require several factors acting in unison and constituting what we might call a constellation. During the course of our discussion we have encountered all types of cases from the simplest one, in which (apparently) only one factor is involved, to the most complex in which a complex constellation is required as in the case of yeast and lactic bacteria. A general observation is that the more thorough the study of the organism in question is, the greater appears to be the complexity of the constellation of essential factors. By comparing the growth factor requirements of several organisms it is possible to establish a general rule concerning the extent to which losses of capacity for synthesis have occurred.

Thiamin seems to be required more frequently than any of the other growth factors. This is due to the fact that this vitamin has been studied more extensively than the others. None of the latter has been studied sufficiently (with enough organisms) to permit any definite conclusions concerning the relative importance of the various factors. It is actually impossible to say with certainty that one factor is required more often than others.

Certain factors occupy a special position in our scheme of classification. For example, pimelic acid is required only by *Corynebacterium diphtheriae;* hemin is needed only by certain hemophilic organisms, as far as is known at present; riboflavin is required chiefly by certain lactic bacteria; cholesterol, which is probably not a true growth factor, is required only by certain flagellates. The other factors, thiamin, pyridoxine, nicotinic acid, and the other bios substances are frequently required in various combinations by widely different organisms.

The following table gives some typical examples of the growth factor requirements of a number of organisms. At present, 17 chemically identified growth factors can be considered. Several of the organisms listed require additional factors which remain to be identified.

TABLE XVI. *Action of 17 chemically known growth factors on various microorganisms (modified from* SCHOPFER, 1939):-

	Pyrimidine	Thiazole	Thiamin	Inositol (Bios I)	Biotin (Bios IIb)	β-Alanine (Bios IIa)	Pantothenic acid	Riboflavin	Nicotinic acid	Cozymase	Adenine	Uracil	Pyridoxine (vitamin B_6)	Hemin	Pimelic acid	Ascorbic acid	Cholesterol	
a: *Rhodotorula rubra* (I)	x																	
b: *Mucor Ramannianus* (P)		x																
c: *Phycomyces Blakesleeanus* (P)	x	x																
d: *Phytophthora cinnamomi* (O)			x															I.
e: *Melanconium betulinum* (A)			x	x	x													
f: *Trichophyton interdigitale* (I)			x	x			x	x										
g: *B. proteus* (B)									x									
h: Lactic bacteria (B)*			x		x		x		x	x	x		x					
i: *Staphylococcus aureus* (B)**	x	x			x				x			x						
k: *Corynebacterium diphtheriae* (B)						x			(x)						(x)			IV.
l: *Saccharomyces cerevisiae* (A)			x	x	x	x	x						x					
m: *Hemophilus canis* (B)														x				
n: *Hemophilus parainfluenzae* (B)										x								
o: *Hemophilus influenzae* (B)										x				x				
p: *Strigomonas oncopelti* (F)			x															II.
q: *Strigomonas fasciculata* (F)			x											x				
r: *Schyzotrypanum cruzi* (F)														x		x		
s: *Trichomonas foetus* (F)																x	x	III.

I — Fungi Imperfecti; P — Phycomycete; O — Oomycete; A — Ascomycete; B — Bacteria; F — Flagellate. The parentheses in the columns under nicotinic acid and pimelic acid signify that the factor is not absolutely necessary and that it functions as a co-factor of growth.

*Lactic bacteria = *Bacterium acetylcholini.*

**Uracil is the only growth factor required by *Staphylococcus aureus* under anaerobic conditions.

This table indicates neither the relative activity of the factors nor the differences in the requirements of the various strains in each species. The requirements given for certain species are those of the average strain. It must be remembered, however, that various strains differ in their requirements; this is true particularly of the lactic bacteria.

It is impossible to establish groups which are clearly separated because they overlap. However, it is possible to distinguish four groups: one group requiring the simple bios substances - thiamin, inositol, β-alanine (group I in the table); another group with a complex constellation requiring at least one of the simple bios substances plus one or more of the following: Pyridoxine, pantothenic acid, and nicotinic acid (group IV in the table). *Corynebacterium diphtheriae* which requires pimelic acid, stands somewhat apart, although it is linked through β-alanine with the group requiring bios substances. Still another group requires hemin with or without other factors (group II in the table). The last group (group III in the table) requires cholesterol. It can be seen that there is overlapping between groups I and II because of the fact that *Strigomonas oncopelti* requires only thiamin and is able to synthesize hemin. Overlapping also occurs between group II (with hemin) and group III (with cholesterol) since there are organisms in both groups requiring ascorbic acid. (Fig. 15).

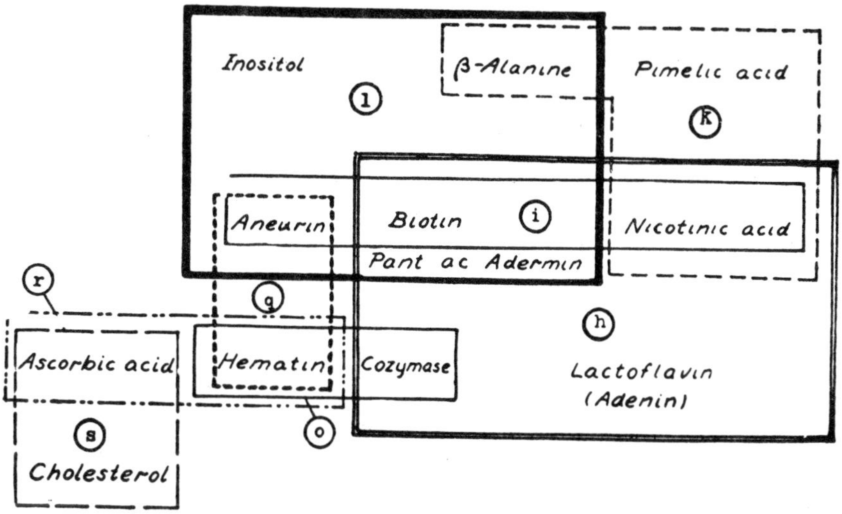

FIG. 15. — Relationships among the principal constellations of growth factors required by certain types of microorganisms (the encircled letters indicate organisms listed in table XVI; a=*Rhodotorula rubra*, b=*Mucor Ramannianus*, etc.

The names in this figure have the following American equivalents:
Hematin = hemin
Aneurin = thiamin
Pant ac = pantothenic acid
Adermin = pyridoxine
Lactoflavin = riboflavin
Adenin = adenine

A more logical scheme of classification would be one based on the functions of the growth factors and on their relationships with enzyme systems. On this basis we might establish three groups: (1) a group concerned with cocarboxylase (requiring thiamin); (2) a group concerned with codehydrogenases (requiring cozymase and nicotinic acid); and (3) a group concerned with cytochrome (requiring hemin). These groups likewise overlap because members of different groups may require other factors whose functions are still unknown and because of organisms belonging to two groups, *e.g.*, requiring both thiamin and nicotinic acid - *B. acetylcholini* (see Fig. 15 representing graphically the table on p. 206).

Subordination of the factors of a constellation. Co-factors. — When a constellation of factors is required, as a rule the factors do not act with equal intensity. Furthermore the activity of each factor may vary as shown in the following cases: (1) The factors when *used individually* have practically no effect; they act only when used in combination. The mechanism of this phenomenon of synergism escapes us completely. (2) On the other hand, a factor when used alone may have a marked effect but its action is increased by the presence of other factors. In the latter case we may speak of a subordination of factors, the principal factor being the one which acts when alone. This case also involves the criterion, "intensity of action". The principal factor is active even in very weak dosage. The other factors with lesser activity can be considered as co-factors. For example, the principal factor for the growth of yeast is unquestionably biotin because of its intensity of action; inositol is a co-factor and acts in higher dosage (this is true of the races studied). In the case of *Staphylococcus aureus,* biotin, which accelerates the effect of the combination thiamin + nicotinic acid by 600 per cent, is the principal factor whereas nicotinic acid, which is required in higher dosage, can be considered as a co-factor. In the case of *Corynebacterium diphtheriae,* beta-alanine is the essential factor because of its action in weak dosage; nicotinic acid and particularly pimelic acid are co-factors.

A factor may be the essential and the principal member of one constellation and secondary in another. This is not true of biotin, since it seems to be the essential factor in all constellations studied in which it is a member.

An accurate view of these phenomena will be possible only when complete information is available concerning the functions of each factor in each organism studied. Unfortunately this is not the case at present.

Replacement of factors. — When a more or less complex group of factors is required by an organism, it is sometimes found that reciprocal quantitative variations of two factors lead to the same amount of growth. This phenomenon is exemplified by *Staphylococcus aureus* which requires biotin, thiamin (or pyrimidine + thiazole) and nicotinic acid. The same effect is obtained with 5γ thiamin + 5γ nicotinic acid as with 0.05γ thiamin + 0.05γ nicotinic acid + 0.05γ *biotin.* In both cases the increase in growth is 670 per cent (KÖGL and v. WAGTENDONK, 1938). This behavior has two possible explanations; either the presence of a small quantity of biotin permits the synthesis of thiamin and nicotinic acid, or, conversely, the presence of a supraoptimal dosage of thiamin and nicotinic acid renders biotin superfluous or capable of being synthesized. By supplying a trace of biotin (suboptimal) it is possible to effect a saving of thiamin.

The same phenomenon is exhibited by *Ustilago violacea* and *U. scabiosae.* These fungi require pyrimidine and thiazole and grow very satisfactorily when supplied with an equimolar and optimal dosage of these factors, but they grow equally well when given suboptimal doses of pyrimidine and supraoptimal ones of thiazole, or vice versa. If we assume that the molecule of thiamin must be synthesized from pyrimidine and thiazole, it is necessary to assume that, directly or indirectly, the supraoptimal dose of pyrimidine favors the synthesis of thiazole which is deficient, or vice versa. It is even conceivable that there is a true replacement of one of the components by the other or a formation of the one at the expense of the other (BLUMER and SCHOPFER, 1940). Similar phenomena of replacement are found with *Trichophyton interdigitale* which requires thiamin, pantothenic acid, inositol and riboflavin. In this case, however, the replacements are not so easily observed (MOSHER, SAUNDERS, KINGERY and WILLIAMS, 1936). Another case in which a replacement is also possible is represented by yeast.

These phenomena of replacement are still poorly understood. They may be interpreted in various ways. One possible explanation is that the function carried out by the lacking factor has become superfluous or that this function can be accomplished by another factor. Another possibility is that of a transformation of the excess of the one factor into the other factor which is deficient. In the case of *Staphylococcus aureus,* either the excess thiamin performs the same function as the biotin would, if present, or a certain quantity of thiamin is transformed into biotin. It is impossible to formulate an opinion when no information is available concerning the chemical relationships existing between factors such as biotin and thiamin. We have already pointed out that the two factors are probably formed from the same precursor, but this has not been demonstrated experimentally (SCHOPFER, 1935; GREWE, 1937).

The possibility of a substitution of functions is not precluded. This is better appreciated when we consider the manner in which growth factors act. They function as coenzymes, and when one of these is absent, an enzymatic reaction is blocked. It is not impossible that when one course is closed, another opens by virtue of a related factor, and the reaction proceeds. These are merely speculations. This problem remains to be worked out experimentally.

The question of replacement of factors touches on that of specificity. This phenomenon of replacement operates in such a manner that it counteracts specificity. We have already pointed out that when compounds normally lacking vitamin activity are employed in concentrations far exceeding the optimum, the specificity of the vitamin action is decreased. From this we concluded that there occurs either a transformation of a normally inactive constituent into the specifically active substance, or a replacement of function. When the normally inactive substance is very closely related chemically to the active one, it can be assumed that a transformation from the former to the latter occurs. We have shown this to be the case with thiamin and its substitution products. When the structure of the normally inactive substance is very different (heterovitamins, for example) the explanation becomes difficult. A similar difficulty is encountered in the hormone field where substances which are very different chemically, such as auxin and heteroauxin, carry on the same functions.

A case of replacement involving excessive dosages was demonstrated by ROBBINS (1940) in his work on *Phycomyces.* This fungus requires pyrimidine and thiazole, and when supplied with

0.5γ of each component attains the same development as that reached with 1γ of thiamin. When 10γ of pyrimidine and 1000γ of beta-alanine are employed the same development is obtained as with 1γ of thiamin, the beta-alanine apparently replacing thiazole* (ROBBINS, 1940).

The author observed another case of replacement in *Ustilago violacea* which requires thiamin (or pyrimidine + thiazole). The maximal growth is attained under a given set of conditions, with 0.01γ of thiamin but the same development is obtained with 40γ of a heterovitamin, 2-methyl-3-α-hydroxyethyl-N-[2-methyl-4-amino-pyrimidyl-(5))-methyl]-pyridine (bromide). In this case thiazole is replaced by a *pyridine* (SCHOPFER, unpublished). Although we have not investigated *Phycomyces* concerning its response to these compounds, it is possible that this fungus might be able to transform beta-alanine (in supraoptimal dosage) into thiazole and also to transform the heterovitamin just mentioned into the active vitamin. Furthermore, the analogies existing between the *thiazole* group and the *pyridine* group are known (HANTZSCH, OCHIAI and NAGASAWA). In the two cases just described the replacement must be a transformation made possible by the large dosages employed.

In brief, replacement may be due to:

(1) A substitution of function. This is a theoretical case not actually demonstrated, but for which analogies in enzymology are found (*e.g.*, respiratory enzymes).

(2) a transformation of an inactive factor into another active factor. This case has been demonstrated experimentally (SCHOPFER). A prerequisite is that the inactive substance to be transformed be present in large dosage. This transformation is dependent on conditions of the medium (among other things the presence of mineral catalyzers) and is effected by enzymes.

Replacement can occur between:

(1) factors which differ greatly. This type is difficult to explain.

(2) a specifically active factor and its products of substitution, *i.e.*, homologues or analogues. This type can be demonstrated experimentally. Since this case is possible and explainable, the same can be assumed for case 1). The substitution of beta-alanine in place of thiazole for *Phycomyces* is of the same nature as that of thiamin in place of biotin.

It may be possible in both cases, but especially in the first, that one of the factors is the natural precursor of the other and that the two substances are related despite the dissimilarity in chemical structure.

Gradient leading to heterotrophism. — The steps leading to an ever increasing degree of heterotrophism can be observed by considering the various organisms belonging to the group requiring one or more bios substances. For example, *Phycomyces* requires thiamin; *Nematospora Gossypii* must have inositol, thiamin and biotin, while *Saccharomyces cerevisiae* needs inositol, biotin, thiamin, pantothenic acid and other factors. We recognize the fact that the losses of capacity for synthesis did not appear simultane-

*The root of the tomato is able to utilize a thiazole having, in position 5, a beta-alanine group in place of the beta-hydroxyethyl.

ously, but successively. The losses of capacity for synthesis, moreover, are dependent on the medium and external factors.

This stepwise progression toward heterotrophism can be shown still more clearly by limiting our description to a single substance such as thiamin. Accordingly the following stages may be recognized: loss of capacity to synthesize pyrimidine, thiazole or its components, both pyrimidine and thiazole, and finally the loss of the ability to condense pyrimidine and thiazole into a molecule of thiamin. The gradient leading to heterotrophism becomes even clearer when we consider the intermediate cases involving the partial loss of the capacity to synthesize thiazole (*Parasitella simplex*), thereby conditioning the utilization of pyrimidine (see p. 115). This partial loss of synthesis, bearing on the same substances in widely differing groups of plants, is a general characteristic of all living matter. It is possible to speak of a strictly physiological evolution leading to degeneration and heterotrophism (*cf.* chap. II) which can be followed in cells and which occurs in all the groups of organisms (see LWOFF, 1936; KNIGHT; FILDES).

Irreversibility of the loss of capacity for synthesis. — A return toward auxo-autotrophism appears to occur when the capacity for synthesis has not been completely lost, but has simply been much reduced (*Rhodotorula rubra* and pyrimidine, see p. 199). In cases where organisms appear able to reestablish their ability to synthesize a vitamin, is not a question of an actual reversibility but simply of a quantitative variation in capacity for synthesis. It is possible to conceive of all degrees of variation extending from the normal capacity for synthesis to one much reduced but still genetically present. In these cases the medium has a strong influence on the capacity for synthesis. If the capacity for synthesis is genetically lost (see p. 214), there can be no recovery of the lost faculty.

On the other hand, it is possible to observe, under certain experimental conditions, a progression toward auxo-heterotrophism, expressed by the appearance of new needs. For example, *Ustilago violacea* possesses the ability to condense pyrimidine and thiazole into a molecule of thiamin. However, during the course of our experiments we have found strains which no longer have this capacity in full and, as a result, their growth with pyrimidine + thiazole is less than that with thiamin. During the course of prolonged experiments, it has been possible to assist in the progressive loss of this capacity with *Ustilago violacea* (BLUMER). *Ustilago scabiosae* (the strain employed) normally possesses a low capacity for this condensation (SCHOPFER and BLUMER).

The progressive losses of capacity for synthesis by a sort of orthogenesis, bring about new needs. These render the organism more and more dependent—heterotrophic. This irreversibility in the evolution of the capacity for synthesis corresponds to that known for evolution in general.

The phenomena involved in the origin of heterotrophism have thus been reduced to hypotheses. In cases where the capacity for synthesis have been genetically lost, a physiological mutation may be envisaged. These may originate under the influence of the medium. Similar cases are known in microbiology. For example, *Bacillus anthracis,* cultivated on dichromate bouillon, occasionally loses the ability to produce spores. This loss of function becomes noticeable during the course of several years. Another case is that of the pyocyanic bacillus (*Bacillus pyocyaneus*), race A, which loses the ability to produce a green fluorescent pigment when it has been cultivated for one year on liquid albumin. This new race (1931) had been stabilized by 39 years of culture previous to this experimentally induced alteration (*see* LWOFF, 1932). The loss of capacity to synthesize a pigment such as pyocyanin, which participates in oxidation-reduction, strangely resembles the loss of capacity to synthesize a vitamin growth factor.

These observations concerning the influence of the medium on the capacity for synthesis and concerning the dependence of the latter on external conditions have forced us to conclude that this genetic mutation is induced by a chemical agent in the medium. To strengthen this conviction we have a similar case for the higher plants in which grafting can modify, by chemical means, the genes of the stock (chimeras).

This concept is in accord with all known relationships between genes and hormones.

Mechanism of the loss of capacity for synthesis. — Information on this subject is so meagre that it is difficult to arrive at any definite conclusions. It is probable that the synthesis of a growth factor is dependent on the action of specific enzymes. For example concerning thiamin, we speak of: (1) a thiaminase (aneurinase) which brings about the condensation of pyrimidine and thiazole (SCHOPFER, BONNER and BUCHMAN), and (2) thiazolase which permits the condensation of thioformamide and acetopropyl alcohol into a molecule of thiazole (BONNER and BUCHMAN). There must also exist an enzyme which hydrolyzes thiamin to pyrimidine and thiazole (perhaps a reversible action of thiaminase). The existence of this enzyme is necessary for the explanation of the absorption of thiamin by organisms which require only half of the molecule. Such an enzyme must also be required for the utilization of analogues of thiamin in which only a portion of the molecule is active.

The author recognizes the existence of an enzyme carotinase, producing a hydrolytic splitting of beta-carotene into two molecules of vitamin A.

The internal or external factors which affect the capacity to synthesize a growth factor of vitamin nature must therefore act on these specific enzymes. Hence, if an enzyme such as carotinase or thiaminase is absent, the corresponding vitamin action can not become manifest.

All experimental results indicate that we are dealing with a category of fundamental enzymes which control the formation of vitamin factors functioning as coenzymes. These enzymes occupy a central position in this problem. Very probably they are immediately dependent upon genes; hence this type of enzyme acts as an intermediary between the gene and the growth factor. The chain of processes connecting the gene with the growth factor and the subsequent action on metabolism is necessarily very complex. Only a few such chains are known.

The inhibition of the gene of which we have spoken (p. 203) may well be, in the final analysis, an inhibition of one of these fundamental enzymes upon which the production of the growth factor depends.

We shall limit our discussion on the loss of synthesis to the well known vitamin thiamin whose function as a codehydrogenase and as cocarboxylase has been mentioned in chapter XVIII and whose biosynthesis has, for the most part, been elucidated.

A general principle which the author proposes is that all the constituents (precursors) of thiamin are dependent on genes which condition their formation. This mechanism may be illustrated as follows:

Genes → fundamental enzymes ⟨ thiazole / pyrimidine ⟩ (1) thiamin (2) ⟨ cocarboxylase / codehydrogenase

↑ (acting on simple substrates containing the ingredients necessary for the formation of vitamin factors)

These coenzymes act in the presence of suitable apoenzymes and specific substrates.

When an organism is unable to exhibit *cocarboxylase* and *codehydrogenase* activity we may conclude that the organism requires thiamin as a growth factor. We may even trace the cause for this avitaminosis to its actual source. According to the highly schematic chain of reactions shown above the following possibilities may very likely exist: (1) inhibition of the gene, (2) inhibition of the fundamental enzymes, (3) absence of the substrate necessary for the action of these enzymes, (4) inhibition of the action of the cocarboxylase or the codehydrogenase, or (5) an inhibition of the relationships of these two substances with other enzyme systems. This interpretation is justified by the existence of intermediate types of heterotrophism in which the ability to synthesize a certain vitamin is only partially lost.

To determine the exact reasons why an organism has become heterotrophic for thiamin is an extremely difficult problem. One of the main reasons is that the action of the fundamental enzymes is lacking. This is the step on which our attention must be focused.

If the loss of genes or their inhibition is immediate and complete, we might speak of it as a physiological mutation; if the

(1) thiazolase (2) thiaminase

medium is responsible for a definitive irreversible inhibition, we might call it an induced mutation; if the medium brings about a temporary inhibition we might refer to it as a reversible "modification".

This discussion remains theoretical and speculative. Experimental work in this domain has only recently commenced. This type of research would be of interest to the general problem of evolution because it may demonstrate the fundamental chemical mechanism of the origin of physiological mutations, and of mutations in general.

The irreversibility of the loss of capacity for synthesis is in complete accord with the law of irreversibility as applied to evolution (DOLLO).

REFERENCES:-

1932: A. LWOFF. — Recherches sur la nutrition des Protozoaires. Monographies de l'Institut Pasteur, Paris, Masson & Cie.

1935: W. H. SCHOPFER. — Recherches sur l'utilisation des facteurs de croissance. La synthèse biologique des facteurs de croissance. Arch f. Mikrobiol. 6:196.

1936: A. LWOFF. — Etudes sur les fonctions perdues. Ann. Fermentations 2:419.

1936: W. A. MOSHER, D. H. SAUNDERS, L. B. KINGERY and R. J. WILLIAMS. — Nutritional requirements of the pathogenic mold *Trichophyton interdigitale.* Plant Physiol. 11:795.

1937: R. GREWE. — Das Aneurin. Erg. Physiol. 39:352.

1937: G. P. GLADSTONE. — The nutrition of *Staphylococcus aureus*: nitrogen requirements. Brit. J. Exper. Path. 18:322.

1938: H. KARSTRÖM. — Enzymatische Adaptation bei Mikroorganismen. Erg. d. Enzymforschung 7:350.

1938: F. KÖGL und W. J. VAN WAGTENDONK. — Über die Bedeutung von Biotin für das Wachstum von *Staphylococcus pyogenes aureus.* Rec. Trav. chim. Pays-Bas (Amsterd.) 57:747.

1938: N. FRIES. — Über die Bedeutung von Wuchsstoffen für das Wachstum verschiedener Pilze. Symbolae botanicae Upsalienses 3(2):1-188.

1938: C. FROMAGEOT et J. L. TCHANG. — Sur la synthèse des pigments caroténoïdes par *Rhodotorula sanniei.* Arch. f. Mikrobiol. 9:434.

1938: W. J. ROBBINS and F. KAVANAGH. — Thiamin and growth of *Pythium Butleri.* Bull. Torrey Botan. Club 65:453.

1939: A. LANCZOS. — Über Gewöhnung an Vitaminmangel. Klin. Wochenschrift 18:244.

1939: W. H. SCHOPFER et S. BLUMER. — Recherches sur l'action de la vitamine E sur les plantes. Zeitschr. f. Vitaminforschung 9:344.

1939: M. SILVERMAN and C. H. WERKMAN. — Thiamin effects in bacterial metabolism. Iowa State College J. of Sc. 13:365.

1939: A. JUNG und W. H. SCHOPFER. — Adermingehalt und Aderminwirkung bei *Phycomyces.* Verh. Ver. Schweiz. Physiol., February 1939.

1940: W. J. ROBBINS. — Response of excised tomato roots to β(-4-methylthiazolyl-5-)-alanine. Plant Physiol. 15:547.

1940: S. BLUMER. — Über die Aneurinheterotrophie bei einigen antherenbewohnenden Brandpilzen. Mitt. Naturforsch. Ges. Bern, p. 19.

1940: S. BLUMER und W. H. SCHOPFER. — Beiträge zur Biologie und Wirkstoffphysiologie von *Ustilago scabiosae* (Sow.) Winter. Ber. d. Schweiz. bot. Ges. 50:248.

1941: G. W. BEADLE and E. L. TATUM. — Genetic control of biochemical reactions in *Neurospora.* Proc. Nat. Acad. Sci. U.S.A. 27:499-506.

1942: E. L. TATUM and G. W. BEADLE. — Genetic control of biochemical reactions in *Neurospora*: an "aminobenzoicless" mutant. Proc. Nat. Acad. Sci. U.S.A. 28:234-243.

Chapter XX.

VITAMINS IN RELATION TO OTHER ACTIVE SUBSTANCES.

In accordance with purely arbitrary convention, we have studied vitamins and their action as if they existed alone, assuming that all other active factors were present. But we know perfectly well that vitamins exercise their effect only in conjunction with numerous other active substances of hormonal nature. Since a study of the latter is beyond the scope of this book, we shall merely enumerate them.

If we consider the phenomena which contribute to the formation of a fully developed plant we find the following schematic arrangement: (1) assimilation, resulting in an increase in the amount of living matter; (2) cell divisions correlated with the increase in living matter and contributing to the distribution of the latter; (3) development and differentiation of tissues; (4) formation of organs; (5) movements and curvatures whose biological significance is obvious. Growth results from a combination of these processes which in turn are interdependent.

In an even more schematic manner it is possible to distinguish the fundamental phenomenon of assimilation from that of growth and differentiation.

The more complicated the organism becomes morphologically, the more complex will be the chemical conditioning of these phenomena.

We have attempted to distinguish vitamins from hormones without considering whether the mode of formation of the active substance is endogenous or exogenous. Vitamins are concerned with assimilation, whereas hormones have a specific effect on morphology. This distinction is justified from what has been said concerning vitamins. Because of their rôle as coenzymes or as fragments of coenzymes, the vitamins are the typical catalyzers of assimilation.

The other phenomena, cell division, cell elongation and growth, and the formation and differentiation of organs depend on specific hormones, of which auxin is a typical example. Hormones are the instruments of differentiation and morphogenesis.

Hormones of cell division. — At the time of the early investigations on biotin, this substance was considered as a hormone of cell division because its action caused an increase in the number of cells. This concept had to be abandoned; instead biotin must be considered as a substance of vitamin nature concerned with assimilation; this is undoubtedly its function despite the fact that the proof is still lacking and the chemical mechanism of its action is unknown.

The typical hormone of cell division is the necrohormone (wound hormone) which participates in the restitution of an injured tissue. This was described in the classical investigations of HABERLANDT as the substance producing meristems on the surface of unwashed sections of a potato tuber. The same substance was obtained in a chemically pure crystalline form by ENGLISH, BONNER and HAAGEN SMIT (1939) from fruits of *Phaseolus vulgaris* and called traumatin. It is known chemically as 1-decene-1,10-dicarboxylic acid, has the empirical formula $C_{12}H_{20}O_4$, and has been synthesized. It is capable, even in very small dosage, of initiating cell division in the parenchymatous mesocarp lining the seed chambers of fruits of *Phaseolus vulgaris.* Glutamic acid plays the rôle of a co-factor and enhances the activity of the hormone almost ten times. The crystalline factor at a concentration of 10 mg. per liter, and in the presence of 0.25 per cent glutamic acid, exhibits an effect which can be detected. A drop, 0.01 cc. in size, contains 0.1γ and acts effectively.

Traumatin is apparently the same as the necrohormone of HABERLANDT since it causes the formation of a periderm on the washed surface of a disc of potato tuber. It is also capable of partially replacing tomato juice in its inhibiting action on the germination of tomato seeds. Thus this substance acts primarily and specifically on cell division but its action may be of a more general nature.

An effect on cell division can be obtained by means of other substances, *e.g.*, heteroauxin exercises such an effect on cambial meristems. The action of the heteroauxin is of an indirect nature, probably accelerating the production of the cell division hormone. Thiamin likewise (ADDICOTT) tends to accelerate cell divisions in meristems of roots. Such non-specific indirect actions can be observed with several other substances. However, nothing is known concerning the relationships between the substances of cell division and the vitamins present in all cells.

Hormones of cell elongation; hormones of root formation. — The hormones concerned in these processes are of the auxin and heteroauxin type. These substances will be treated briefly here since they are discussed in detail in the monographs of F. W. WENT and THIMANN (1937), and of BOYSEN-JENSEN (1936*b*), translated by AVERY and BURKHOLDER (1936).

These hormones are present not only in the higher plants but also in microorganisms, for example, *Aspergillus niger* (BOYSEN-JENSEN, BÜNNING, and other authors). Heteroauxin is present in yeast (KÖGL and KOSTERMAN); it corresponds to the rhizopin of NIELSEN, an ether-soluble substance contained in the culture medium of *Rhizopus suinus.* The hormone found in microorganisms is not auxin, but instead heteroauxin. They are able to synthesize it from a medium based on tryptophane which plays the rôle of precursor of heteroauxin and which is chemically related to it. Heteroauxin can also be formed in the presence of other amino acids, asparagine or KNO_3 by *Aerobacter aerogenes* and *Escherichia coli* (BURKHOLDER). One of the most interesting organisms in this connection is *Rhizobium* which is also an active producer of heteroauxin (THIMANN), thereby explaining the hypertrophying morphogenic effect of these bacteria and of their extracts on the roots of legumes. All of these microorganisms produce one or several substances of the hormone type, acting on the coleoptile of *Avena*. On the other hand, we are entirely uninformed concerning their possible action on the microorganisms producing them. Despite an extensive literature on this subject the results are, with few exceptions, unconvincing and indecisive; although these hormones sometimes act as stimulants, they do not appear to be indispensable to the life of the organism.

KÖGL and TÖNNIS, working with *Saccharomyces cerevisiae* found heteroauxin inactive, but their observations do not agree with those of other authors. *B. coli* is strongly stimulated by this hormone, the optimal dosage being 10^{-4} to 10^{-5} percent. The growth of *B. auxinophilum* is likewise favored by heteroauxin (TINCKER and JACOB) but it is not known whether the action is of a hormonal nature or simply nutritive. Certain of the fungi respond strongly to heteroauxin. For example, the growth of *Aspergillus niger* and *Rhizopus suinus* is accelerated by this hormone (THIMANN and DOLK). Furthermore the formation of conidia of *Aspergillus niger* (BÜNNING) and the formation of sporangia of *Rhizopus suinus* are favored by this substance (THIMANN and DOLK). The most interesting observations are those of KERL concerning the ascomycete *Pyronema confluens*. He found that the growth of the mycelium and the formation of apothecia is accelerated by biotin and thiamin and also by auxin, but that the greatest effect is produced by heteroauxin. He observed that the latter, in a dosage of 200AE, increases the formation of perithecia 4037 percent. In this relatively complex organism we recognize two actions: (1) on assimilation, by thiamin and biotin, and (2) on the formation of apothecia; the latter action is of formative nature and simulates the action of this hormone on the higher plants.

It has been suggested that heteroauxin probably acts as a primary excitant of respiration in *Aspergillus niger*, the energy produced accelerating the assimilation of nitrates (BONNER).

The conclusion reached is *a priori*, that heteroauxin, since it is produced by the metabolism of numerous lower organisms, very probably exerts an effect on the organisms themselves. The conflicting observations concerning the activity of this hormone may be due to the fact that the organisms studied differ in their ability to synthesize it.

Heteroauxin has been found to exert a favorable action on the development of unicellular green algae: *Chlorella* (YIN, PRATT, BRANNON) and *Oocystis* (BRANNON).

The fundamental action of auxin in the higher plants is exerted on the elasticity of the cell wall and consequently upon the growth of the shoot and the root. Auxin plays a part in correlations and, along with other factors, induces root formation and is also responsible for the inhibition of lateral buds on the shoot. The effect of auxin and heteroauxin on root formation on cuttings is sometimes augmented by factors concerned with assimilation such as biotin and thiamin. As a root inducing substance, heteroauxin is not very active and must be employed in relatively high dosage.

The rôle played by heteroauxin in the formation of organs is manifested in phytopathology by the formation of galls. The hormone supplied artificially is able to replace the effect of a microorganism such as *Pseudomonas tumefaciens* which must be regarded as a producer of hormones (BROWN and GARDNER; LINK; WILCOX and LINK).

Because of their action on the elasticity of the cell membrane, the hormones of the auxin type take part in tropistic curvatures. Certain movements, such as those of *Mimosa,* are controlled by another specific substance (UMRATH and SOLTYS) which has likewise been isolated. Cytoplasmic streaming (cyclosis) in the epidermal cells of *Vallisneria* is dependent on histidine and its derivatives (FITTING).

This discussion of hormones need not be extended further since these substances are adequately treated in other works. We merely wish to point out that the physiological relations between hormones and vitamins are little known.

Action of animal hormones on plants. — The term animal hormone is reserved for hormones normally produced in animals. This definition must be modified in view of the fact that the "female sex hormone", estrone (theelin), or a substance closely related physiologically, is also produced by plants, *e.g.*, estrone in palm kernels, and estriol in pussy willow.

A vast literature exists concerning the action of animal hormones on plants. All the work concerning microorganisms and higher plants carried out with uncrystallized products extracted from glands is of no value, or at most is of value only in orientation. We shall not discuss such results here (see BOYSEN-JENSEN, 1936*a*).

The action of numerous hormones, particularly estrone, has been tested on microorganisms. The results are extremely variable. This female hormone is inactive on *Aspergillus* and *Saccharomyces cerevisiae* according to A. LUND. It, likewise, has no influence on the formation of zygotes of *Absidia coerulea* or *Phycomyces.* A certain number of yeasts have been studied by WEBER using several pure hormones. A positive effect was produced only on *Rhodotorula Suganii* and *R. glutinis* var. *Satoi;* the hormones arranged according to decreasing activity are as follows: insulin (impure preparation) > estrone > benzoate of dihydrotheelin >

testosterone > androsterone > thyroxine > adrenaline > and heteroauxin. The action is probably related to the composition of the medium; the mechanism of action, although actually unknown, perhaps resides in the rapid establishment of a favorable redox potential.

The information concerning the action of animal hormones on microorganisms is as fragmentary as that concerning the plant hormones which the microorganisms produce.

The flowering plants, on the other hand, have been studied extensively with regard to animal hormones. Prolan A (the gonadotropic hormone produced by the anterior lobe of the pituitary) produces parthenocarpy in *Antirrhinum,* and in certain cases stimulates root formation. In the case of rye, a favorable action on the growth of the shoot (in length) and on the mineral content has been found (HERBST).

The effect of estrone on the phanerogams is of particular interest. The first work of SCHOELLER and GOEBEL was carried out in the following manner: potted plants were sprinkled with water containing progesterone of technical grade (contaminated by auxin). Almost at once the results became readily visible. A positive action by the hormone was manifested by an acceleration of growth and flowering of hyacinths, fuchsias and primroses. In the case of tomatoes, larger fruits were produced. Other investigators were unable to verify the results completely and were inclined to attribute the action to the auxin contaminating the hormone preparation. However, this criticism has broken down because other workers have observed the same positive effects with the absolutely pure crystalline hormone.

The hormone exerts three effects which must be discussed: (1) on growth in general and on the opening of the flowers, (2) on the speed of flowering and on the number of flowers, (3) on the production of living matter. (Literature in BOYSEN-JENSEN, 1936*a*).

The contradictory effects observed by numerous investigators suggest the view that the action of the hormone depends on internal and external conditions which were not the same in all experiments.

The effect on flowering has stirred up a lively interest. It is thought, on the basis of the observations of CHOUARD regarding the relation between the hormone and photoperiodism, that estrone is one of the hormones of flowering, the probable existence of this substance in higher plants has been pointed out by CAJLACHJAN and MELCHERS. This hypothesis has not yet received definite experimental confirmation.

The effect on the production of dry matter appears more definite. Positive results have been obtained with the following: barley (SCHARRER and SCHROPP), rye (SCHARRER and SCHROPP), tomato (SCHOELLER and GOEBEL, HARDER and STÖRMER), *Vicia Faba* and *Triticum* (SCHARRER and SCHROPP), and *Zea Mays* (ORTH, HARDER and STÖRMER). Similar findings have at times been reported from experiments with cultures of embryos deprived of their cotyledons where the hormone acts in conjunction with thiamin and biotin.

C. ZOLLIKOFER (1938) made some very interesting observations with a grass, *Poa alpina.* This species has three different forms, one fertile with normal flowers and seeds (*fructifera*), and two sterile which propagate by bulbils developed in the inflorescences (*vivipara* and *intermedia*). All three forms respond favorably when the hormone is sprinkled on the plants, *viz.*, there occurs an increase in growth and the production of dry matter. A relationship exists between the increase in dry matter and the daily dosages of hormone. The maximal dosage effects a 96 per cent increase, which is much greater than the experimental error. This work is interesting because of the fact that the hormone does not bring about an increase in the number of flowers on the form *fructifera,* but it does increase the number of bulbils for vegetative reproduction on the form *vivipara.* Hence, the effect of the hormone is not exercised on flowering and the reproductive system of any of the forms of *Poa alpina;* instead it is on the vegetative system and on the production of dry matter.

The female sex hormone (estrone) is a typical growth factor of vitamin nature, as we have defined it because of, (1) its great activity (a fraction of a gamma exerts an effect on cultures of embryos), and (2) its effect on assimilation. It can therefore be considered as a vitamin since it occurs in plants. It is found in seeds but in smaller quantities than the optimal dosage necessary for the culture of an embryo. Chemically it is related to the sterols which are equally common in plants.

This is the only hormone for which activity has been claimed; however, it does not appear to be indispensable to growth.

Other hormones have been studied, namely, thyroxine and adrenaline, but have given contradictory results and have not aroused the same interest as estrone.

Aside from the female sex hormone, animal hormones have not yet been found in plants. They have been tested for activity and, in some cases, have been found to have a stimulating effect; nevertheless, with the exception of estrone, they must be considered as non-specific. The observed stimulations have nothing in common with the action of a true growth factor as defined by the author.

Actions of various substances. — Colchicine, whose action on nuclear division is the topic of the moment (EIGSTI, DUSTIN, BLAKESLEE), has a non-specific action on phytocarcinomas produced experimentally on tomatoes by *Bacterium tumefaciens.* Its action in this case is that of an inhibition on tumor development (HAVAS). In wheat, on the other hand, it causes the production of hyperplastic malformations at the tips of the roots. This action, which must be related to that of growth hormones, has not been explained. Colchicine, by virtue of its action on nuclear division, makes it possible to obtain plant forms frequently (but not always) characterized by a more rapid vegetative development and a larger size. This might be spoken of as an indirect action of a growth factor.

Carcinogenic substances are chemically related to sexual hormones and are of considerable interest. Attempts to produce an action on plants simulating that observed in cancerous tissue of animals has, in the case of certain microorganisms, yielded positive results. For example, *Escherichia communior* undergoes a 50 per cent increase in growth under the influence of 1, 2, 5, 6-dibenzanthracene and methylcholanthrene (GOLDSTEIN, 1937). These substances also stimulate growth and fermentation in yeast (C. W. DODGE and B. S. DODGE, COOK, HART and JOLY). Nothing is known, however, concerning the specificity of these actions. They must, for the present, be considered as purely stimulatory.

The results obtained from the use of carcinogenic substances on flowering plants are less clear. A slight stimulation has been observed in the case of *Pisum;* root tips in aseptic culture are stimulated somewhat by 1, 2-benzpyrene, methylchloranthrene, and 1, 2, 5, 6-dibenzanthracene. In the case of *Helianthus,* the hypocotyl portion of the axis when treated with 1, 2-benzpyrene develops small tumefactions (BERTHELOT and AMOUREUX). LEVINE (1939) in his work on *Kalanchoe Daigremontiana* was able to demonstrate a specific morphogenic action with carcinogenic substances. These substances, especially 1, 2-benzpyrene, have been studied regarding their action on *Solanum lycopersicum, Nicotiana suaveolens, Pelargonium zonale,* and *Sambucus nigra* (KISSER and LINDENBERG, 1940). The substance is applied as a paste to the stump of the stem of the plant being studied, the application being repeated for a prolonged period. There occurs a stimulating action which is expressed as an intensification of cellular activity. This action is non-specific, and is of no greater validity than the action of animal hormones for showing relationships between plants and animals. The carcinogenic substances applied to plants do not produce tumors comparable with those produced by *B. tumefaciens.* This organism acts by virtue of hormones which it produces. The slight effect occasionally produced by carcinogenic substances must be attributed to a local stimulus which in turn causes the formation of hormones of cell division and organ formation.

This short examination of these widely diverse substances leads to the conclusion that all these compounds are linked physiologically in their action. Absolutely nothing is known, however, concerning the nature of these relationships. When we observe a particular effect of a substance we do not necessarily regard it as a result of a primary action. Other possibilities exist; for example, a substance acting externally on the organism may intervene at a given place in the chain of reactions, which unites various substances, thereby controlling their action and their production. Hence, an effect which one might attribute to the action of a specific factor may be merely the result of one or several previous actions, the observed effect being indirect.

When we are better informed concerning the physiological and chemical relationships by which these factors are united their effects probably will appear in an entirely different light.

REFERENCES:-

1936*a*: P. BOYSEN-JENSEN. — Die Bedeutung der Wuchsstoffe für die Pflanzenproduktion. Z. f. Pflanzenernährung, Düngung und Bodenkunde 43:142.

1936*b*: P. BOYSEN-JENSEN. — Growth hormones in plants. Translated and revised by G. S. AVERY and P. BURKHOLDER. McGraw-Hill, New York and London, 268 pp.

1937: F. W. WENT and K. V. THIMANN. — Phytohormones. Experimental Biology Monographs, The Macmillan Company, New York, 294 pp.

1937: J. ENGLISH and J. BONNER. — The wound hormones of Plants, I. Traumatin, the active principle of the bean test. J. Biol. Chem. 121:791.

1938: C. ZOLLIKOFER. — De l'influence des hormones animales sur les plantes. — Scientia, August 1938, p. 66.

1939: J. ENGLISH, J. BONNER and A. J. HAAGEN SMIT. — The wound hormones of plants, II. The isolation of a crystalline active substance. Proc. Nat. Acad. U.S.A. 25:323.

1937: S. GOLDSTEIN. — A microbiological test for carcinogenic hydrocarbons. Science 86:176.

1939: M. LEVINE. — Crown gall-like tumors induced with scharlach red on the plant, *Kalanchoe*. Proc. Soc. Exper. Biol. and Med. 40:599.

1940: J. KISSER und L. LINDENBERG. — Untersuchungen über die Wirkung karzinogener Substanzen auf höhere Pflanzen. Jahrb. f. wiss. Bot. 89:89.

Part 3.

GENERAL PROBLEMS INVOLVING VITAMINS.

Chapter XXI.

VITAMINS IN NATURE. THEIR ROLE IN AGRICULTURE AND HORTICULTURE. VITAMIN CYCLES.

Plants possess the power of synthesizing their own vitamins, whereas animals receive their vitamins from plants. Both contain many of them. As the substance of these organisms is returned to the soil, there remains, from a theoretical and *practical* point of view, the very important question: are vitamins lost when an organism dies?

At first glance such a loss would be contrary to all that is known about element cycles; thanks to these, the elements, carbon, nitrogen, sulfur, calcium, etc. pass through many groups of organisms. Similarly, such a loss would be contrary to another equally valid principle of biology, *i.e.*, "nothing is lost and nothing is created".

The question then, is simply this: are vitamins present in the soil?

Vitamins in the soil. — According to the old and classic researches of BOTTOMLEY and MOCKERIDGE, bacterial infected peat, having undergone fermentation, contains substances which have an auxogenic action similar to that of vitamins. These "auximones" act on the growth of higher plants as well as on the fixation of nitrogen by *Azotobacter*. These results, obtained between 1914 and 1918, have been confirmed by recent experiments, which have demonstrated the sensitivity of higher plants to vitamins.

LWOFF and LEDERER (1935) have shown that a flagellate, *Polytomella agilis,* is aided in its development by soil extracts. Contrary to expectation, growth factors, rather than humus, are the active substances. The same year, PRINGSHEIM (1935) realized the importance of soil in this connection and raised the question whether the soil really contained growth factors.

The problem was taken up again in 1937 by the author's collaborator, W. F. MÜLLER (1937, 1941), with *Mucor Ramannianus.* This fungus is a typical member of the *Mucoraceae* found in the soil; it is auxo-heterotrophic and requires thiamin as an essential growth factor; however, it is satisfied with the thiazole half of this vitamin (MÜLLER and SCHOPFER). MÜLLER (1941) made an extract of soil from a locality which had never received manure. This soil was taken from a coniferous forest, in which *Mucor Ramannianus* finds favorable conditions. It was passed through a sieve, the stones and fragments of roots were removed, and then dried at

100°C. to determine its dry weight. Later, enough water was added to produce a semi-liquid decoction, which was heated to 100°C. and then filtered. A liquid was obtained which contained the dissolved water-soluble growth factors from the soil.

With the optimal dose of vitamin B_1, 1.5γ under the conditions of our experiments, a dry weight of about 200 mg. was obtained. With the soil extracts, results were obtained as follows: to each experimental flask was added a quantity of soil extract corresponding to 40 g. of soil: soil taken from 1-3 cm. below a layer of conifer needles, 28.5 mg. of dry weight; soil taken from 6-8 cm., 18 mg. of dry weight; soil taken from 12-15 cm., 4.5 mg. of dry weight; control, 0.

There can be no doubt, therefore, that the soil studied contained a growth factor active on *Mucor Ramannianus.* The effect observed with the soil taken from 1-3 cm. corresponds with 0.05γ of thiamin. Experiments carried out with soils taken from other forests, from prairies, and from various fields led to the same results. For *Mucor Ramannianus,* it is not the amount of growth factor contained in the soil which acts as the limiting factor, but the quantity of assimilable nitrogen which is usually too small (MÜLLER, 1941). The factor present in the soil can be adsorbed by animal charcoal and eluted with a mixture of alcohol and N/10 HCl. Experiments performed with the product of the elution show that the product is soluble in water, and in alcohol, but not in chloroform; thus it can not be preformed thiazole which is strongly soluble in alcohol. The presence of thiamin is not impossible.

LILLY and LEONIAN (1939) made similar observations. They found sufficient thiamin in the soil to produce a good development of *Phytophthora erythroseptica* (requiring the complete molecule of thiamin), *Phycomyces Blakesleeanus* (pyrimidine + thiazole), *Pythiomorpha gonapodioides* (pyrimidine) and *Mucor Ramannianus* (thiazole), as well as *Sordaria fimicola,* which, according to LEONIAN, requires biotin. They also observed that soil taken close to the surface is richer than subsoil in growth factors.

It is known that thiamin is strongly adsorbed by Fuller's earth. It is quite possible that thiamin may be likewise adsorbed by the fine particles of soil, and that it may be held fast in the upper layers of the latter; it is found near living matter, being liberated upon the decomposition of this matter. Thus it can readily be expected that soil acts on plants not only through its plastic foods and minerals, but also through its growth factors.*

Origin of Growth Factors Contained in the Soil. — One of the possible sources of the growth factors in the soil is traceable to the microorganisms living in the soil and giving off growth factors to the soil. They include all the (totally or partially) auxo-autotrophic organisms: bacteria, flagellates, and fungi. This

---

***RUSSEL had already demonstrated the fact that soil must act because of substances other than those of inorganic nature.**

liberation of growth factors permits the development of other organisms which are auxo-heterotrophic. Experiments made with culture media, in which auxo-autotrophic bacteria or fungi have lived, show that the growth factors have diffused into the medium and are again utilizable.

Certain organisms are particularly rich in growth factors. BONNER and GREENE (1938) showed that *Azotobacter* contained 140 mg. of thiamin per kg. (this species is auxo-autotrophic and synthesizes its thiamin when cultured on a mineral medium). Other organisms certainly belong in this class. The organs of higher plants permit the diffusion of vitamin factors into humid soil. We know that under certain conditions, roots allow amino acids to diffuse out of them (VIRTANEN). It is not surprising that vitamins likewise diffuse into the soil. *Mucor Ramannianus* is greatly activated in its development by the natural dialyzates from roots. Dialyzates from conifer seeds have the same effect (MELIN). We have shown that the natural dialyzates of pollen and of floral organs, as well as of leaves of numerous plants, are extremely active on *Phycomyces* (SCHOPFER, 1936). One may attribute to the rhizosphere a certain richness in vitamin growth factors. As a result, auxo-heterotrophic organisms are likely to aggregate there since conditions favorable to them are usually found in the neighborhood of the sources of growth factors.

Furthermore, auxo-autotrophic and auxo-heterotrophic organisms die and their matter returns to the soil. Their products of decomposition, the seat of multiple bacterial activities, do not necessarily lose all their vitamin growth factors; a certain number of these bacterial activities may lead even to the formation of growth factors which will pass into the upper layers of the soil.

The same is true with reference to the decomposition of higher plants. The roots of decomposing plants give up their growth factors to the soil just as does the decaying foliage. Some of the experiments with *Mucor Ramannianus* are conclusive: a very active aqueous extract is obtained from decomposing leaves found on the ground.

Likewise the action of animal excreta can not be overlooked. Because of their high vitamin content the feces of horses and other large animals serve as excellent media for the culture of numerous fungi. Cat feces, according to BONNER, contain 0.1γ of thiamin per gram of dry matter. The author has found large quantities in human and dog excreta. A man with normal nutrition eliminates through the urine, every 24 hours, from 30-90γ of thiamin (WESTENBRINK and GOUDSMIT), 100γ (W. KARRER). Decomposing animal organs must also give up their vitamins to the soil.

Finally, manure applied to the soil represents another source of growth factors. BONNER and GREENE (1938) found the following figures for a few manures: Arizona steer manure 0.13 mg., local steer manure 0.08 mg., dairy manure 0.13 mg. of thiamin per kilogram. The presence of growth factors in natural manures must contribute to their superiority as compared with artificial fertilizers. This superiority has been established for a long time (RUSSEL).

Action of vitamins supplied by nature and by fertilizers. — It is easy to understand that these vitamin factors of the soil act on the microorganisms of the soil, a situation which needs no further discussion; but there remains the question whether higher plants profit equally from them. Although higher plants are auxo-autotrophic and synthesize their own growth factors, it has been demonstrated that this capacity for synthesis is not always sufficient. *Camellia japonica* studied by BONNER and GREENE (1938-1939) is an example of this. When seedlings are cultivated in sand culture they do not develop until vitamin B_1 is

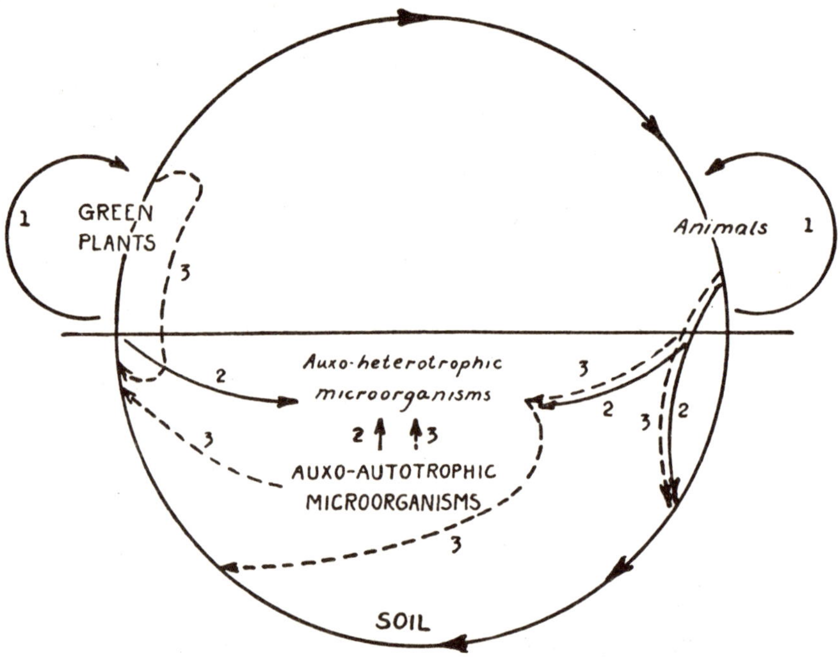

FIG. 16. — Cycle illustrating the direct and indirect relationships among the various groups of organisms. Capital letters designate auxo-autotrophic organisms; small letters, auxo-heterotrophic organisms. 1: symbiotic microorganisms; 2: products secreted and excreted by the living organisms; 3: products released by the disintegration of dead organisms containing vitamins.

added to the culture. Other plants (see chapter 7, page 76) grown from seeds (*Bougainvillea glabra, Arbutus Unedo, Eucalyptus ficifolia*), or started as seedlings with roots (*Aleurites Fordii*), develop without vitamins, but their development is considerably accelerated when thiamin is supplied*. After a period

*Other workers have been unable to obtain any improvement in the growth of intact green plants upon the addition of thiamin (ARNON; MINNUM; GORHAM; C. L. HAMNER; SWARTZ).

of one year, the growth of *Ceratonia siliqua* is increased about 100% by the addition of thiamin to a sand culture (0.01 mg. per liter) (BONNER and GREENE, 1939). The same condition applies to plants cultivated in soil (*Arbutus Unedo, Prunus ilicifolia, Bryophyllum* sp.) (BONNER and GREENE). The vitamin accumulates in the leaves of the treated plants.

Obviously not all plants have the same capacity for synthesizing thiamin. Some of them must suffer from a natural hypovitaminosis; these are the ones that will react strongly to thiamin added to the medium and will be influenced by the natural growth factors of the soil. Plants which are sufficiently rich in thiamin will not react to exogenous growth factors.

It is probable that growth factors of the soil must also be considered as ecological and sociological factors. Their presence may determine a particular population of certain localities, because of inadequate synthesis by some plants.

What has been said in regard to thiamin may be applied to any other growth factor, provided it is sufficiently stabile to remain in the soil (Fig. 16).

Cycles of growth factors. — Definite relations among the various groups of living organisms of the soil and sub-soil as based on various growth factor requirements are conceivable. They may possibly lead to the establishment of a cycle, analogous to the nitrogen and carbon cycles, but of a more flexible nature.

The higher plants represent the starting point; they carry out the initial synthesis and furnish their vitamins to animals. Wastes and feces restore to the soil a certain quantity of vitamins. In the soil the auxo-heterotrophic microorganisms profit by the vitamins produced by the auxo-autotrophic forms, diffusing from the latter or liberated by their decomposition. They likewise profit from the factors liberated from the roots of higher plants and from the products of decomposition of plants and animals. Those higher plants which have an inadequate synthesis, also profit by the factors of various origin in the soil.

This cycle is not completely closed; nevertheless it clearly shows the relations which unite the various groups of living organisms, autotrophic or heterotrophic to thiamin. What is possible for thiamin is equally possible for other growth factors (KÖGL and HAAGEN SMIT, BOAS).

A complete cycle, just as complex, is conceivable for vitamin A and carotene. The latter is synthesized by marine algae, which in turn furnish it to microplankton, *e.g.*, minute crustaceans. Vitamin A is formed from carotene in the animal organism; it is produced in small and large fishes and is found in the oil extracted from the liver. This is utilized by man. For vitamin A, animals and man are totally dependent on the plant's power of synthesis which alone manufactures enough endogenous carotene. It is not a true cycle which can be directly closed; there is simply a passage from plants to animals. When our knowledge becomes more advanced concerning the other vitamins, we will undoubtedly be able to construct other cycles.

This information may have no great practical importance; however, in the field of horticulture and in intensive cultivation, it may be possible to accelerate growth by adding pure vitamins, *e.g.*, thiamin; even this procedure is hardly practicable. An indirect interest lies in the fact that the development of seedlings and the production of roots can be stimulated by the action of thiamin, together with that of heteroauxin.

The principal interest lies in a better and more complete comprehension of the action of natural fertilizers. Natural fertilizers can act by virtue of substances other than typical vitamins. Urine is rich in auxin and it is not impossible that by its presence in excreta this powerful substance acts on the plant. PRÉVOT (1939) pointed out that pigeon dung possesses a remarkable action on the formation of adventitious buds on the leaf of *Begonia rex*.

Effect of the external environment on the vitamin content of food plants. — This aspect of the problem is really the most important. Is it possible, in a practical way, to increase the vitamin content of a food plant?

In chapters 8 and 9 we have shown that the biosynthesis of any vitamin is not solely dependent on internal factors, but also on external conditions. The production of vitamin B_1 in young leaves is dependent upon light, whereas that of vitamin C is influenced by temperature and metallic catalyzers. These results are particularly significant for laboratory experiments. In agricultural practice, it is of interest to determine whether the vitamin content of the plant can be indirectly influenced by a particular fertilizer acting on the general metabolism. The question is especially important for wheat (vitamin B_1 content of grain, flour, and bread), for legumes in general, cabbage, spinach, tomatoes, etc. (vitamins A, B_1 and C), for carrots and tomatoes (vitamin A, carotinoids), and has already engrossed numerous investigators. ROWLANDS and WILKINSON (1930) proved that the thiamin content of grass seeds is lower when grown with an artificial fertilizer than with a natural one.

Vitamin B_1. — We know that this vitamin is widely distributed in the whole plant kingdom. It is found not only in food plants, but also in a great number of species not utilizable for food. The *Phycomyces* test gives a positive reaction with 134 species belonging to 71 different families of monocotyledons and dicotyledons (see SCHOPFER, 1936; and WILLIAMS and SPIES, 1938, p. 376). In the wheat grain, thiamin is found mainly in its peripheral part. It is also abundant in the germ (which is an excellent source of growth factors for microorganisms). Grinding with too much pressure, and using too small a proportion of the whole grain, yields a flour with a low vitamin B_1 content, insufficient to prevent beri-beri in animals. The proportion of kernel going into the flour must not be below 72%, an appreciable portion of vitamin B_1 thus being conserved. This was approximately the portion in flour (in Switzerland) during peace time. To lose the least possible vitamin B_1, the proportion of kernel milled into flour must be above 80% (82-85%); the flour is then of a gray tint which corresponds well to that of our sad times (see the graphs of SCHEUNERT).

The vitamin B_1 content depends also on the type of wheat. The whole grain of wheat contains, on the average, 3-5γ of thiamin per gram; the germ of wheat, 15-20γ per gram; an extract of wheat germ, studied with an animal test and with the *Phycomyces* test, gives us a content of 20γ per gram (SCHOPFER and JUNG, 1936). Wheat flour made from 94% of the kernel contains 3.5γ; with 75%, the content falls to 1.2γ and with 60% (very white flour) to 0.5γ per gram. There is practically as much vitamin B_1 in wheat bran as in the entire grain. Barley and rice germs are very rich: 35γ per gram; rye germ contains only 10γ and corn germ only 12γ per gram. Oat flour is one of the richest, containing 8-10γ per gram which is almost three times that of wheat flour with 94% of the kernel present.

SCHEUNERT and SCHIEBLICH (1936) used three methods of fertilizing wheat: (1) natural fertilizer, composed of stable manure prepared in a special manner; (2) commercial fertilizers; (3) alternating fertilizers, the first year natural manure in conjunction with commercial fertilizer, then the second and third years artificial commercial fertilizers alone. The wheat studied was a winter variety and the experiments extended over a three year period. The differences observed (with rats as test animals) were small, the wheat obtained with the natural fertilizer being a little richer than the others.

The experiments were extended to barley (SCHEUNERT and WAGNER, 1939a). The experimental fields were well known from records available for 60 years. The results of the experiments show that the variations are small; the vitamin content varies from 4.5 to 6γ per gram and bears no relationship to the nature of the fertilizer. The plants grown without fertilizer have the highest vitamin B_1 content. Rye gave the same results.

The following vegetables were studied with a complete fertilizer (NPK), a natural fertilizer (stable manure), and a mixture of the two: Brussels sprouts, white cabbage, carrots, spinach, green cabbage, celery, onion, and red cabbage. The results are identical, the figures showing, at most, insignificant variations in thiamin content. Brussels sprouts contain 3γ per gram, celery 3γ per gram; the thiamin content of the other vegetables is between 0.9 and 1.8γ per gram (SCHEUNERT and WAGNER).

Vitamin A. — The carrot (*Daucus Carota*) has been studied in detail (SCHEUNERT and WAGNER, 1939 *b,c*). The biological determination of the vitamin A value in carrots by means of an animal test proves that the variations due to fertilizers are insignificant. The same is true for Brussels sprouts, green cabbage, and spinach.

Vitamin C. — Researches on this subject are of very great practical importance because of the instability of ascorbic acid.

The potato has been studied most extensively. The ascorbic acid was determined by the usual method (2,6-dichlorophenolindophenol, in acid solution). The essential fact demonstrated is that storage notably diminishes their ascorbic acid content. SCHEUNERT and WAGNER (1939*d*, 1940) made analyses at the end of October, and in the following January and October; between the first and the last analysis, the content of titratable ascorbic acid became two to three times weaker (with three varieties of potatoes). OLLIVER, in a detailed study, proved that potatoes, after 10 days of storage, already lost a good part of their vitamin C. The destruction, which is brought about enzymatically, is less rapid at 0°C. than at room temperature. The content may change from 25 to 7.5 milligrams. Various authors, FUHRMEISTER, KRÖNER and SREINHOFF, WACHHOLDER, etc. obtained similar results (see LUNDE, 1940).

The rapid and progressive diminution in the content of ascorbic acid was also proved in spinach; after two days of storage at room temperature the decrease is about 80% (OLLIVER). VAN EEKELEN states that the loss is 60% after three days at room temperature and 90% after five days; on ice after five days the loss is 65%.

Other vegetables and fruits have been studied with similar results (see the excellent book by LUNDE, 1940 (Stavanger), published by Springer).

What is the influence of environment and fertilizers on the formation of vitamin C? The researches of VON HAUSEN (see p. 95) have shown that, within the scope of laboratory experiments, the vitamin C content is dependent on the chemical composition of the culture medium (peas). The greatest production of ascorbic acid is obtained when $Ca(NO_3)_2$ is supplied to the growing plants, but the content may vary a great deal. Studies conducted in the field concerning the function of fertilizers have not led to the same results (SCHEUNERT and WAGNER, 1936; SCHEUNERT, RESCHKE, and KOHLEMANN, 1940). Lettuce, spinach, green cabbage, white cabbage, red cabbage, Brussels sprouts, carrots, onions, and potatoes have been studied with stable manure plus artificial fertilizer (NPK), and with stable manure alone. The results concerning ascorbic acid are similar to those obtained by studies of other vitamins, *viz.*, there are no appreciable variations (animal and chemical tests).

The abundant data of practical nature can be summarized in the following way: the vitamin content (especially A, B_1, C) of a plant is a function of its development, at a given stage of growth the content being at a maximum. This content represents a biological constant characterizing a given race. In a large measure it is practically independent of the nature of the fertilizer utilized. Under extreme conditions a diminution can occur, but this is the result of a disturbed metabolism.

The data obtained in the course of limited laboratory experiments do not always correspond to the results furnished by actual practice and by cultures on a large scale. It is not impossible that someone will succeed in altering the vitamin content of food plants; however, no suitable method has as yet been found. The only available method is the selection and the hybridization of races particularly rich in vitamins. The vitamin content of a plant is a character which is determined genetically. It can more easily be influenced by genetical means than by chemical ones such as fertilizers.

REFERENCES:-

1930: M. J. ROWLANDS and B. WILKINSON. — Vitamin B_1 content of grass seeds in relationship to manures . . . Biochem. J., 24.199.

1935: A. LWOFF et E. LEDERER. — Remarque sur l'extrait de terre envisagé comme facteur de croissance pour les Flagellés. C. r. Soc. Biol. Paris, 119:971.

1935: E. G. PRINGSHEIM. — Wuchsstoffe im Erdboden? Naturwiss., 23: 197.

1936: W. H. SCHOPFER. — Vitamines et facteurs de croissance chez les plantes. Recherches sur l'action de divers extraits végétaux sur le développement de *Phycomyces*. Arch. f. Mikrobiol., 7:156.

1936: A. SCHEUNERT und M. SCHIEBLICH. — Über den Einfluss der Düngung auf den Vitamin B_1-Gehalt von Weizen. Tierernährung, no. 8:120. See also: Die Ernährung der Pflanze, 1936, 32:288.

1937: W. F. MÜLLER. — Zur Physiologie von *Mucor Ramannianus*. Ber. d. schweiz. Bot. Ges., 47:277.

1938: R. R. WILLIAMS and T. D. SPIES. — Vitamin B_1 (thiamin) and its use in medicine. New York, Macmillan, 411 p.

1938: J. BONNER and J. GREENE. — Vitamin B_1 and the growth of green plants. Botan. Gaz., 100:226.

1939: J. BONNER and J. GREENE. — Further experiments on the relation of vitamin B_1 to the growth of green plants. Botan. Gaz., 101:491.

1939: P. PRÉVOT. — Production expérimentale de bourgeons sur des feuilles de *Begonia rex* non isolés de la plante: action de la colombine. Bull. Soc. roy. Sc. Liège, 1939, no. 2, p. 136.

1939: V. G. LILLY and L. H. LEONIAN. — Vitamin B_1 in soil. Science, 89:292.

1939*a*: A. SCHEUNERT und K. H. WAGNER. — Weitere Untersuchungen an Gerste über die Frage ob Düngungseinflüsse auf den Vitamin B-Gehalt bestehen. Biochem. Z., 303:200.

1939*b*: A. SCHEUNERT und K. H. WAGNER. — Über den Einfluss der Düngung auf den Vitamin-Gehalt (mit quantitativen Vitamin A-Bestimmungen in Grünkohl, Spinat, Rosenkohl und Tomaten). Biochem. Z., 303:266.

1939*c*: A. SCHEUNERT und K. H. WAGNER. — Über die Vitamin A-Wirkung von Möhren (*Daucus carota*) unter dem Einfluss verschiedener Düngungsweisen. Biochem. Z., 303:208.

1939*d*: A. SCHEUNERT und K. H. WAGNER. — Die Konstanz des Vitamingehalts in frischen Vegetabilien. Ber. d. Mathem.-Physischen Klasse der sächsischen Akad. der Wissenschaften, Leipzig, 91:307.

1940: A. SCHEUNERT, J. RESCHKE, und E. KOHLEMANN. — Über den Vitamin C-Gehalt der Kartoffeln, IV. Mitteilung. Biochem. Z., 305:1.

1940: G. LUNDE. — Vitamine in frischen und konservierten Nahrungsmitteln. Berlin, Julius Springer, 272 p. (This contains the most recent data concerning the vitamin content of foods).

1941: W. F. MÜLLER. — Zur Wirkstoffphysiologie des Bodenpilzes *Mucor Ramannianus*. Ber. d. schweiz. Bot. Ges., 51:165.

Chapter XXII.

GROWTH FACTORS, VITAMINS, AND SEXUALITY.

The phenomena of sexuality occur normally in the life of plants. For a long time they were regarded as a mere continuation of vegetative development which influenced them without any particular factors coming into play.

Higher plants. — In the higher plants the formation of the flower and the sexual organs as well as gametogenesis are extremely complex phenomena. One can distinguish: (1) a state of maturity of the plant preparatory to the development of the flower; (2) the appearance of the flower primordia; (3) the development of the flower and the appearance of special reproductive organs; (4) haplosis, terminating in the formation of the haploid male and female gametophytes; (5) syngamy leading to the formation of the zygote and endosperm; (6) the growth of the zygote, the development of the embryo, and the formation of the mature seed and fruit.

The preliminary stages of reproduction, which are of a nonsexual nature, have been the subject of numerous investigations concerning the nutrition and metabolism of the plant. KLEBS, using *Sempervivum* and *Sedum,* proposed the theory that the production of flowers is directly related to the carbon/nitrogen ratio. Although it can sometimes be observed that the first period is associated with a high nitrogen content (MURNEEK, ASAMI and ITO), there is no direct and constant relation between the formation of primordia and a particular carbon/nitrogen ratio. At any rate, this factor is not primary and determinative (MURNEEK, 1939).

Today the initiation of reproduction is regarded as the result of hormone action. SACHS had already advanced the hypothesis that the development of the flower depends on "specific floral substances" (*Blühstoffe*) developed in the leaves and diffused up to the meristem where they initiate the differentiation of the primordium. Such substances are neither known chemically, nor have they been isolated. CAJLACHJAN has revived this hypothesis and has modernized it by giving the name *florigene* to the substance which causes the appearance of the floral primordia and *vernaline* to that responsible for the subsequent development.

The formation of these substances is related to photoperiodism. In plants with long photoperiods, *i.e.,* long-day plants, the floral substances develop only after long daily periods of illumination; in plants with short photoperiods, *i.e.,* short-day plants, a short daily period of illumination is required to produce these substances. HAMNER and BONNER (1938), working with *Xanthium pennsylvanicum* (with a short photoperiod), discovered that the light stimulus is perceived by a single leaf subjected to a short photoperiod. The result of this stimulus is the rapid biosynthesis of specific floral substances, which are transmitted to the meristem through other parts of the plant, even if these are subjected subsequently to a long photoperiod.

Despite our ignorance of the chemical nature of these substances, their existence is quite probable in view of the fact that in grafts they can pass from one partner to the other (MELCHERS, KUIJPER and WIERSUM). There is no doubt that we are dealing with a typical hormone phenomenon (according to our definition), which we shall not discuss in greater detail here. We simply follow HAMNER and BONNER in stating that the hormone responsible for floral initiation cannot be replaced by the known vitamin factors: vitamins B_1, B_2, B_6, ascorbic acid, nicotinic acid, pantothenic acid, *i*-inositol, nor by the hormone heteroauxin, nor by such hormones as theelin (estrone) or theelol (estriol). Some investigators think that the action of the "female hormone" (theelin) offers a clue to the understanding of photoperiodism. CHOUARD observed that after eight hours of daily exposure to light, *Calisthephus sinensis* reacts positively to the addition of 25γ of dihydrotheelin per week, *i.e.*, flowering is advanced. There is no advance when the daily exposure to light is 12 hours. Flowering is poor and dihydrotheelin is ineffective if the expose to light is 18 or 20 hours. It seems, therefore, that this hormone "can partially replace a deficit in illumination", and that it "behaves like the hypothetical hormones of flowering invoked in the mechanism of photoperiodism." These experiments have not been confirmed and the rôle of the "female hormone", present also in the plant kingdom, remains obscure. The whole phenomenon is very complex; it is probable that certain vitamins may also play an indirect rôle.

Various lower microorganisms. — A study of the nature of sexuality must include microorganisms. In these the morphological phenomena of sexuality are definitely less complicated and the physiological aspects are more accessible to experimentation. In the lower microorganisms the expression of sexuality is closely dependent on nutrition and metabolism. Yet specific factors of sexuality have recently been demonstrated, as for example in *Phytophthora* (LEONIAN).

Schizophyllum commune (Basidiomycete) requires thiamin as a growth factor, but it can be satisfied with pyrimidine (ROBBINS, SCHOPFER and BLUMER). The haploid mycelia of both sexes have the same requirements, thiamin being the essential growth factor. With this factor alone the diploid fruit bodies, produced after plasmogamy, develop poorly. No doubt other essential factors must be involved (SCHOPFER and BLUMER, 1940). THREN (1941) found that the haploid mycelium of *Ustilago nuda* differs physiologically from the mycelium of the dicaryon phase, the former being auxo-autotrophic while the latter requires the pyrimidine component of thiamin.

The telemorphic chemical influences, which were regarded as being of hormonal nature, have been studied in various fungi: the formation of progametangia in the *Mucoraceae* (BURGEFF); the formation of perithecia in *Neurospora* (MOREAU); the formation of oogonia and antheridia in *Achlya* (RAPER, 1939*a* and *b*). In the last case RAPER demonstrated the existence of four different sexual hormones: one secreted by the female mycelium inducing the formation of antheridia; a second secreted by antheridial branches causing the appearance of oogonial initials; a third secreted by the oogonial initials producing a directive growth on the part of the antheridial hyphae; and a fourth secreted by the antheridia, causing a wall to be formed at the base of the oogonium. The existence of these various substances is proved biologically but not chemically. An attempt by RAPER (1942) to determine the chemical nature of the substance secreted by the female mycelium inducing the formation of antheridia (hormone A), in which he tested the hormonal activity of various known substances, revealed two groups of compounds with positive effect. The first group is represented by the saturated dicarboxylic acids. Three of these, glutaric, malonic, and pimelic acid, were found to induce the formation of a limited number of antheridial hyphae when applied to the male plants. These acids are not as effective on the male plants as they are in causing the female plants to secrete the hormone. The second group contains barbituric acid and certain of its substitution products. Some of these are inactive while others are more active than the unsubstituted barbituric acid. These compounds are active only when present in relatively high concentration (10^{-5}). The barbiturates, unlike the active dicarboxylic acids, have no apparent effect on the female plant and the production of hormone A.

This problem can be studied in greater detail in the *Mucoraceae* since they produce known specific sexuality factors.

In *Mucor hiemalis* the influence of the nutritive medium on the production of zygotes (therefore on the origin of the gametangia and on syngamy) is well known (SCHOPFER, 1928). Definite quantities of a carbon (maltose) and of a nitrogen source (asparagine) are necessary, as well as a definite ratio between these two constituents of the medium. Above or below a certain ratio, zygotes develop poorly or not at all. *Mucor hiemalis* is an auxo-autotrophic organism.

In *Phycomyces Blakesleeanus* (SCHOPFER, 1931-32) the natural medium is indispensable for the formation of zygotes. In a liquid synthetic medium with vitamin B_1, the vegetative development is excellent but the zygotes do not form readily. Thus other specific conditions for sexuality, besides thiamin, must be involved. We know that certain general conditions of nutrition have specific effects on sexuality. On a natural medium, with a malt extract base, vegetative development and the formation of zygotes occur.

If the medium is enriched with gelatin, the vegetative development is luxuriant but zygotes fail to develop. The inhibition is complète. The best medium for *Phycomyces* is composed of maltose (containing an active vitamin as an impurity) and of 0.1% asparagine. A synthetic medium with a glucose base induces the formation of zygotes only if the impurity contained in maltose is added to it. The zygotes do not form as well when the amount of asparagine is increased. This partial inhibition can be compensated for by the addition of a very large quantity of the vitamin impurity found in maltose. Development proceeds as if the latter had overcome the inhibition produced by too large an amount of asparagine (SCHOPFER, 1931-32).

At the time when the action of pure vitamin B_1 on *Phycomyces* was observed, the author assumed that it replaces the vitamin impurity in maltose (SCHOPFER, 1934). Thus, without actual proof, the possibility of the presence of two substances was assumed: one acting on the vegetative development, and the other on sexuality (SCHOPFER, 1931-32). Later ROBBINS (1939) took up the problem on an experimental basis. He proved that in a liquid medium vitamin B_1 does not suffice to assure the formation of zygotes, despite the excellent growth of the mycelium. On the other hand, if a solid agar medium is employed, as the author did in his first experiments, the zygotes develop in the presence of vitamin B_1 alone. Thus, the agar must contain a particular substance whose action, in conjunction with that of thiamin, induces sexuality. The agar can be purified by subjecting it to and extraction by methyl alcohol or aqueous pyridine which removes the active substance. This purified agar, even in the presence of thiamin, is no longer suited for the formation of zygotes. The substance contained in the agar is not indispensable for vegetative growth; however, under certain conditions it has a beneficial effect on growth.

Under normal conditions of culture on a liquid synthetic medium containing thiamin, *Phycomyces* is incapable of synthesizing one substance necessary for reproduction. Temperature plays an important rôle in the synthesis of this substance. At 20°C. the synthesis of this factor is possible and, in a liquid medium with vitamin B_1, zygotes are produced. Conversely, at 25°C. the synthesis is sufficient to meet the needs of vegetative development but not those for the formation of zygotes. This substance is not identical with the known vitamins: riboflavin, pantothenic acid, nicotinamide, pyridoxine (vitamin B_6), carotene and vitamin E. The substance in question is present in various plant extracts. Potatoes are rich in it and can serve as a source for experiments. Attempts at purification and identification (ROBBINS and HAMNER, 1940) lead to the fact that two factors are present in potato extract: one is adsorbed on animal charcoal and is eluted with an ammoniacal-acetone mixture (factor Z_1); the other is only slightly adsorbed and is present in the filtrate (Z_2). Each of the two factors acts alone but the best effect is produced by the combination of the two in the presence of vitamin B_1. Factor Z_1 has been identified as hypoxanthine. The same activity is exhibited by guanine (see ROBBINS and KAVANAGH, 1942, concerning the specificity of action of this factor). Factor Z_2 remains unidentified. Apparently mineral catalysts are not involved since they were supplied in the medium employed by ROBBINS.

The author has been able to confirm the results of ROBBINS in so far as the effect of agar on sexuality is concerned. But in the presence of a sufficient quantity of agar (supplying a supraoptimum of factor Z), thiamin plays the rôle of a limiting factor and the number of zygotes (already formed) to reach maturity depends on the amount of thiamin supplied. With a medium containing 2% crude agar in 20 cc. of the usual nutritive solution, the maximal number of zygotes is formed with 0.4-0.5γ of thiamin.

Factor Z_1 (hypoxanthine) is highly thermostabile. The author considers it as one of the constituents of factor M (1935), including the components of thiamin, acting eventually on *Phycomyces* and other *Mucoraceae* (SCHOPFER, 1940). LEONIAN and GREENE (1940) assert that a complete medium is indispensable for a normal development of zygotes but they were unable to demonstrate a growth factor with a specifically sexual action.

The problem of *Phycomyces* which seemed simple in the beginning becomes surprisingly complicated.

These experiments show us that sexuality in general is dependent on specific factors but they do not enable us to penetrate further into the problems of sex determination and of its chemical mechanism. However, experiments with flagellates show us a different aspect of the problem.

Flagellates. — The remarkable work of KUHN and MOEVUS (summarized by THIMANN, 1940 and by MURNEEK, 1941), on the sexuality of *Chlamydomonas eugametos* has exposed the problem of sexuality in an entirely different light. In these organisms under certain conditions, the entire flagellated and motile cell functions as a gamete.

The starting point of these experiments was the following observation: *non-motile* female gametes become motile under the influence of a filtrate from a culture of motile female gametes, the same filtrate induces motility in male gametes. Obviously the filtrate containing the substances secreted by the gametes produces a chemical effect, whose origin and nature should be determined.

The phenomena of motility, of gametic union, and of sex determination are intimately related; they depend on the same category of substances. However, it is better to study them separately.

Motility (MOEVUS, 1938). — The cells from an agar culture are non-motile and without flagella. If they are placed in the light and in water, they become motile but after a few moments the flagella become paralyzed (*Geisselstarre*). The duration of their motility is directly proportional to the exposure to light. It is equal to one minute and three seconds for an exposure of five minutes; 59 minutes and 18 seconds for an exposure of three hours. Longer exposure does not increase the duration of motility; the effect is quantitative. Red, yellow, green, blue, and violet rays have the same effect. In water and in darkness motility does not appear. In the dark with 1% glucose and without oxygen, the cells remain without flagella. In the dark with 1% glucose but with oxygen, the cells become flagellated and motile.

There is no doubt that the effect of light may be replaced by sugars. These sugars in order of decreasing activity are: *gentiobiose, d*-glucose, cellobiose, lactose, saccharose and raffinose.

Finally, the effect of light, or of darkness plus oxygen plus glucose, can be obtained with a filtrate from motile cells. The latter have formed the specific motility substance which has diffused into the medium and acted on the non-motile cells.

An analysis of the filtrate (200 liters) leads to the conclusion that the substance has the characteristics of a carotinoid; the two absorption bands at 4670 and 4380 Å are those of *trans*-crocetin. Hydrolysis with HCl yields glucose in a quantity similar to that furnished by crocin. Therefore, it is certain that the motility substance is represented or replaced by *crocin.* This carotinoid, $C_{44}H_{64}O_{24}$, is a glucoside - the digentiobiose of crocetin, $C_{20}H_{24}O_4$. Acid hydrolysis of the glucoside furnishes only *glucose;* hydrolysis with alcoholic ammonia, *gentiobiose.* It should be remembered that in darkness glucose and gentiobiose were the most active sugars.

The activity of crocin is still evident at a dilution of 1:250,-000,000,000,000. At this dilution, there are approximately 2.4×10^6 molecules per cubic centimeter; thus a small number of molecules suffices to make a cell motile (five molecules for four cells).

Gametic union (MOEVUS, 1938). — *Chlamydomonas eugametos* is heterothallic and isogamous; sex determination is genotypic. The sex of the gametes can be ascertained, however, by growing *C. eugametos* with *C. Braunii* which is anisogamous.

The cells of *C. eugametos* having become motile in darkness in the presence of oxygen and glucose or in the presence of the filtrate from an irradiated culture are unable to undergo gametic union (to copulate). A specific copulation substance is necessary. This is formed by a photochemical reaction which is dependent on blue or violet light. It is not necessary to irradiate the organisms themselves, for exposure of the filtrate from cell cultures which had been previously made motile (with darkness + glucose) will produce this copulation substance. A study of this filtrate permits an analysis of the substance inducing gametic union.

If the male and female filtrates are exposed to blue or violet monochromatic light, they become inactive. If they are irradiated with red or green light, they retain their activity. Therefore, the active substances for gametic union are *formed* in blue light and are also *destroyed* by blue light.

Experimentation with progressive irradiation will furnish us a definitive explanation. Active male and female filtrates are subjected to blue light; at regular intervals they are tested with motile male and female cells (having undergone the darkness-glucose treatment). After 30 minutes of irradiation, the *female* filtrate is still active on *female* cells. After 75 minutes and up to 105 minutes of irradiation, the *female* filtrate initiates the union of *male* cells. After 105 minutes of irradiation, the female filtrate has become definitely inactive (on males and females). The conclusion drawn is that by proper timing of irradiation it is possible to transform the *female* filtrate into a *male* filtrate.

The male filtrates behave differently: after 30 minutes of irradiation with blue light, they are still active on male cells. A prolongation of irradiation to 12 hours renders them definitively inactive.

If the cells given the darkness-glucose treatment are irradiated with red light, they do not become capable of uniting. If after an hour of such irradiation, a filtrate is prepared which is absolutely free of cells, and if this filtrate, irradiated with red light, as in the preceding experiment, is subjected to progressive irradiation with blue and violet light, the female filtrate is at first active on female cells, then on male cells; with the same time of irradiation, the male filtrate is active at first on female cells, then on male ones. Therefore, under the influence of red light, the motile cells (darkness + glucose) have synthesized a *precursor* without acquiring the ability to effect a union; at the expense of the precursor the female filtrate is formed first, then the male, and finally the inactive stage appears:

red light
↓
motile cells from darkness + glucose treatment
↓
filtrate
↓
precursor (inactive) → female substances (30 minutes) → male substances (60 minutes) → final substance, inactive (12 hours)
(increasing irradiation with blue-violet light)

The filtrate with female activity is composed of three parts of the precursor plus one part of the final substance. The filtrate with male activity is composed of one part of the precursor plus three parts of the final substance. Mixtures in other proportions are inactive.

The following experiments prove that the two important substances, the precursor and the final one, are carotinoids derived from crocin; the precursor is replaceable by the *dimethyl ester of cis-crocetin,* the final substance by the *dimethyl ester of trans-crocetin.* These two substances behave like the female and male substances in so far as red and blue light are concerned. The same mixtures are active: three parts *cis-* plus one part *trans-* equal female action; one part *cis-* plus three parts *trans-* equal male action. The two substances are active at a dilution 1:33,000,000. They are less active than the source substance, crocin (motility substance).

Biological experiments teach us that the two sexual potentialities, male and female, are the extremes of a series composed of numerous intermediates. Thus we see how *one* substance, through successive chemical transformations, acquires first a female effect and then a male effect. *Sex is not determined by any one substance, but rather by a ratio between two forms of the same substance.* As a result, intermediates may exist. By varying the time of exposure one obtains (MOEVUS, 1939), with eight types of gametes for the two sexes, four sexual intensities: very strong (4), strong (3), medium (2), and weak (1). Each type of gamete produces a specific substance, *i.e.*, a specific mixture of the precursor and of the final substance. By irradiating the eight types of filtrates with blue-violet light they are inactivated for some types of gametes but activated for others.

By treating the eight types of gametes with mixtures of *cis*- and *trans*-crocetin in varying proportions, it is possible to determine the mixture in which each type of gamete becomes active; an astonishing agreement is established:

Intensities obtained by the irradiation of filtrates

female →				male →			
(4)	(3)	(2)	(1)	(1)	(2)	(3)	(4)
95/5	85/15	75/15	65/35	35/65	25/75	15/85	5/95

Corresponding *cis/trans* ratios

The transformation from one stage to the following requires 10 minutes, except that from female (1) to male (1) which requires 30 minutes. One minute is necessary to transform 1 percent of *cis*- into *trans*-crocetin (methylester). Thus relative sexuality and sexual valence, expressed in *Chlamydomonas eugametos* by genetically determined valences, have been demonstrated chemically.

Sex Determination (KUHN, MOEVUS and WENDT, 1939). — *Chlamydomonas eugametos* f. *synoïca* is homothallic (♀♂) and the gametes have genotypically undetermined sex (determination is phenotypic). If the cells are treated with a typical female filtrate, all of them become females, but if treated with a male filtrate, all of them become males. The filtrates, therefore, must contain substances capable of *determining* sex. These substances are related to those just described; the female-determining substance is a glucoside, *picrocrocin* (gentiobioside of *safranal*), whereas the male-determining one is *safranal*. Their action is observed only in homothallic forms. In heterothallic forms no reaction to picrocrocin and to safranal takes place.

The properties of the female-determining (replaceable by picrocrocin) and male-determining (replaceable by safranal) substances are very different. The male substance is soluble in ether and can be extracted with them. The female substance lacks these properties; it is destroyed by heating with dilute H_2SO_4 or dilute $Ba(OH)_2$, and as a result of this inactivation acquires male activity.

That picrocrocin and safranal are really the sex determining substances can be proved in the following way: assuming that this fact is correct, the cells of the homothallic f. *synoïca* must, under the action of picrocrocin and safranal, produce and secrete the dimethyl esters of *cis-* and *trans-*crocetin in the proper proportions. Experiments reveal that under the influence of picrocrocin (female-determining substance) the cells produce the *cis/trans* ratio 65/35; whereas, under the influence of safranal they produce the *cis/trans* ratio 35/65, according to expectations.

Picrocrocin and safranal are really hormones which determine *primary sexual characters.* These two hormones determine sex because they control gametic union by means of the *cis/trans* ratio.

The three groups of substances with which we are dealing are the following:

(1) The motility substance, replaceable by crocin and digentiobiose.

(2) The copulation-inducing substances, called gamones, replaceable by the dimethyl esters of *cis-* and *trans-*crocetin in definite proportions (andro- and gynogamones).

(3) Sex determining substances called termones, replaceable by picrocrocin and safranal (andro- and gynotermones).

These three groups are closely related chemically and are derived from one another as shown in the following table. We assume the existence of a common starting substance for the two sexes, and for all observed phenomena: *protocrocin,* composed of digentiobiose, of crocetin, and of picrocrocin.

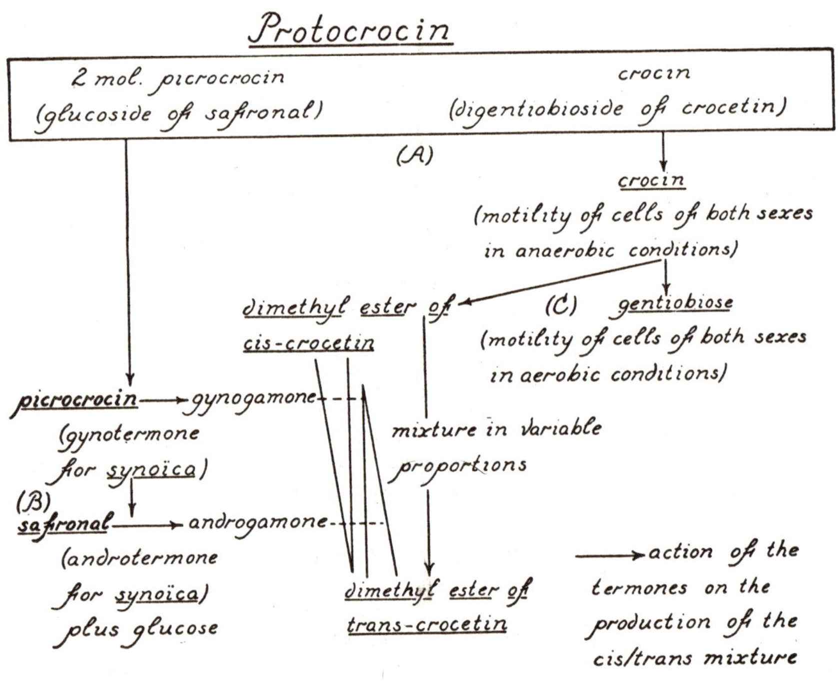

SEX DETERMINATION SYNGAMY MOTILITY

The activity of safranal is such that one can make the following theoretical calculation: a 10^{-1} molar solution of picrocrocin is dissociated to the extent of 50% by 1.7×10^8 cells after 30,000 years.

Summary. — *Chlamydomonas* contains in its plastids a carotinoid which brings about the formation of crocin and picrocrocin. This organism decomposes picrocrocin which supplies safranal and sex determination results. Under the influence of safranal, *cis*-crocetin (produced from crocin) and *trans*-crocetin are secreted in variable proportions and cause gametic union. This is possible because of the motility initiated by the gentiobiose furnished by crocin.

By disintegration, the following substances are formed: picrocrocin and crocin from protocrocin (A), safranal from picrocrocin (B), the esters of crocetin from crocin (C). The B and C reactions are of an enzymatic nature, the enzymes being dependent on a gene (KUHN and MOEVUS, 1939).

Reaction A (Motility gene). — The substances for motility (crocin) and sex determination (picrocrocin) always appear together; this substantiates the preceding table.

Reaction B (Gathe gene). — The male-determined cells, producing safranal, are capable of hydrolyzing picrocrocin. The enzyme, detected biologically, has its maximum activity at *p*H 7 and at a temperature of 26°C.

Reaction C (M_D gene). — All cells producing gamones (esters of crocetin) possess an esterifying enzyme, permitting the substitution of the gentiobiose residue by a CH_3 group. In many species of *Chlamydomonas,* this enzyme has its maximal activity at *p*H 7. The optimal temperature varies, according to the species, from 12° to 27°C.

These three reactions show us how these substances originate from one another.

It is not yet understood what effect the termones, picrocrocin and safranal, exert on the *cis/trans*-crocetin ratio. The termones are the substances that govern this ratio and bring about the formation of specific gamones. It is proven experimentally that the cells of f. *synoïca,* when treated with picrocrocin, produce more *cis*- than *trans*-crocetin (dimethyl ester), and that they have female properties. Conversely, those treated with safranal produce more *trans*- than *cis*-crocetin. Thus picrocrocin accelerates and safranal inhibits the formation of the dimethyl ester of *cis*-crocetin; in other words, the variations of the *cis/trans* ratio are produced by an action of the termones on the esterifying enzymes, on which the "formation" of the gamones depends. Obviously, this action of the termones on the gamones can not be demonstrated in homothallic cells (as in f. *synoïca*). In the heterothallic forms (*C. eugametos* type), the *cis/trans*-crocetin ratios are determined genetically.

Such experiments lead us to a thorough understanding of the mechanism of genotypic and phenotypic sex determination and of its chemical basis. They also touch on the important problem of the relations between genes and active substances. In the author's opinion this is the fundamental part of the problem. We have many examples in genetics in which the quantitative variations of a factor correspond to equivalent variations in the production of a vitamin; JOHNSON and MILLER found that in *Zea Mays* the β-carotene content of the endosperm corresponds to the number of dominant Y genes. The same correspondence is observed with vitamin A activity.

The experiments of MOEVUS and KUHN are remarkably precise. They were conducted in a field in which all expectations were exact and where everything took place according to the wishes of the investigator. There is no reason for doubting these experiments. Nevertheless, it would be comforting to see them confirmed.

The flagellates are at present the only organisms in which such observations have been made. The mechanisms demonstrated must be of general occurrence: firstly, because of the general distribution of carotinoids in the whole plant kingdom; secondly, a remark of KUHN renders this generalization still more probable: the β-carotene of leaves can be oxidized into two molecules of safranal and one of picrocrocin.

Picrocrocin (a colorless glucoside) and crocin are found in saffron, and are produced by the stigmas of *Crocus*. These substances will probably be found in the sexual organs, stigmas, and pollen of other flowers. If they are not found there preformed, they may originate there.

Actually, by the use of the *Chlamydomonas* test for motility, it is possible to prove biologically that crocin is present in the pollen and the stigmas of the following plants: *Allium ursinum, Tulipa australis, Narcissus poeticus, N. odorus, Iris Reichenbachii, Convallaria majalis, Lilium candidum, Anemone pulsatilla, Cheiranthus Cheiri, Wistaria sinensis, Aesculus Hippocastanum, Syringa vulgaris, Bellis perennis, Borrago officinalis, Ribes aureum, Potentilla fruticosa.*

Various observations have shown that an increased quantity of carotinoids is present in the reproductive organs at the time of reproduction. CHODAT and SCHOPFER reported a large quantity of carotinoids in the progametangia of *Mucor hiemalis*. MURNEEK (1939) analyzed the vegetative and reproductive organs of *Cosmos, Salvia*, and *Soja* at the time of anthesis and found that the reproductive organs contain a much greater amount of carotinoids than the vegetative ones. The difference is also clear for xanthophyll. The abundance of carotinoids at the time of flowering has likewise been recorded by VIRTANEN and VON HAUSEN (see p. 84), and by NAGASIMA. It is certain that these substances are connected with the sexual phenomena in plants as well as in animals. These reports suggest the possibility of generalizing on the theories of KUHN and MOEVUS.

In conclusion, attention is called to the fact that in many different phenomena where the effect of a light stimulus is evident, carotinoids are found interposed as receptors. Carotinoids absorb light in the blue and violet wave lengths. In *Phycomyces* large quantities of carotinoids are formed preferably under the influence of blue and violet rays (SCHOPFER). These are the same rays which, absorbed by the carotinoids of the *Phycomyces* sporangiophore and by the *Avena* coleoptile, initiate the greatest phototropic curvatures (BÜNNING). Finally, under normal conditions, the blue rays are responsible for the appearance of the gamones (esters of crocetin) in *Chlamydomonas*. This effect is not explained by the experiments of MOEVUS.

The cases in which specific substances for sexuality have been demonstrated, are not very numerous; however, they suffice to illustrate the full importance of the problem. The particular state of development of an organism, during which sexuality must become expressed, is definitely under the control of hormone and vitamin factors, specific in their effects but general in their distribution.

REFERENCES:-

1928: W. H. SCHOPFER. — Recherches sur la sexualité des champignons. Le problème de la biochimie comparée du sexe. Bull. Soc. bot. Genève, 2ème série 20:149.

1931-32: W. H. SCHOPFER. — Recherches expérimentales sur la formation des zygotes chez *Phycomyes blakesleeanus*. Influence des substances vitaminiques. Bull. Soc. bot. suisse, 40:87, 41:73.

1934: W. H. SCHOPFER. — Les vitamines cristallisées B comme hormones de croissance chez un microorganisme (*Phycomyces*). Arch f. Mikrobiol., 5:513.

1938: F. MOEVUS. — Carotinoïde als Sexualstoffe von Algen. Pringsheim Jahrb. f. Bot., 86:753.

1938: K. C. HAMNER and J. BONNER. — Photoperiodism in relation to hormones as factors in floral initiation and development. Bot. Gaz., 100:388.

1939*a*: JOHN R. RAPER. — Sexual Hormones in *Achlya*. Third Intern. Congress for Microbiology, Report of Proceedings, p. 494.

1939*b*: JOHN R. RAPER. — Sexual hormones in *Achlya*, II. Distance reactions, conclusive evidence for a hormonal coördinating mechanism. Amer. J. Bot., 26:162.

1939: A. E. MURNEEK. — Physiological factors in reproduction of plants. Growth, 3:295. (Bibliography!).

1939: W. J. ROBBINS. — Growth substances and gametic reproduction by *Phycomyces*. Bot. Gaz., 101:428.

1939: F. MOEVUS. — Untersuchungen über die relative Sexualität von Algen. Biol. Zentralbl., 59:40.

1939: R. KUHN, F. MOEVUS und G. WENDT. — Über die geschlechtsbestimmenden Stoffe einer Grünalge. Ber. d. d. chem. Ges., 72:1702.

1939: R. KUHN und F. MOEVUS. — Über die chemische Wirkungsweise der Gene Mot, Md und Gathe bei *Chlamydomonas*. Ber. d. d. chem. Ges. 73:547.

1940: W. J. ROBBINS and K. C. HAMNER. — Effect of potato extracts on growth of *Phycomyces*. Bot. Gaz., 101:912.

1940: W. H. SCHOPFER. — Facteurs de croissance et sexualité chez *Phycomyces*. C. r. Soc. Phys. Hist. nat. Genève, 57:45.

1940: W. H. SCHOPFER et S. BLUMER. — Le pouvoir de synthèse d'un facteur de croissance par *Schyzophyllum commune* (haplontes et diplontes). Protoplasma, 34:524.

1940: L. H. LEONIAN and V. G. LILLY. — Studies on the nutrition of fungi, V. Factors affecting zygospores formation in *Phycomyces blakesleeanus*. Amer. J. Bot., 27:670.

1940: K. V. THIMANN. — Sexual substances in the algae. Chronica Botanica, 6:31-32.

1941: A. E. MURNEEK. — Sexual reproduction and the carotenoid pigments in plants. The Amer. Nat., 75:614-620.

1941: R. THREN. — Zur Entwicklungsphysiologie der Dicaryophase von *Ustilago nuda* (Jensen) Kellerm. und Sw. Die Bedeutung des Aneurins und seiner Komponenten. Arch. f. Mikrobiol., 12:192.

1942: W. J. ROBBINS and F. KAVANAGH. — Hypoxanthine, a growth substance for *Phycomyces*. Proc. Nat. Acad. Sci. U.S.A., 28:65-69.

1942: J. R. RAPER. — Sexual hormones in *Achlya*, III. Hormone A and the initial male reaction. Amer. J. Bot., 29:159-166.

Chapter XXIII.

SYMBIOSIS, PARASITISM, AND VITAMINS.

Important discoveries were made when the available data on vitamin growth factors were used toward the explanation of symbiosis and related phenomena.

Symbiosis. — *Microorganisms.* — The mutual relationship of microorganisms in mixed cultures has been studied for a long time. It has been observed that sometimes two organisms grown in the same medium cause an inhibition of growth; either one of the two organisms does not develop or the growth of both is inhibited. In this case one speaks of unilateral or reciprocal *antibiosis.* Conversely it is possible that one of the two organisms is favored in its development by the presence of the other, or in the same manner, they aid each other reciprocally. This case is known as unilateral or reciprocal *stimulation.*

These phenomena occur in culture media, but they may also appear in the hosts of microorganisms. This case is called either unilateral or reciprocal *synergism* (corresponding to stimulation) or *antagonism* (corresponding to antibiosis). The latter is of less interest to us (see PAPACOSTAS and GATÉ, 1928; WEINDLING, 1938; WAKSMAN, 1940, 1941).

Unilateral and reciprocal stimulation. Satellitism. — Satellitism has already been described with reference to hemophilic bacteria. A Pfeiffer bacillus culture is more or less strongly stimulated by the presence of contaminating organisms (see p. 168), particularly by *Staphylococcus aureus.* The phenomenon of satellitism (DAVIS) corresponds to the giant colonies of GRASSBERGER and to the "nurse colonies" of NEISSER. This stimulation, which is here unilateral, may be explained in various ways. It might be assumed that the growth of *B. influenzae* is favored by the direct action on the medium, due to the production of an acid or alkali by the organism. However, this explanation had to be abandoned when it became known that bacterial extracts exercise the same effect as the living colonies. Instead it must be a matter of the production of certain substances which the Pfeiffer bacillus is perhaps no longer capable of synthesizing and which are now furnished by the nurse colony. These beneficial factors, produced by the satellite (nurse colony), diffuse into the medium and permit the growth of the organism.

When considered in the light of actual findings, most cases undoubtedly involve vitamin action. We encounter again the general phenomenon described many times in preceding chapters:

extracts of an autotrophic (with respect to a given factor) species activate the growth of a heterotrophic (with respect to the same factor) species (as long as other inhibiting factors do not enter into play). Unilateral stimulation, or satellitism *in vivo* under normal biological conditions, can be produced artificially by applying the above-mentioned principle.

Not every case of satellitism can be traced to a vitamin effect. A satellite microorganism can modify the physico-chemical properties of the medium, *p*H, *r*H, etc. By enzymatic action it can render the medium more assimilable. A typical case is represented by the association of *Mucor Rouxi* Wehmer and *Micrococcus roseus.* The former is able to convert starch into sugar and under certain conditions grows on potato producing yellow-orange pigments, whereas the latter does not grow on potato and requires maltose as a source of carbon. If they are inoculated together, *Mucor* grows normally by converting starch into sugar and *Micrococcus* also grows due to the maltose produced by the action of *Mucor* (Vuillemin). Despite these exceptions the conclusion is that the stimulation is frequently due to a vitamin action.

The contamination of a culture of an auxo-heterotrophic fungus by a yeast (rich in vitamins), another mold, or a bacterium often results in a good growth of the fungus. Since the medium contains only assimilable substances and since only the absence of a growth factor prevents the growth of an auxoheterotrophic fungus, it must be concluded that the contaminant acts by producing a vitamin.

In cultures of *Phycomyces* on synthetic media made without vitamins, *Penicillium* may appear at times as contaminants. The latter is auxo-autotrophic and grows well without vitamins. Whereever the *Penicillium* colony grows, the mycelium of *Phycomyces* also grows as vigorously as in the presence of vitamins. *Penicillium* has acted by furnishing *Phycomyces* the missing growth factors (thiamin among others). This fact, which has frequently been observed by the author since 1933, has been confirmed by Robbins. This does not constitute a phenomenon of symbiosis. A similar example is that of an auxo-heterotrophic *Polyporus* whose growth in nature is dependent on the presence of living bacteria which act through their extracts, the active substance being thiamin.

Cultures of *Strigomonas* (flagellate) sometimes become contaminated by yeasts, by fungi such as *Aspergillus niger,* or by bacteria such as *Staphylococcus, Sarcina,* etc. These organisms enable a culture of *Strigomonas* to thrive even if the medium is not provided with blood, which we know is indispensable to this flagellate. The latter is dependent on the presence of hemin. It must be admitted, therefore, that the contaminating fungi and bacteria stimulate the flagellate unilaterally by furnishing it the small quantity of hemin required.

Numerous reports occur in the literature of cases of unilateral stimulation between aerobic microbes, anaerobic microbes, and between fungi and microbes. In many instances the introduction of the concept of growth factors would undoubtedly provide the explanation of the mechanism of stimulation. Moreover, GOY by observing the favorable action of a *Mucor* filtrate on the growth of microbe cultures has attempted to determine the exact mechanism of stimulation. He succeeded in isolating a substance active in very small dosages and attempted to explain this stimulation by the action of vitamins.

The cases cited here concern the action of vitamin B_1, of bios, and of hemin produced by the stimulating organisms. No doubt other cases involving other growth factors and vitamins will be found.

Metabiosis can also be explained in the same way. This phenomenon represents a particular kind of stimulation according to which a microorganism does not grow on a given medium unless a different microorganism has previously grown on it. A classical case of polymetabiosis is represented by the formation of nitrates at the expense of organic nitrogen substances. The two nitrifying bacteria, *Nitrosomonas* and *Nitrobacter*, inoculated together or separately do not grow. With the introduction into the medium of a third bacterium endowed with proteolytic properties (*Bacillus ramosus* for example), growth of the three organisms is made possible and the whole cycle of nitrification can be accomplished. *B. ramosus* produces ammonia which permits the development of *Nitrosomonas*, and the nitrous acid produced by the latter is oxodized to nitric acid by *Nitrobacter*.

The production of "sake" (Japanese alcoholic beverage prepared from rice) may likewise be described as an example of metabiosis in which the primary action of a fungus, *Aspergillus oryzae*, produces sugar from starch, and the subsequent action of yeast produces alcohol. These and other cases are explained apparently without reference to vitamin growth factors. It is simply a preparation of a medium by one organism which elaborates assimilable substances for the next organism which follows. On the other hand, there are cases in which metabiosis depends on the production of vitamins functioning as growth factors. *Streptothrix corallinus* (*Actinomyces corallinus*) requires a growth factor which seems to be thiamin and which is present in a preparation of toruline by PETERS (READER). This fungus does not grow in a purely synthetic medium; but if the medium is first inoculated with a *Meningococcus*, the fungus is able to grow. No doubt the bacterium has produced growth factors favorable to the fungus. It is also known that *Meningococcus* cultivated in association with the Pfeiffer bacillus stimulates the latter (WOLF, DAVIS).

Obviously the cases in which the culture of a microorganism is favored or inhibited by the extracts of another microorganism may be regarded as being related to metabiosis (see above).

Generally speaking, unilateral stimulation acts in a vital manner by assisting growth, *i.e.*, the production of cells, and by favoring the production of toxins by pathogenic microbes. It may similarly influence the production of enzymes and pigments. One may associate with typical metabiosis those cases in which the growth of an organism is favored by, or is dependent upon, extracts of other microorganisms. When these extracts are sufficiently concentrated and purified and when they act in very small amounts, the phenomenon can safely be explained by the action of vitamin factors. The three phenomena, unilateral stimulation *in vivo,* metabiosis, and stimulation through an extract of a microorganism are three forms of one and the same manifestation, and all of them may occur in the same organism.

We do not consider as a stimulation the case in which bacterial cells serve directly as a source of food for another microorganism.

Reciprocal (bilateral) stimulation. — Cases of reciprocal stimulation are of the greatest interest to us, although they are much rarer than those of unilateral stimulation and of metabiosis.

Hemophilus canis requires factor X (hemin) for its development but it synthesizes cozymase (factor V). Conversely, *Hemophilus parainfluenzae* synthesizes hemin but requires cozymase. When grown together (RIVERS) both develop, since they exchange their growth factors with the result that each has the necessary substance at its disposal. This case can be explained only by the action of growth factors.

Analogous cases can be found in the fungi. *Nematospora Gossypii* requires biotin as a growth factor but synthesizes thiamin. *Polyporus adustus* synthesizes biotin, but requires thiamin (in reality only pyrimidine). When cultured together without growth factors, the two fungi develop normally (KÖGL and FRIES, 1937). The most typical case is represented by the association of *Mucor Ramannianus* and *Rhodotorula rubra.* It is explained by means of a single growth factor, which consists of the two components of thiamin. We have seen that *Mucor Ramannianus* is the thiazole type which can synthesize pyrimidine. *Rhodotorula rubra,* conversely, requires pyrimidine but synthesizes thiazole. When cultured separately on a synthetic medium without growth factors, *Mucor* does not grow and *Rhodotorula* shows only a trace of growth, but when cultured together on the same medium they grow as well as they do when supplied with thiamin. The two components of the thiamin molecule have been exchanged, giving each partner of this mixed culture the complete molecule of this vitamin (MÜLLER and SCHOPFER, 1937; MÜLLER, 1940).

Mucor Ramannianus		*Rhodotorula rubra*
synthesizes: pyrimidine	⟶	requires: pyrimidine
requires: thiazole	⟵	synthesizes: thiazole

When the mixed culture is grown on a solid agar medium, the two organisms intertwine, forming a consortium whose morphological aspect is different in certain cases from the cultures of the two partners when isolated. We have here an actual case of *artificial symbiosis* which is ideal. Furthermore, it demonstrates irrefutably that the organism requires only one-half of the molecule because it is capable of synthesizing the other.

Although unilateral stimulation can never be considered as symbiosis, all well-established cases of reciprocal stimulation can be regarded as artificial symbioses, particularly that of the association *Mucor Ramannianus-Rhodotorula rubra.* However, a single example like this might prove misleading because it might lead to the assumption that symbioses among microorganisms depend solely on an exchange of growth factors. Actually a symbiosis is determined by a complex group of various kinds of factors, among which vitamins play a rôle. It is conceivable that the lichen symbiosis can be explained partially in this manner but this has not been proved as yet.

We must again call attention to the fact that, in the majority of cases, symbioses are made up of an autotrophic and a heterotrophic partner—heterotrophic in the broad sense of the term. In the case of *Mucor-Rhodotorula* it is a matter of an artificial symbiosis between two heterotrophic organisms cultivated in the presence of sugar. In nature, it is hardly probable that these organisms encounter each other, and even if they did so, it is not likely that they could live together. Moreover, all the cultural conditions necessary for the symbiosis of *Rhodotorula* and *Mucor* are not yet known in detail. The recent investigations of UTIGER and SCHOPFER have shown that the symbiotic condition develops well in the presence of asparagine. If the latter is replaced by ammonium sulphate or another inorganic nitrogen source, development occurs only upon the addition of a mixture of cations containing at least Li, Cs, Sn, Mo, Hg, U, and Al. The author considers this case a *model* particularly adaptable for teaching purposes. It shows us how nice it would be if natural symbioses could also be explained as simply.

Higher Plants. — It is primarily the endotrophic and exotrophic mycorrhizas which interest us as far as vitamins are concerned. The fungi of exotrophic mycorrhizas are the most accessible. From 1925 on, E. MELIN, studying the development of fungi of this type *in vitro,* has shown that special conditions are required for their culture. He attempted to culture *Boletus luteus* and *B. variegatus* as well as mycorrhizal *Rhizoctonia silvestris* β, *R. silvestris* γ, and *R. Abietis* (the latter from *Pinus silvestris* and *Picea excelsa*) on a synthetic medium, and found that their growth on it was improved by introducing conifer seeds or young conifer plants. He assumed that the substances which diffused from the seeds and plants into the medium were vitamins analogous to growth factors, but since HANSTEEN-CRANNER had just shown that soluble phosphatides diffused from living roots and seeds, particularly those of conifers, into the water in which they had been placed, he promptly made comparisons and found his hypothesis untenable (see SCHOPFER, 1932). The substances which exercise an auxogenic action are not specific, since they can be obtained from seeds other than those of conifers. Furthermore, they are present, not only in seeds, but also in many other types of tissues. Moreover, it has been demonstrated since that time that phosphatides are not the only substances diffused from the seeds, but that vitamin B_1 contained in the seeds also passes into the water. F. W. MÜLLER (1941) proved that the dialyzates of many seeds and of various tissues are a source of growth factors for a soil organism, *Mucor Ramannianus* (which does not form a true mycorrhizal association with roots, but which has likewise been studied by MELIN). The factor in question is vitamin B_1 or one of its constituents, thiazole.

Encouraged by numerous positive results obtained with vitamin B_1 and bios by a number of authors, MELIN and LINDEBERG (1939) resumed their investigations with other fungi: *Boletus elegans* (Schum.) Fr., *Clitopilus prunulus* (Scop.) Fr., *Lactarius deliciosus* (L.) Fr., *Rhizopogon roseolus* (Corda) Th. Fr., *Tricholoma albobrunneum* (Pers.) Fr., *T. imbricatum* Fr., *T. pessundatum* Fr. *Boletus elegans* forms a mycorrhiza with *Larix; Lactarius deliciosus* with *Pinus silvestris*, and *P. strobus; Clitopilus prunulus* and *Rhizopogon roseolus* with *Pinus silvestris;* the three *Tricholoma* species with *Pinus silvestris, P. montana* and *Picea abies.*

Thiamin exercised a very favorable effect on these fungi, when grown on synthetic media and studied according to the usual methods. Although biotin by itself had no effect, it acted favorably with thiamin on *Rhizopogon roseolus.* MELIN and NYMAN (1940) carried out supplementary experiments with other fungi, employing N. NIELSEN's method of suboptimal doses, and found that thiamin is required by *Boletus luteus* (L.) Fr., *B. variegatus* (Sw.) Fr., *B. viscidus* (L.) Fr., but not by *B. granulatus* (L.) Fr., *B. piperatus* (Bull.) Fr., and *Cenococcum graniforme* (Sow.) Ferd. and Winge. Biotin (bios II*b*) stimulates the growth of *Boletus piperatus* and *Cenococcum graniforme* but has no effect whatever on the other species.

Thus it was clearly demonstrated, in accord with our knowledge of growth factors, that mycorrhizal fungi are frequently auxo-heterotrophic. They must receive from the root, which constitutes their substrate, the thiamin contained in it. But it is of great interest to learn that not all are auxo-heterotrophic. In no case is the need for thiamin the only cause for the formation of a mycorrhizal association nor does it explain the specificity of the substrate for a given fungus.

The concept of vitamin growth factors has been even more fruitful for the explanation of endotrophic mycorrhizas. We must restrict ourselves to a few details concerning orchid symbionts. Since the classical researches of NOËL BERNARD we know that orchids require the presence of a specific fungus for germination and development. The fungus forms a mycorrhiza within the cells of the embryo. Since this phenomenon has been studied cytologically this aspect of the problem can be omitted here. Instead we shall consider the part played by the fungus and the mechanism of its action.

Under certain conditions the fungus is indispensable for orchids growing on synthetic media solidified with agar. However, it has been proven that germination and growth *in the absence* of a fungus are possible in certain cases. KNUDSON reported the development and flowering of a *Cattleya* in the absence of a fungus (culture in light on an agar mineral medium in a 12-liter flask). Apparently there exist cultural conditions under which certain species do not require a fungus, *e.g.*, a high concentration of sugar will permit a *non-symbiotic* culture of some species. On the other hand, it is sufficient to know that conditions exist under which the symbiotic fungus is absolutely indispensable, in conformity with the data of N. BERNARD and of BURGEFF. Moreover, since the "synthetic" medium of KNUDSON contains potato starch as the source of carbon, the presence of vitamin impurities in the starch is not excluded. Natural potato starch, untreated and unpurified, certainly contains some such impurities.

How can we represent the mode of action of the fungus? When the medium contains insoluble carbohydrates and these are not directly utilizable by the orchid seedling, the *living* fungus acts by digesting the starch. If the sugar is not all used up by the fungus, the orchid plant has some at its disposal.

However, a living fungus is not necessary under certain cultural conditions. BURGEFF (1936) obtained germination and development of seedlings of *Phalaenopsis* and *Euanthe* by replacing the living fungus with an extract from it. The growth was proportional to the amount of extract added. At this point the apparent specificity of the symbiotic fungus disappeared. Any of the symbiotic fungi are able to bring about the germination of seeds. With an extract of the mycelium of the mycorrhizal fungus *Rhizoctonia Stuarti*, BURGEFF was able to obtain germination of seeds of 16 different hybrids of *Phalaenopsis*. Thus the action is not that of a specific substance as was believed originally. Activation may likewise be effected by extracts of wheat germ, or better yet of yeast.

Thus the problem is placed in its true perspective when we say that, as in the case of auxo-heterotrophic microorganisms, the orchid embryo suffers an avitaminosis which can be compensated by supplying the medium with the missing factor which the plant is no longer able to synthesize. On a medium without vitamin growth factors, cell divisions are arrested from the beginning.

Apparently the mycorrhizal fungus serves as a source of vitamin factors. From this point of view it is a matter of auxo-autotrophic fungi. Actually, the author has experienced no difficulty in culturing different species of *Rhizoctonia* on synthetic media without the addition of any growth factors. Although the symbiotic fungus is not the only source of vitamins, this action is the most important one.

These new results caution us in interpreting culture experiments on a so-called synthetic agar medium. Impurities of a vitamin nature can be supplied not only by the starch but also by the agar (ROBBINS). The most important problem now is that of identifying the active substance. BURGEFF ascribed to it the following general properties: soluble in water and in absolute alcohol, insoluble in ether, thermostabile, resistant to oxidation by H_2O_2, resistant to the action of ultra-violet rays, capable of being adsorbed by animal charcoal but not by filter paper or Fuller's earth. These are general properties, which indicate that this growth factor differs from all other known factors. Still it must be a growth factor since BURGEFF, by concentrating the active substance obtained a product (beginning with yeast) active at a concentration of 300 γ per cc. of medium. With this concentration it is possible to prevent symptoms of avitaminosis in *Vandae.*

SCHAFFSTEIN (1938), a student of BURGEFF, again took up the problem from the point of view of growth factors. As might be foreseen, the development of *Phalaenopsis* seedlings can be stimulated by extracts of various tissues: stem, leaf, root of *Vicia Faba,* coleoptile and vegetative tip of *Avena,* as well as the extracts of *Rhizoctonia mucoroïdes* and those of baker's and wine yeasts. It is sufficient that the favorable dose be found. A very small dosage is inadequate, whereas too large a dose can induce phenomena of inhibition.

Comparative experiments with the concentrations varying from 0.02 to 2.5 g. of extract per 50 cc. of medium show that the development of orchids occurs with the following extracts: pith of *Aesculus,* bark of *Aesculus,* stem of *Asparagus,* root disk of *Allium cepa,* epicotyl of *Pisum,* stem of *Cucurbita,* seeds and stem of *Vicia.* The extract from the seeds of *Vicia* are much more active than that from *Rhizoctonia mucoroïdes.* Attempts to concentrate the extract were carried out particularly with the seeds of *Vicia.* An aqueous extract is prepared; a treatment with animal charcoal adsorbs the active substance and separates it from the inhibitory substances contained in the crude extract. An elution can be effected with aqueous alcohol. The active factor can not be precipitated by lead acetate in an acid or alkaline medium. As had already been observed by BURGEFF, the orchid factor can not be replaced by any other known vitamins, such as B_1, B_2, A (Vogan), or D (Vigantol). The female hormone is likewise inactive. The same is true of preparations of biotin, extracts of egg yolk, and of lecithin. It is apparently unnecessary to include the auxins as a possible cause. It is true, many vitamins, identified and obtained in the crystalline state since 1938 (B_6, pantothenic acid, etc), have not been tested. The extract purified by animal charcoal is still active in the concentration of 1:1,000,000, although the optimum concentration is 1:100,000. With 160γ of extract from seeds of *Vicia* added to the basal medium, the percentage of development reaches 50 in the hybrid *Phalaenopsis Schilleriana* × *P. amabilis* and 90 in *Dendrobium Phalaenopsis.*

SCHAFFSTEIN succeeded in demonstrating that the seed of *Phalaenopsis* contains very little vitamin, as would be expected; on the other hand, the green parts of a full-grown plant are relatively rich in vitamin. It is thus during the early development, which is of a saprophytic nature, that the avitaminosis appears. At this time the seedling is dependent on the fungus or on the extract added to the medium.

The problem now is to determine the mode of action of this unknown vitamin. BURGEFF had previously observed that absorption of nutritive materials is still possible in the absence of growth factors but that no cell divisions occur. Cytological observation shows that the cells of the growing point of the undifferentiated embryo die of an avitaminosis. Because of the failure of development, an accumulation of starch occurs in the other cells and nuclear hypertrophy appears. Despite the fact that it would be natural to suspect the absence of a substance specific for cell division, it is probable that the principal active substance is a vitamin (according to our definition) acting on the assimilatory power. If such a substance is lacking, cell division can no longer take place. At present it is difficult to draw any conclusions regarding the nature of the action since the substance or substances are not identified. Possibly a combination of factors is involved, some of which are vitamins (affecting assimilation) whereas the others are specific hormones (affecting structure and cell division; substances of the traumatin type).

In conclusion it may be pointed out that because of these experiments the orchids have lost the exceptional place that they once occupied. In the last analysis, the fundamental phenomena of the physiology of germination are the same in orchids as in *Pisum.* If the embryo of *Pisum* is rendered partially heterotrophic by premature removal of its cotyledons, it will likewise incur an avitaminosis. To a certain degree it is also dependent on vitamins which must be added to the culture medium. Orchid seeds are characterized by a normal physiological avitaminosis, brought on because of the absence of reserves, and by the ingenious method devised by them to utilize the contribution of the auxoautotrophic fungus that has become symbiotic. This unusual case of symbiosis has been investigated very thoroughly.

The case of bacterial nodules, produced on legumes by *Rhizobium,* may likewise give way to a similar interpretation. The growth factor requirements of *Rhizobium* have been studied (p. 160). We have recognized the necessity of "coenzyme R" which may be replaced by biotin, and under certain conditions thiamin may likewise come into play. These growth factors are undoubtedly furnished by the host plant, which is amply supplied with them. MCBURNEY, BOLLEN, and R. J. WILLIAMS (1935) pointed out that *Rhizobium meliloti* produces appreciable quantities of pantothenic acid. Cultures of alfalfa on a synthetic medium without bacteria are very much benefitted by the addition of this acid. It is tempting to conclude from this that the pantothenic acid produced by *Rhizobium* is utilized by legumes. There would thus be an exchange of growth factors, a character of true symbiosis: the bacterium receives biotin and liberates pantothenic acid. This exchange is entirely possible. However, the pantothenic acid employed in the experiments on alfalfa culture was only a concentrate (1935) and it is not certain whether the action observed was specific for pantothenic acid or was due to impurities. At any rate these results indicate the direction that future experiments should assume.

Symbionts of Insects. — This subject is outside the scope of our book but the parallelism between the results obtained with symbionts of orchids and those of insects is so remarkable that we can not overlook it. The investigations of KOCH as well as those of W. SCHWARTZ (1939) and his collaborators (BODE and RESÜHR) have contributed much to the clarification of this subject.

KOCH (1933) was the first to introduce the concept of vitamins in this field. The symbionts of insect larvae are principally bacteria and saprophytic fungi, a large number of which are known to be auxo-autotrophic. The larvae of *Sitodrepa* (a lignivorous beetle) contain chiefly yeasts as symbionts. In aseptic cultures these larvae do not develop without yeasts. The living symbionts can be replaced by extracts of concentrated yeast (Cenovitan), as well as by extracts of wheat germ, whose richness in vitamins is known. As is the case of orchids, no specific substance is involved, since the substance or substances responsible for the observed effect are widely distributed in the plant kingdom.

The attempts at identification of the active substances did not succeed immediately. An absorbate of vitamin B_1 (on Fuller's earth) does not replace yeast extract. Vitamin A (Vogan) and D (Vigantol) have no effect. The thermostability of the active factors is remarkable; heating at 170°C. for two hours has no effect on the strength of the vitamin. It is certain that we are dealing with a constellation of vitamin factors. In 1933 KOCH pointed out the existence of at least two factors differing in their solubility and recently he stated that this symbiosis involves vitamins B_1, B_2, B_6, bios, and a factor in the B group not yet identified.

It has been repeatedly demonstrated that vitamins are required by insects when grown aseptically. Yeast extract is indispensable for *Drosophila* (larvae) cultures (GUYÉNOT). VAN'T HOOG has determined precisely the required culture conditions and has shown that the yeast extract, replacing the bacteria, acts through vitamins B_1 and B_2 which it contains, as well as through sterols (as originally demonstrated by GUYÉNOT). *Drosophila* larvae in aseptic culture may thus serve as test objects for the quantitative determination of thiamin. It is also possible to follow the thiamin metabolism in the rat by means of this method (VAN'T HOOG, 1935).

Nicotinic acid is another growth factor in insects (RUBENSTEIN and SHEKUN. Thiamin and riboflavin are required by mosquito larvae (TRAGER and SUBBAROW, SERGENT). (For literature on the vitamin requirements of insects see FRÖBRICH, 1939, OFFHAUS, and STEINHAUS, 1940).

With insect symbionts we meet the same phenomena and explanations as for the symbionts of higher plants, since the same growth factors as those encountered to date in higher plants and in microorganisms come into play here.

It must be pointed out, however, that many cases of symbiosis in insects can not be explained by a vitamin action of the symbiont.

Parasitism. — In this field the concept of growth factors did not lead to any new interpretations. We have established the fact that general heterotrophism and auxo-heterotrophism do not always run parallel. Many saprophytes and parasites synthesize their essential growth factors, hence it is not necessarily the need of growth factors that has brought about parasitism. In the *Mucoraceae,* some saprophytic species such as *Pilaira anomala* and *Phycomyces Blakesleeanus* require thiamin as a growth factor and are similar in this respect to such parasitic species as *Parasitella simplex.* Other parasitic species such as *Piptocephalis Fresenianus* can be grown successfully on a synthetic medium lacking growth factors. *Mycobacterium tuberculosis* var. *hominis* is capable of synthesizing growth factors, whereas *M. tuberculosis* var. *bovis* is no longer able to do so. Certain cases are more definite: hemophilic organisms are parasitic on blood because they require hemin and cozymase, but again the parasitic specificity is not explained. Furthermore, it might be asked whether it is the need of hemin which brought on parasitism, or whether the latter has arisen from other causes after the power of synthesizing this factor had been lost because it became useless in the new habitat. But here we reach the domain of speculation which holds no practical interest for our problem.

REFERENCES:-

1925: E. MELIN. — Untersuchungen über die Bedeutung der Baummykorrhiza. — Jena, Gustav Fischer.

1925: B. HANSTEEN-CRANNER. — Weitere Beiträge zur Biochemie und Physiologie der pflanzlichen Zellphosphatide. Meld. fra Norges Landbrukshøiskole, 48 pages.

1928: G. PAPACOSTAS et GATÉ. — Les associations microbiennes et leurs associations thérapeutiques. Paris, Doin, Encyclopédie scientifique.

1932: W. H. SCHOPFER. — Certains phosphatides peuvent-ils se substituer au facteur de croissance des Mucorinées? C. r. Soc. Phys. hist. nat. Genève, 49:155.

1933: A. KOCH. — Symbionten und Vitamine. Naturwiss., 21:543.

1933: W. H. SCHOPFER. — Etude d'un cas de stimulation unilatérale et d'un cas d'inhibition chez un microorganisme. C. r. Soc. Phys. hist. nat. Genève, 50:152.

1935: C. H. MCBURNEY, W. B. BOLLEN and R. J. WILLIAMS. — Pantothenic acid and the nodule bacteria-legume symbiosis. Proc. Nat. Acad. Sc. U.S.A. 21:301.

1935: E. G. VAN'T HOOG. — Aseptic culture of insects in vitamin research. Zeitschr. f. Vitaminforschung, 4:300.

1936: H. BURGEFF. — Samenkeimung der Orchideen und Entwicklung ihrer Keimpflanzen. Jena, Gustav Fischer.

1937: W. F. MÜLLER et W. H. SCHOPFER. — L'action de l'aneurine et de ses constituants sur *Mucor Ramannianus* Möll. C. R. Acad. Sc. Paris, 205: 687.

1937: F. KÖGL und N. FRIES. — Ueber den Einfluss von Biotin, Aneurin und Mesoinosit auf das Wachstum verschiedener Pilzarten. H. S. Zeitschr. f. physiol. Chem., 249:93.

1938: G. SCHAFFSTEIN. — Untersuchungen über die Avitaminose der Orchideenkeimlinge. Jahrb. f. Bot., 86:720.

1938: R. WEINDLING. — Association effects of fungi. Bot. Rev., 4:475-496.

1939: W. SCHWARTZ. — Die physiologischen Grundlagen der Symbiose von Tieren mit Pilzen und Bakterien. VII. Intern. Kongress für Entomologie, Berlin, p. 916.

1939: G. FRÖBRICH. — Untersuchungen über Vitaminbedarf und Wachstumsfaktoren bei Insekten. — Zeitschr. vergl. Physiologie, 27:335.

1939: K. OFFHAUS. — Der Einfluss von Wachstumsfördernden Faktoren unter besonderer Berücksichtigung der Phytohormone. Zeitschr. vergl. Physiologie, 27:384.

1939: E. MELIN und G. LINDEBERG. — Ueber den Einfluss von Aneurin und Biotin auf das Wachstum einiger Mykorrhizenpilze. Botaniska Notiser, p. 241.

1940: E. MELIN und B. NYMAN. — Weitere Untersuchungen über die Wirkung von Aneurin und Biotin auf das Wachstum von Wurzelpilzen. Arch. f. Mikrobiol. 11:318.

1940: E. A. STEINHAUS. — The microbiology of insects. Bact. Rev. 4:17.

1940: S. A. WAKSMAN. — Antagonistic interrelationships among microorganisms. Chronica Botanica, 6:145-148.

1941: F. W. MÜLLER. — Zur Wirkstoffphysiologie des Bodenpilzes *Mucor Ramannianus*. Ber. d. schweiz. bot. Ges., 51:165.

1941: S. A. WAKSMAN. — Antagonistic relations of microorganisms. Bact. Rev., 5:231.

Chapter XXIV.

MICROORGANISMS AS TEST OBJECTS FOR VITAMINS.

The quantitative determination of vitamins, in the pure state or in raw products, is made by means of bio-assays, *e.g.*, rat, pigeon, guinea pig, etc., and by means of chemical tests whenever possible (see the excellent book by GSTIRNER, 1940).

The fundamental methods are always those which utilize animals as test objects. Before being identified as chemical substances, vitamins were recognized biologically by their action on animals which react very specifically and must always be taken as a basis of reference. Animal tests have the disadvantage of taking a great deal of time, whereas chemical tests are very rapid but often lack specificity.

The fact that microorganisms react more or less specifically to a vitamin and that, within certain limits, the vitamin action is quantitative has induced biologists to use them as test objects. Some of them have actually attained a certain degree of popularity. We shall consider them briefly.

To be useful a test object must have the following characteristics:

(1) The microorganism must be easily cultured and must present a certain constancy in its cultural characteristics.

(2) It must be able to grow on a relatively simple medium, and it must require only a single growth substance or a simple group of substances.

(3) The duration of growth must be short.

(4) It must be possible to measure the growth quantitatively (gravimetrically, nephelometrically, etc.).

(5) It must react specifically to the vitamin studied.

(6) It must be utilizable also with impure or crude products, in which the vitamin in question is found only in a very small amount.

(7) It must not react too strongly to the presence of other substances contained in the impure products.

Vitamin A. — There is no microbiological assay for this vitamin nor for carotinoids. *Chlamydomonas eugametos,* which reacts to the presence of very small quantities of carotinoids, can not be used in actual practice.

Vitamin B_1. — This vitamin may be assayed biologically by several different procedures. We will not discuss the animal tests here except to mention, in passing, the interesting test of VAN'T HOOG, based on thiamin requirements of *Drosophila* larvae. When the other factors required by this organism are present in the medium (sterols in particular), a quantitative determination of the vitamin can be made.

Chemical tests applicable to the quantitative analysis of thiamin are numerous: the thiochrome test of JANSEN (oxidation of thiamin to thiochrome and quantitative analysis of this by determining its fluorescence in ultra-violet light) ; the formaldehyde-azo test of KINNERSLEY and PETERS (reaction of thiamin with diazotized sulfanilic acid in an alkaline medium) ; the test of PREBLUDA and McCOLLUM with diazotized para-aminoacetophenone and para-aminoacetanilide is based on the same principle. In all these tests the vitamin must first of all be extracted. WILLSTAEDT and BARANY report a reaction of thiamin with diazotized 2,4-dichloraniline, the color obtained being determined with a Pulfrich photometer. Various colorimetric reactions are known, none of which are very practicable.

The precipitation reaction with potassium tetranitritodiaminocobaltate and Reinecke's salt (ROSENTHALER) permits the qualitative detection of thiamin. This test permits the detection of a very small quantity of pure thiamin (less than 5γ) (SCHOPFER). Another precipitation test employs bismuth potassium iodide (NAIMAN) which gives a characteristic orange-red precipitate. Neither of these tests is specific and therefore cannot be used for quantitative analyses. We will not mention the other precipitation reactions, which are interesting from the chemical point of view, but are not suitable for our purpose.

Among the biological tests, we mention first of all that proposed at an early date by R. J. WILLIAMS in which yeast is utilized. This test can not be employed in its original form because it was based on the belief that bios and vitamin B_1 are identical. Recently yeast was taken up again by SCHULTZ, ATKIN and FREY (1937, 1942) and employed in a fermentation method. Under a given set of conditions, thiamin produces a marked acceleration of alcoholic fermentation, the intensity of which permits a quantitative determination of as little as 0.5-4γ of thiamin. While a control produces only 185 cc. of CO_2 in three hours, the addition of thiamin furnishes 325 cc. Although a hot alkaline treatment destroys the activity on animals, it leaves intact its effect on fermentation. An active product of disintegration must thus be formed, proving that the method is not specific. Actually 2-methyl-5-ethoxy-methyl-6-amino-pyrimidine acts as strongly as thiamin, whereas thiazole is without effect. Nicotinic acid (the same authors, 1938*a*) exercises a weak stimulation in the presence of thiamin, while adenylic acid is without effect. These authors describe a technique (1938*b*) which permits the elimination of the error due to pyrimidine. The total fermentation is first of all determined (due to pyrimidine plus thiamin), the thiamin is then oxidized into thiochrome which has no effect on the fermentation and the rest of the fermentation is due to the pyrimidine and is deducted from the total value. The best results are obtained with this test when the thiamin content is between 2 and 5γ. The method, when accompanied by proper attention to several important details, is applicable to numerous raw materials (HEYNS).

In regard to yeast, it is necessary to mention the interesting observation of HAAG and DALPHIN (1940). Mycolevure of DUCLAUX as well as other yeasts (*Hansenula anomala* var. *productiva* for example) accumulate pyruvic acid on a synthetic medium with a saccharose base neutralized with $CaCO_3$. HAAG and DALPHIN observed that the addition of a trace of cocarboxylase reduces the pyruvic acid content of the medium; even 0.00003γ is active and with 0.0003γ the acid is no longer formed. It is probably possible and practical to determine the cocarboxylase contained in a natural product.

A *Staphylococcus* test has been proposed by WEST and WILSON (1938). It will be recalled that *S. aureus* (KNIGHT, KÖGL and v. WAGTENDONK) requires thiamin, nicotinic acid, and especially biotin. WEST and WILSON made use of this test to detect the presence of thiamin in *Rhizobium trifolii.* The basal medium must naturally contain nicotinic acid and biotin (the authors used hydrolyzed casein). Under these conditions the bacterial growth, measured with an electrophotometer, is proportional to the quantity of vitamin. This test is very sensitive since as little as 0.5 microgamma can be detected by it. The growth curves as a function of the quantity of vitamin obtained with pure thiamin and with an extract of hog liver, are satisfactory. The method must, however, be applied carefully with regard to the constancy and stability of the bacterial culture and to the constitution of the basal medium with hydrolyzed casein. It has the advantage of being rapid (36 hours). Up to the present it has not been employed extensively and no definite statement can be made regarding the possibilities of its general use. Furthermore, *S. aureus* reacts also to pyrimidine plus thiazole.

A *Propionibacterium pentosaceum* test has been considered by SILVERMAN and WERKMAN (1939). It has been demonstrated with this bacterium that thiamin accelerates the anaerobic metabolism of pyruvic acid when the cells are deficient in this vitamin. The measurement of the metabolism and of its variations under the influence of the added vitamin should permit the quantitative determination of the latter. The test has not received sufficient experimental application to prove its merits.

Protozoa, heterotrophic for thiamin, offer some interesting possibilities. *Glaucoma piriformis,* under a definite set of cultural conditions, requires thiamin (whole molecule) as the only growth factor. The number of cells formed is, within the limits defined, proportional to the thiamin dosage (A. and M. LWOFF, 1937). The advantage of this test lies in the fact that the specificity of this ciliate has been adequately studied (A. and M. LWOFF, 1938). The authors have not described in detail a quantitative method. We know that MINZ has been able to prove with this very sensitive test that the excitation of a nerve causes a liberation of thiamin.

Although it is not known to what extent this test is utilizable with impure raw products, it is certainly suitable for numerous applications.

Fungi which are easy to culture, especially the Phycomycetes, promise numerous possibilities. The *Phytophthora* test has been proposed by ROBBINS (1938). This Oomycete, like *Glaucoma piriformis,* requires the whole molecule of thiamin but nothing is known about the possible necessity of other growth factors. Thiamin seems to be the essential growth factor for *Phytophthora cinnamomi.* Unfortunately this organism must be inoculated by means of mycelial transfer and its growth is subject to much variation. By making use of quadruple series we have been able to obtain a growth curve showing the thiamin required for optimal growth to be 0.8γ (112 mg. of dry yield) for 25 cc. of our usual medium; up to 40γ the yield remains unchanged. Although it is not practicable to use this fungus for quantitative determination, it may be of service for a semi-quantitative determination or simply to detect thiamin qualitatively (ROBBINS, J. BONNER). It has served to demonstrate the presence of the whole molecule of thiamin in *Rhodotorula rubra* grown on pyrimidine. Theoretically, it is not important whether a test fungus requires the whole molecule of thiamin or is able to utilize the two components (pyrimidine plus thiazole).

So far *Phycomyces Blakesleeanus* has given the best results. Because of its great regularity of growth, as well as the exact proportionality between the vitamin dosage and the weight of the yield, the author proposed it as a test for thiamin (SCHOPFER, 1935).

In the original method, 25 cc. of medium (glucose 3 per cent, asparagine 1 per cent, magnesium sulphate 0.5 per cent, potassium acid phosphate 1.5 per cent) were used, sterilized at 115°C. for 20 minutes, with a *p*H of 4 to 6. At first an attempt was made to construct a standard curve, obtained with various amounts of vitamin B_1, to serve as a comparison for the curves obtained with unknown products. This procedure was unsatisfactory in view of the many uncontrollable influences which cause variation in the rate of growth from one experiment to another. It was found necessary to work out a series of standards for each experiment, with the pure vitamin paralleling the series of the unknown product.

The best method consists in making many series of basal medium with increasing volumes of the pure vitamin solutions. The latter are of increasing concentrations: 0.05, 0.1, 0.2, 0.3, 0.4γ of thiamin per cc.; each of these concentrations is supplied in volumes ranging from 0.05 to 4 cc. For each of the above concentrations a growth curve (based on the *dry weight* of the fungus mycelium) is obtained. Five growth curves are thus available and can be superimposed regularly over those obtained from the substance being assayed. (Only one of these curves is shown in the figure below — the one obtained with the standard solution having a concentration of 0.2 of vitamin B_1 per cc.). Solutions of varying concentrations are prepared from the substance to be analyzed; each of these is supplied in increasing volumes (0.05 to 4 cc.) giving the same series as above. From the "vitamin" curves and the "unknown-product" curves are chosen two which are the nearest to being superimposed, in order to calculate the vitamin B_1 content of a given volume of the unknown substance (SCHOPFER and JUNG, 1936) (Fig. 17).

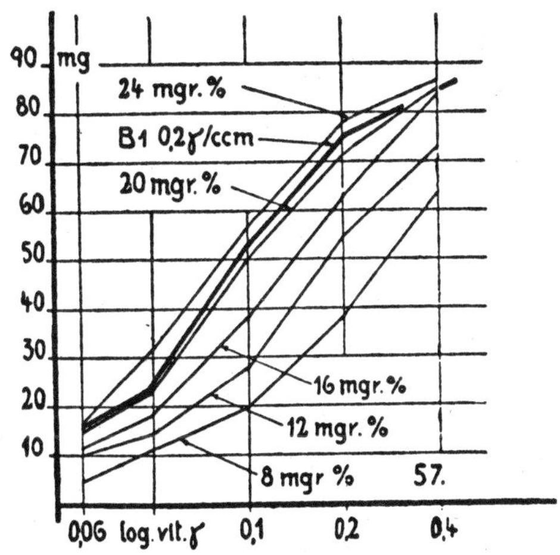

FIG. 17. — Example of a thiamin assay made by means of the *Phycomyces* test. On the abscisse are the doses of thiamin in γ (0.001 milligram). The heavy line (curve) represents the amount of growth (dry weight of mycelium in mg.) of *Phycomyces* produced by the respective doses of thiamin. The fine lines (curves) represent the amount of growth produced by 8, 12, 16, 20, and 24 mg. of the substance being tested. The vitamin curve corresponds with the curve obtained with 20-22 mg. of the substance being tested; thus 20 mg. contains 20γ of thiamin or 1 gram contains 1,000γ.

Numerous variations of the method can be employed, especially in regard to the manner of making the calculations (BONNER and ERICKSON, 1938). The following points are important: all the media of an experiment of one series must be prepared at the same time; the inoculation must be made with the same spore suspension in distilled water from a well-developed, but not too old, stock culture; if need be, the inoculated spores can be counted; the number of spores can vary within rather wide limits without influencing the final weight of the culture; only the rate of growth is affected. Furthermore, after inoculation (on the second or third day) it is necessary to agitate the culture in order to obtain an even distribution of the germ tubes; the surface pellicle must be uniform and must cover the whole surface; phototropic curvatures must be avoided; the cultures should be incubated at about 22°Centigrade. After about 10 days, when the sporangiophores have attained their maximum length, the cultures should be collected on weighed filters, dried at 100°C., and weighed on an analytic balance. With these conditions of culture it is possible to detect even 0.01 to 0.05γ of the vitamin in 25 cc. of medium.

A *Phycomyces* micromethod (SCHOPFER and JUNG, 1938) utilizes 1 cc. of medium. The length of the hyphae grown in the small experimental tube is measured with the aid of a microscope and compared with that of the pure vitamin series. The method is less precise but permits detection of as little as 0.0004γ of thiamin. By this test we have been able to show the differences existing in the vitamin content of organs of normal rats, and of those suffering from avitaminosis.

The *Phycomyces* method was adopted after a detailed comparison with the animal test (SCHOPFER and JUNG, 1936). The following table gives the results obtained with a number of substances varying considerably in vitamin content. The agreements are generally satisfactory, even when the vitamin content is low (for example, 1 international unit).

TABLE XVII. — Thiamin content of various products determined by the *Phycomyces* test and by the animal test (from SCHOPFER and JUNG, 1936).

Product	International units per gram	
	Rat	*Phycomyces*
Pure crystalline:		
(2) Crystalline vitamin (VAN ROCHE)	200,000 250- 300,000	200,000 333,000
(2) Crystalline vitamin (VAN VEEN)	200,000	200,000
(3) Standard = one International unit (U.I)	1 U.I = 2-3 micrograms	1 U.I = 2-3 micrograms
Concentrates:		
(4) 2909/4 (ROCHE)	625	500
(5) 3163/4 (*id.*)	2,050	2,000
(6) 3163/6 (*id.*)	10,000	11,400
(7) 3163/8 (*id.*)	30,000	27,000
Extracts of yeast:		
(8) N.C. 2 (WANDER)	32	35–40
(9) A.C. 2 (*id.*)	7	4
(10) N.C. 4 (*id.*)	1.6	1–1.5
(11) A.C. 4 (*id.*)	5–7	10–12
(12) Yeast extract, autoclaved and dried	0	10
Extract from wheat germ:		
(13) Concentrated extract, strictly from germs	5.3–6	6.8
Malt extract:		
(14) Malt extract (WANDER)	1.2 ± 0.15	1.5
Extracts from polishings of rice:		
(15) R.K.E. (WANDER)	6	7
(16) *id.*	12	14
(17) *id.*	15	16–17
(18) *id.*	19; 19.6	18–19
(19) *id.*	15.4; 17.5	25
(20) *id.*	15; 16	18–19
(21) *id.*	14; 16	16
Malt preparations:		
(22) N. (WANDER)	1	2
id.	1; 1.3	2.5–3
id.	1; 1.5	3–5

(23)	C. (*id.*)	1.4	2–2.5
	id.	1.5–2	3–5
(24)	L. (*id.*)	1.5–2	3–5
(25)	Cc.	1.5–2	3–4
(26)	Ch.	1.8–2.2	2.5
(27)	M.C.	2	2.5
(28)	H.M.M.	1	1
(29)	B.V.	1–1.5	1.5–2
Mixtures of extracts from malt and yeast:			
(30)	H.M. (WANDER)	10–13	10
(31)	H.M. (*id.*)	——	——
(32)	One year old	10–12	10
(33)	He. *id.*	8–9	4–6
Mixtures of yeast extracts and malt preparations:			
(34)	H.50 (WANDER)	4–5	8
(35)	H.33 (*id.*)	2.8	4–5
		3.2	5
(36)	H.1 (*id.*)	1.1	4
(37)	H.2 (*id.*)	1.5–1.8	5–6

If necessary, an adsorption with Fuller's earth can easily be made; only the eluted fraction is assayed (FAGUET, 1937). This eliminates the disturbances caused by the other substances of the crude product.

Causes of Errors. — The most important error is caused by assimilable nitrogen, carried possibly by the raw material. We have seen (p. 103) that by raising the asparagine content of the medium the effect of a given thiamin dosage is augmented. Therefore the "unknown substance"- and the "vitamin"- curves will no longer be comparable. It is possible to compensate for this source of error by comparing the curves only in their lower ascending portion, where the volumes of the substances utilized are small. Furthermore, it is possible to reduce the effect of the nitrogen in crude products by using a medium very rich in nitrogen (with an increased asparagine content).

Another cause of error lies in the fact that not all the thiamin of a product is always available to the microorganism. It may be masked (adsorbed). In this case, it is necessary to liberate it. This can generally be partially accomplished by autoclaving. *Phycomyces* has the great advantage of being able to utilize the products made available by this treatment. The reaction of *Phycomyces* to blood is a great deal more marked when the blood is added to the medium before sterilization than when added subsequently.

One possible cause of error is due to prolonged autoclaving. This can result in thermal decomposition and even complete destruction of the thiamin. An exhaustive study with the formaldehyde-azo and the thiochrome tests have shown that autoclaving for 20 minutes at 115°C. does not modify perceptibly the *pure* vitamin, if the medium is slightly acid or neutral. The nature of the effect of autoclaving on a crude product is unknown, although experience has shown us that, generally, a destruction of thiamin need not be feared. It is impossible to assay a product which alkalizes the medium or one which through sterilization gives rise to alkalizing substances, since the vitamin is rapidly destroyed.

Failure to observe the directions indicated above is responsible for other errors.

Specificity of the Phycomyces test. — The specificity of the action of thiamin has been studied previously (p. 119). Within the limits of biological quantities, it is evident that the substitution products, the analogues, the homologues and the heterovitamins generally are active only in large amounts. It is uncertain that they occur in nature. On the other hand, it is necessary to recall that thiamin is quantitatively replaceable by the pyrimidine + thiazole combination. Again it is uncertain that the two thiamin components are found in any quantity in nature. We know (SCHOPFER and MÜLLER) that thiamin can undergo thermal disintegration upon prolonged sterilization, which destroys its effect on animals but not on *Phycomyces.* The latter test is therefore useless with a substance that has undergone prolonged autoclaving before testing (see product No. 12 in table XVII), since it does not assay thiamin but instead the disintegration products which are active only on the fungus. If the thiamin is destroyed *completely,* the test will no longer give any reaction. The observation of LEONIAN and GREENE that some organic acids (particularly fumaric acid) and some metals have a marked effect on *Phycomyces* is cautioning. The test is useless if these acids are added to the basal medium. The Z factors of ROBBINS must also be considered although it seems improbable that they should give much trouble in the biological assay of thiamin. This test is certainly capable of modifications and improvements; it can be changed to accommodate each new discovery.

Comparison of the specificities of various tests. — It is known that the results of various tests, *i.e.,* animal and thiochrome, sometimes vary considerably (PYKE). For assaying the complete molecule of thiamin, the *Phycomyces* test has as great a specificity as the thiochrome test. We have shown that many substitute vitamins, still possessing the NH_2 group in position 4 of pyrimidine and H in position 2 of thiazole, react positively to the thiochrome test in the same manner as thiamin does, even though they have little or no effect on *Phycomyces.* Cocarboxylase has the same effect on *Phycomyces* as thiamin. It also reacts to the ordinary thiochrome test. However, one modification of the thiochrome test permits separate titration of cocarboxylase and thiamin (WESTENBRINK).

It would be very interesting to make a systematic and exhaustive comparison of the various tests: *Phycomyces,* animal, thiochrome, and formaldehyde-azo.

Aside from its use in purely scientific research, the *Phycomyces* test has been applied extensively. It has been employed with 0.4 per cent asparagine for the determination of thiamin in the blood by MEIKLEJOHN (1937), SINCLAIR (1939), ROWLANDS and WILKINSON (1938), MORELL (1938), GUHR (1939), and TANGARI (1940).

GUHR points out that it is accurate within ± 8 per cent per 1-2 cc. of blood. FAGUET employed it with success in assaying various substances, after having carried out a preliminary adsorption and suggested that the animal should be replaced by the *Phycomyces* test. By means of the latter test VILLELA (1939) was able to detect thiamin in cerebrospinal fluid as well as in various plant products (Paraguay tea, etc.) (1938 *a, b*). The *Phycomyces* test is applicable to urine only with precautions. MARCQ and MANIL (1939) found good agreement between the chicken and *Phycomyces* tests as far as yeast extract is concerned.

Recently, R. THREN (1941) has made a critical and detailed study of the *Phycomyces* test from the data of SCHOPFER and his collaborators. By carefully applying the precautions prescribed, he concludes that the test is perfectly utilizable and gives precise results. He is one of the first biologists who recognizes, in accordance with the findings of SCHOPFER, the great importance of the nitrogen content of the medium to the action of thiamin.

Still other microorganisms are utilizable. *Ustilago violacea* would be the most delicate test for thiamin were it not so sensitive to numerous factors (for example, cotton falling into the medium is a source of growth factors for it). This organism is not recommended because of the multiple precautions that it necessitates.

Rhodotorula rubra permits the detection of pyrimidine, and *Mucor Ramannianus* of thiazole, providing preformed thiamin can be eliminated.

Riboflavin. — We have seen that the specificity of the action of flavines on the lactic acid bacteria (p. 155) is rather definite and approaches that on animal cells (SNELL and STRONG, 1939). Good agreement between the lactic bacteria assay and the rat growth method has been reported by numerous workers (ELVEHJEM and coworkers, MICKELSON *et al.*, EMMET and coworkers) (*cf.* R. J. WILLIAMS and coworkers, 1941). Although the complex constellation of factors required by the lactic acid bacteria reduces their usefulness for a quantitative analysis of flavines, they are satisfactory for the assay of a wide variety of products. This microbiological method determines not only riboflavin in the free form, but also in the combined forms (coenzymes and enzymes) (SNELL and STRONG, 1941).

Pyridoxine (B_6). — MÖLLER, using *Streptobacterium plantarum*, obtained very regular growth curves as a function of the pyridoxine dosage employed. The specificity of action is sufficiently exact (p. 156).

With the same reservations as for riboflavin, a lactic acid bacterium can serve as an almost quantitative test for this vitamin. A method has recently been announced by LANDY and DICKEN (1942) in which *Lactobacillus casei* is used for the assay of six B vitamins — pyridoxine, riboflavin, biotin, pantothenic acid, folic acid, and nicotinic acid.

R. J. WILLIAMS and coworkers (1941) have proposed a method in which the "Gebrüder Mayer" strain of *Saccharomyces cerevisiae* is used as the best organism for pyridoxine. These authors report that the method gives consistent results and recoveries. They are of the opinion that it acts as a quantitative test for *free* pyridoxine and that it has a greater degree of specificity than the animal test.

Nicotinic acid. — The *Bacterium proteus* test can be used without reservation in assaying this vitamin. This bacterium requires only nicotinic acid for its growth and, within certain limits, reacts quantitatively to it. FRASER, TOPPING and SEBRELL first proposed this organism as a test object for this vitamin. LWOFF and QUERIDO (1938), utilizing the observation of FILDES, point out that *B. proteus* is entirely suitable for the bio-assay of nicotinic acid. The growth is measured with a nephelometer, and the growth curves obtained as a function of the concentration are regular. *B. proteus* can be used not only in the determination of nicotinic acid and its amide but of all the substances acting as antipellagra vitamins for animals. The assay includes not only nicotinic acid and its derivatives, but also the codehydrogenases (nucleotide with a nicotinic acid base). This method has been employed successfully with blood, both normal and pathological (LWOFF, QUERIDO, LATASTE, ALBEAUX-FERNET).

SNELL and WRIGHT (1941*a*) have described a method of assay for nicotinic acid in which *Lactobacillus arabinosus* is used. Results obtained by this method agree reasonably well with those obtained by the dog assay and certain chemical methods (see SNELL and WRIGHT, 1941*b*).

Vitamins C, D, E, and K. — No microbiological test has been found for any of these vitamins. In the case of vitamin C it would not even be of interest since we have already a very practical titration method with the Tillmann reagent.

Biotin. — A very precise quantitative analysis of biotin (Kögl's method) is possible with *Saccharomyces cerevisiae* (Race M). Various methods employing yeast as a test for biotin have been proposed. SNELL, EAKIN, and WILLIAMS (1940) have described a practical procedure whereby biotin can be rapidly and accurately assayed. *Nematospora Gossypii,* requiring biotin and inositol (BUSTON and PRAMANIK, KÖGL and FRIES), can also be used for the detection of biotin (ROBBINS and M. BARTLEY-SCHMIDT, 1939). The basal medium in such a case must contain inositol.

Pantothenic acid. — A biological assay of this substance appears possible since it is required by a number of organisms. R. J. WILLIAMS and his collaborators (1940) employ yeast as well as *Streptococcus lactis, Lactobacillus casei e, Bacillus brassicae,* and *Propionibacterium pentosaceum.* PELCZAR and PORTER (1941) have proposed *Proteus Morganii* as a test object for the assay of this vitamin; it is sensitive to as little as 0.0001 γ per cc.

Inositol. — A microbiological assay method for this factor in which yeast ("Gebrüder Mayer" strain of *Saccharomyces cerevisiae*) is used as the test organism has been described by R. J. WILLIAMS and coworkers (1941).

Folic acid. — Two microbiological assays are now available for this factor (SNELL and PETERSON, 1940; MITCHELL and SNELL, 1941).

Discussion of bio-assays. — As a rule, all organisms requiring growth substances can be used in the qualitative determination of a growth substance, and with a complete basal medium a quantitative determination is theoretically possible. Only practical reasons determine whether to employ or reject a test organism.

Higher plants, because of the slowness of their growth and because of the numerous inherent difficulties in their culture on a sterile synthetic medium, are unsatisfactory as test objects.

Our experience indicates that one should not expect too much from these microbiological tests. Their value must be examined in each particular case and the results must always be studied critically. A microorganism can not be handled in the same way as a chemical substance in a test tube and a biological reaction can not be detected as readily as a colorimetric one. We maintain, nevertheless, that some of them have occasioned notable progress in the knowledge of vitamins. The vitamin B_1 content of the blood of a living individual can be determined only with the *Phycomyces* test. In the same way, the *B. proteus* test discloses the evolution of nicotinic acid in the blood.

REFERENCES:-

1935: W. H. SCHOPFER. — Recherches sur l'emploi possible d'un test végétal pour la vitamine B_1. Essai d'étalonnage. Bull. Soc. chim. biol. Paris, 17:1097.

1936: W. H. SCHOPFER et A. JUNG. — Un test végétal pour l'aneurine. C. r. 5ème Congr. intern. technique et chimique des Industr. agric., p. 22.

1937: A. et M. LWOFF. — L'aneurine facteur de croissance pour le Cilié *Glaucoma piriformis*. C. r. Soc. biol. Paris, 126:644.

1937: A. P. MEIKLEJOHN. — The estimation of vitamin B_1 in blood by a modification of Schopfer's test. Biochem. J., 31:1441.

1937: A. SCHULTZ, L. ATKIN and CH. N. FREY. — A fermentation test for vitamin B_1. J. Amer. chem. Soc., 59:948.

1937: M. FAGUET. — Titrage de la vitamine B_1 par un test végétal. C. r. Soc. biol. Paris, 126:856.

1938: P. M. WEST and P. W. WILSON. — Biological determination of vitamin B_1 (thiamin) in *Rhizobium trifolii*. Science, 88:344.

1938*a*: A. S. SCHULTZ, L. ATKIN and C. N. FREY. — Influence of nicotinic acid on the fermentation method for vitamin B_1. J. Amer. chem. Soc., 60:1514.

1938*b*: A. S. SCHULTZ, L. ATKIN and C. N. FREY. — The specificity of the fermentation test for vitamin B_1. J. Amer. chem. Soc., 60:3084.

1938: A. et M. LWOFF. — La spécificité de l'aneurine facteur de croissance pour le Cilié *Glaucoma piriformis*. C. r. Soc. biol. Paris, 127:1170.

1938: W. H. SCHOPFER et A. JUNG. — Etude des organes d'animaux normaux et en avitaminose B_1 au moyen du test *Phycomyces*. Etablissement d'une microméthode. Zeitschr. f. Vitaminforschung, 7:143.

1938: W. J. ROBBINS. — Thiamin and the growth of species of *Phytophthora.* Bull. Torrey Bot. Club, 65:267.

1938: E. N. ROWLANDS and J. F. WILKINSON. — The clinical significance and estimation of blood vitamin B_1. Brit. med. J., p. 878.

1938*a*: G. G. VILLELA. — Sobre a dosagem da vitamina B_1 (aneurina) pela technica de Schopfer. O Hospital, 13:43.

1938*b*: G. G. VILLELA. — Sur le dosage de l'aneurine du maté par la technique de Schopfer-Jung. C. r. Soc. biol. Paris, 129:987.

1938: A. LWOFF et A. QUERIDO. — Dosage de l'amide de l'acide nicotinique au moyen du test *Proteus.* Principe de la méthode. C. r. Soc. biol. Paris, 129:1039.

1938: J. BONNER and J. ERICKSON. — The *Phycomyces* assay for thiamin (vitamin B_1). The method and its specificity. Amer. J. Bot., 25:685.

1938: TH. MORELL. — Ein neuer biologischer Vitamin B_1-Test. Deutsche Med. Wochenschrift, 64:1722.

1939: G. G. VILLELA. — Sur la teneur en aneurine (vitamine B_1) du liquide céphalorachidien. C. r. Soc. biol. Paris, 130:1493.

1939: H. M. SINCLAIR. — The estimation of vitamin B_1 in blood, II. A further modification of Meiklejohns method. Biochem. J., 33:2027.

1939: E. E. SNELL and F. M. STRONG. — A microbiological assay for riboflavin. Ind. and Eng. Chem. (Anal. Ed.), 11:346.

1939: G. GUHR. — Untersuchungen über den Vitamingehalt des Blutes bei Gesunden und Kranken. Klin. Wochenschrift, 18:1028.

1939: J. MARCQ et P. MANIL. — Détermination comparée de la teneur en vitamine B_1 (aneurine) de levures sèches et non extraites. Bull. Inst. Agron. Gembloux, 8:172.

1939: W. J. ROBBINS and M. BARTLEY-SCHMIDT. — Preliminary experiments on biotin. Bull. Torrey Bot. Club, 66:139.

1940: C. TANGARI. — Il comportamento della vitamine del sangue nella peritoniti acute sperimentali. Archivo italiano Medicina sperimentale, 6:1.

1940: E. HAAG et C. DALPHIN. — Une réaction biochimique de l'aneurine d'une grande sensibilité. C. r. Soc. phys. hist. nat. Genève, 57:76.

1940: F. GSTIRNER. — Chemisch-physikalische Vitamin-Bestimmungsmethoden. F. Enke, Stuttgart, 266 p.

1940: E. E. SNELL and W. H. PETERSON. — Growth factors for bacteria, X. Additional factors required by certain lactic acid bacteria. J. Bact., 39:273.

1940: E. SNELL, R. E. EAKIN and R. J. WILLIAMS. — A quantitative test for biotin and observations regarding its occurrence and properties. J. Amer. chem. Soc., 62:175.

1940: H. K. MITCHELL, H. H. WEINSTOCK, E. E. SNELL, S. R. STANBERRY and R. J. WILLIAMS. — Pantothenic acid, V to IX. J. Amer. chem. Soc., 62:1776, 1779, 1785, 1791.

1941: M. J. PELCZAR and J. R. PORTER. — A microbiological assay technique for pantothenic acid with the use of *Proteus morganii.* J. Biol. Chem., 139:111.

1941: E. E. SNELL and F. M. STRONG. — Assay method for riboflavin. Univ. Texas Publication No. 4137:11-13.

1941: F. M. STRONG, R. E. FEENEY, B. MOORE, and H. T. PARSONS. — The riboflavin content of blood and urine. J. Biol. Chem., 137:363.

1941: R. J. WILLIAMS, R. E. EAKIN and J. R. MCMAHAN. — Assay method for pyridoxin. Univ. Texas Publication No. 4137:24-26.

1941*a*: E. E. SNELL and L. D. WRIGHT. — A microbiological method for the determination of nicotinic acid. J. Biol. Chem. 139:675.

1941*b*: E. E. SNELL and L. D. WRIGHT. — Assay method for nicotinic acid. Univ. Texas Publication No. 4137:22-23.

1941: H. K. MITCHELL and E. E. SNELL. — Assay method for "Folic Acid". Univ. Texas Publication No. 4137:36.

1941: R. J. WILLIAMS, A. K. STOUT, H. K. MITCHELL and J. R. MCMAHAN. — Assay method for inositol. Univ. Texas Publication No. 4137:27-30.

1941: R. THREN. — Kritische Versuche zum biologischen Pilztest (*Phycomyces*-Wachstumstest) auf Vitamin B_1. Vitamine und Hormone, 1:100.

1941: W. H. SCHOPFER. — Bemerkungen zur Arbeit von R. THREN, kritische Versuche zum biologischen Pilztest, avec remarque finale de R. THREN. — Vitamine und Hormone, 1.

1942: M. LANDY and D. M. DICKEN. — A microbiological assay method for six B vitamins using *Lactobacillus casei* and a medium of essentially known composition. J. of Lab. and Clin. Med., 27:1086.

1942: A. S. SCHULTZ, L. ATKIN and C. N. FREY. — Determination of vitamin B_1 by yeast fermentation method. Ind. and Eng. Chem. (Anal. Ed.), 14:35.

CONCLUSION

It can no longer be doubted that vitamins are essential factors for the growth of plants. They are regulators for metabolism and serve to establish chemical relations between the various organs of a complex plant. The need of vitamins is due to a loss of the ability to synthesize them. When this condition prevails the plant reacts like a heterotrophic animal. The reactions of plant and animal cells are similar, for the fundamental functions of vitamins are the same in the two kingdoms. The only aspects that differ are the morphological expressions to which the vitamin deficiency gives rise. In an animal, a complex and highly differentiated organism, an avitaminosis expresses itself in an extremely complex manner. The disturbances of metabolism and of cellular equilibrium, which are the first and most fundamental symptoms of an avitaminosis, affect not only the functions of the various organs but also their structure.

An undifferentiated unicellular microorganism can not display a complex avitaminosis; there is only one possible reaction—it stops growing.

From a biochemical viewpoint there are no differences of reaction between animals and lower plants. At all phylogenetic levels the requirements of living matter are approximately the same regardless of the structure of the organism. By using a vitamin particularly suitable for this demonstration, we are able to establish a line which cuts across all phyla of living organisms, irrespective of their classification, from bacteria to higher vertebrate animals. The same must be true for other vitamins.

It should be emphasized that this field is the meeting ground of specialized sciences whose relation could only be suspected at the end of the last century. Organic chemistry, enzymology, vitaminology, human, animal, and plant physiology, meet again to solve fundamental problems. In order to understand the problem of vitamins in all of its ramifications it is no longer possible to confine oneself to one field. The plant physiologist has learned from his confrère, the human physiologist, what an avitaminosis and a vitamin is. The plant physiologist in turn has shown that plants are the seat of the biosynthesis of vitamins and thus has established a new intimate relationship between two kingdoms. The biochemist, by establishing the chemical structure of vitamins, has been obliged to create new groups of chemical compounds. The enzymologist finds to his surprise that these vitamins are nothing but the active portions of enzymes which have been studied for a long time. The microbiologist, who for years had been trying, without success, to isolate the "growth factors" of his microorganisms, proved that typical animal vitamins were the factors he was looking for. The concept of growth factors (in the exact sense) conforms with that of vitamins (in the strict sense) and is identical with it.

The field is capable of many applications. It may be possible in the near future to grow on synthetic media such organisms as the *Erysiphales* (powdery mildews) and *Uredinales* (rusts), which up till now have resisted all such attempts. In short, the problem treated here is the place for an ideal collaboration of various biologists. More than any other problem, it proves the importance and the necessity of general physiology.

The problem of vitamins started with man and, in the last analysis, it returns to man after an apparent departure from him. All the progress accomplished in this domain contributes to a better understanding of the problem in general and ultimately to human well-being.

To speak of the well-being of Man at this time seems to be macabre humor. Why is it that skilful men from different nations can solve problems in fundamental biology but can not do so when the matter directly concerns themselves? The question remains unanswered.

Additions

Note 1 (*cf.* page 135): — *Rhizopus suinus*, whose growth is inhibited by thiamin (see p. 104), responds to *i*-inositol which functions as a growth factor. The latter brings about a 50 per cent increase in dry matter production in cultures grown without thiamin. In cultures grown with and inhibited by thiamin, *i*-inositol, after a certain period of development, nullifies the inhibiting action of thiamin. The action of *i*-inositol on cultures grown with and without thiamin is dependent upon the age of the culture, the composition of the medium, and the temperature.

The period of inhibition by thiamin is preceded by a short period of acceleration when thiamin actually functions as a growth factor. The passage from the stage of acceleration to that of inhibition is a function of time, being rapid at 36° and proportionately slower at temperatures of 29°, 22-24°, and 18°C. During the period when thiamin exerts a stimulating effect, a positive action of inositol occurs in addition to that of thiamin (synergism). During the period when thiamin has an inhibiting effect on growth, this action is nullified by the stimulating effect of inositol (antagonism) (SCHOPFER, 1942).

The specificity of action of *i*-inositol is very pronounced. This compound cannot be replaced by any of the following substances: scyllitol (which differs from *i*-inositol only by the position of an OH) (POSTERNAK), mytilitol, quercitol, quebrachitol, inosose (obtained biochemically by KLUYVER or chemically by POSTERNAK), or cyclohexanol.

These investigations again demonstrate the relativity of the capacity for synthesis and of the vitamin requirements of the organism.

Note 2 (*cf.* p. 135): — SCHOPFER and BLUMER (1942) have shown that, under certain conditions, purified biotin acts as a growth factor for *Trichophyton album.* The growth factor requirements of this species are conditioned by the composition of the medium and the age of the culture. When the cultures are supplied with asparagine, biotin is required only during the early stages of development. The growth of the controls soon equals that of the cultures supplied with biotin. When the cultures are supplied with ammonium sulfate, biotin is required until an advanced stage of development has been reached. When ammonium citrate is supplied to cultures, the controls remain poorer throughout an incubation period of 24 days than the cultures supplied with biotin. This organism is able to synthesize a certain amount of biotin but not enough for normal development. The capacity for synthesis varies with the age of the culture and the nitrogen source. This characteristic must be looked upon as a function of time and of the composition of the medium. Thus we see that *Trichophyton album*, cultured on the usual medium of the author (glucose, asparagine, magnesium sulfate, and potassium acid phosphate), requires only thiamin as a growth factor. By changing the composition of the medium the following become necessary for maximal dry matter production: thiamin, biotin, *i*-inositol, and pyridoxine (SCHOPFER and BLUMER).

Recent work of ROBBINS and MA (1942*a* and *c*) concerning the vitamin deficiencies of several species of *Ceratostomella* (*Graphium*) and related fungi reveal the following facts: thiamin and pyridoxine are required by *C. pseudo-tsugae, C. pilifera, C. multiannulata, C. pluriannulata, C. fimbriata* (in this case pyridoxine can be replaced by biotin), and *Ophiostoma catonianum;* thiamin, pyridoxine, and biotin are required by *C. picea-perda, C. penicillata, C. microspora, C. montium* and *C. ips;* thiamin and biotin are required by *C. pini* and *C. obscura;* pyridoxine alone is required by *Grosmannia serpens;* thiamin alone is required by *C. stenoceras, Endoconidiophora paradoxa, E. coerulescens.* Several of the species studied require additional factors supplied by malt extract, but the identity of these factors was not investigated.

The specificity of action of pyridoxine on *C. ulmi* was studied by ROBBINS and MA (1942*b*) and found to be very pronounced.

AUTHOR INDEX

GENERAL INDEX

PLATE I.—Increase in growth of excised tomato roots with increasing amounts of thiamin chloride. The figures represent γ per 40 cc. medium. (From ROBBINS and MARY B. SCHMIDT, in WILLIAMS and SPIES, p. 380, Fig. 19, courtesy Macmillan.

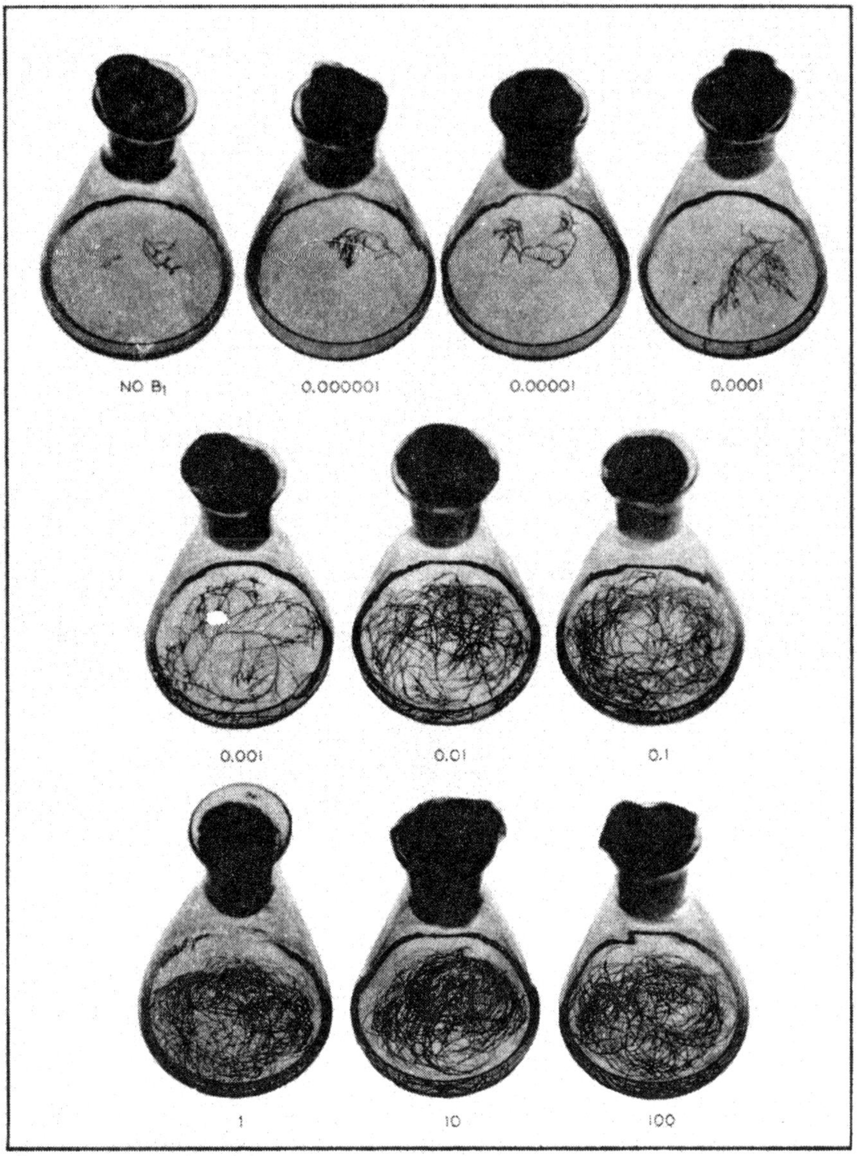
NO B_1
0.000001
0.00001
0.0001
0.001
0.01
0.1
1
10
100

PLATE II.—Action of thiamin and its components on various microorganisms. Top row: *Mucor Ramannianus*, a thiazole requiring fungus. Second row: *Rhodotorula rubra*, a pyrimidine requiring fungus. Third row: *Phycomyces Blakesleeanus*, a pyrimidine + thiazole requiring fungus. Bottom row: *Phytophthora cinnamomi*, a thiamin requiring fungus. (Top row from MÜLLER, unpublished; the remainder from SCHOPFER, Erg. Biol. 1939, 16:66, Springer, Berlin).

From left to right: K=Control. P=Pyrimidine. T=Thiazole. PT=Pyrimidine and Thiazole. B′=Thiamin (whole molecule).

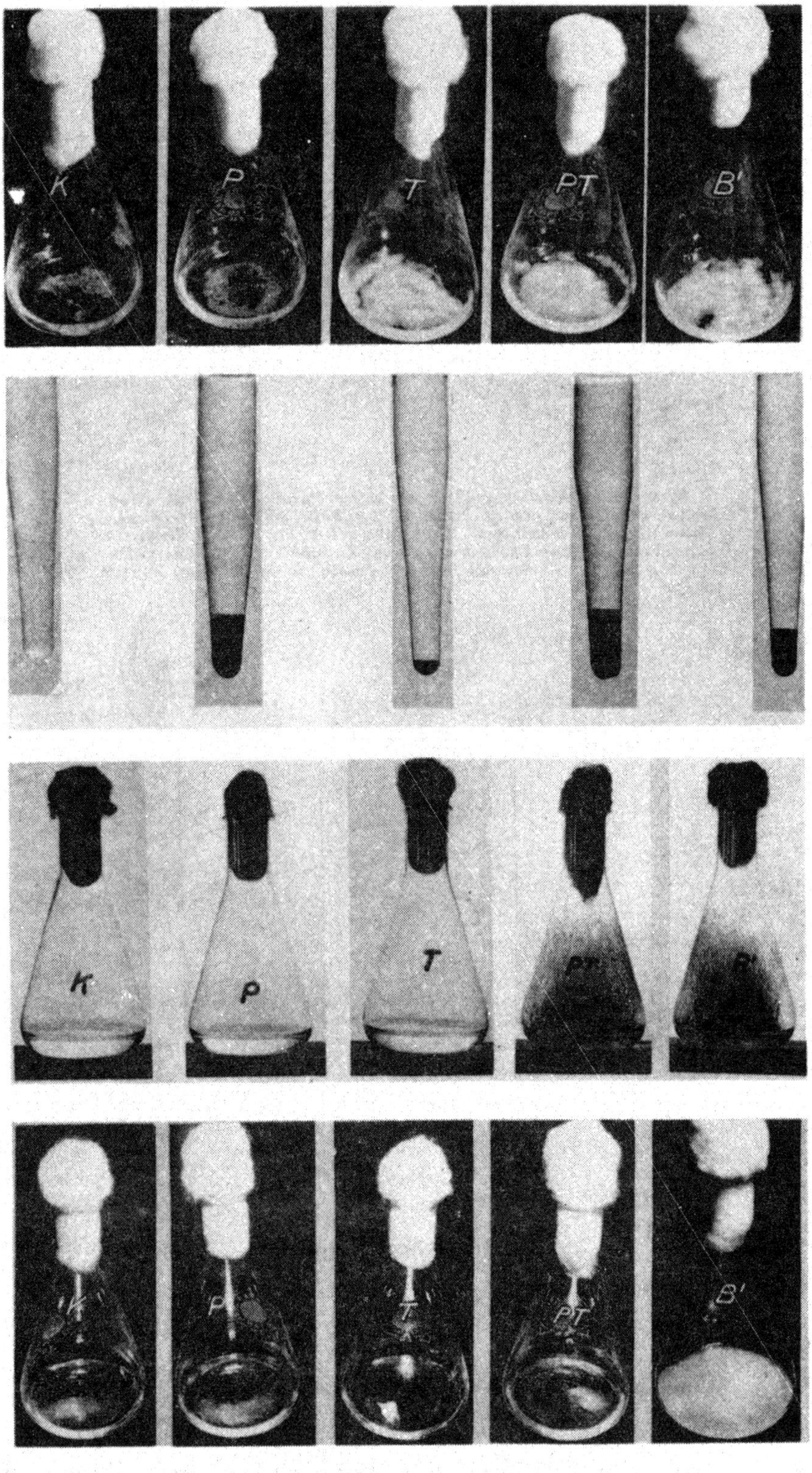
K
P
T
PT
B'
K
P
T
PT
B'
K
P
T
PT
B'

PLATE III.—Artificial symbiosis of *Mucor Ramannianus* with *Rhodotorula rubra*. From left to right: M. I.=*Mucor* with thiazole, M. II.= *Mucor* without growth factor. From right to left: H. I.=*Rhodotorula* with pyrimidine. H. II. + *Rhodotorula* without growth factor. In the middle. H + M.=*Mucor* + *Rhodotorula* without growth factor. (From MÜLLER, 1941).

M
I
M
II
H
II
H
I